FOR USE IN THE LIBRARY ONLY

Programmed Cell Death

PROGRAMMED CELL DEATH:
The Cellular and Molecular Biology of Apoptosis

Edited by

Martin Lavin and Dianne Watters

Queensland Cancer Fund Research Unit
Queensland Institute of Medical Research
The Bancroft Centre, Brisbane, Australia

harwood academic publishers

Australia • China • France • Germany • India • Japan • Luxembourg
Malaysia • The Netherlands • Russia • Singapore • Switzerland • Thailand
United Kingdom • United States

Copyright © 1993 by Harwood Academic Publishers GmbH.

All rights reserved.

First published 1993
Second printing 1995

No part of this book may be reproduced or utilized in any form or by any means, electronic or mechanical, including photocopying and recording, or by any information storage or retrieval system, without permission in writing from the publisher. Printed in Singapore by Kyodo Printing Co (Singapore) Pte Ltd.

Emmaplein 5
1075 AW Amsterdam
The Netherlands

Library of Congress Cataloging-in-Publication Data

Programmed cell death : the cellular and molecular biology of
  apoptosis / edited by Martin Lavin and Dianne Watters.
       p. cm.
  Includes bibliographical references and index.
  ISBN: 3-7186-5461-X
  1. Apoptosis.   I. Lavin, Martin, 1943–   .   II. Watters, Dianne,
1952–   .
QH671.P76   1993
574.87'65–dc20   93-27865
                                                             CIP

FOREWORD

Mitosis was defined last century. In contrast, the distinctive features of normal cell death were described only relatively recently. The long neglect may have been partly due to a human tendency to embrace positive rather than negative concepts. However it was mainly a result of three other factors: the inherently cryptic nature of the morphological manifestations of apoptosis; the superficial resemblance that apoptosis sometimes bears to necrosis at the light microscope level, both being accompanied by nuclear chromatin condensation and the development of cytoplasmic eosinophilia; and the firm entrenchment of a notion that regarded all cell death as a degenerative phenomenon. Thus the tendency for apoptosis to involve scattered single cells, the small size of many apoptotic bodies and the rapid disposal of these bodies by nearby cells without the induction of inflammation would have impeded its detection in most normal tissues, where the rate of cell death is low. It was only in a few locations where dying cells are particularly numerous in healthy animals, such as the centres of reactive lymphoid follicles and sites of programmed tissue deletion occurring during embryonic development, that early histologists explicitly recognized the existence of cell death under physiological conditions. It was, however, difficult for them to appreciate that the death present in these locations is morphologically basically different from necrosis using light microscopy alone. The categorization of apoptosis as a discrete phenomenon had to await the application of electron microscopy to the study of cell death.

Following definition of the features of apoptosis, including the diagnostic cleavage of nuclear DNA at the linker regions between nucleosomes, it became possible to determine its incidence. In addition to subserving essential kinetic, homeostatic and regulatory roles in normal animals, it was found to occur under a wide range of pathological conditions. The most important implication of the apoptosis concept for medicine is clearly the possibility that cell deletion by this means might be selectively and predictably controlled in disease states.

Knowledge of the regulation of apoptosis at the molecular level is now growing rapidly. Significantly, some of the genes recently found to modulate its occurrence are among those previously shown to be involved in controlling mitosis. Many of the exciting new insights into the biochemistry of apoptosis that have been achieved in the past two or three years are drawn together in this book. It is hoped that it will act as a further stimulus to research in the field.

The rewards of such research may be great. There is, for example, evidence that the $CD4^+$ T lymphocyte deletion that is responsible for the development of immunodeficiency in patients with HIV infection is mediated by apoptosis and that infection of these lymphocytes by the virus is not necessary for the triggering of their death. Rather, it has been proposed that much of the lymphocyte apoptosis is a result of defective support by cytokines or an inappropriate response to activation. Detailed understanding of the processes involved may make it possible to prevent or delay the development of immunodeficiency without its being necessary to eliminate the virus from the body. Another potentially important area of application of the currently rapidly evolving knowledge of the regulation of apoptosis is in the development of novel approaches to tumour therapy. Progress in this field is reviewed in several chapters.

John F. R. Kerr

CONTENTS

Foreword		v
Preface		xi
Contributors		xiii
Chapter 1	Introduction: Definition of Apoptosis and Overview of its Incidence *John F. R. Kerr*	1

BIOCHEMICAL MECHANISMS

Chapter 2	Signal Transduction in Thymocyte Apoptosis *David J. McConkey, Sten Orrenius and Mikael Jondal*	19
Chapter 3	Regulation of Thymocyte Apoptosis: Glucocorticoid-Induced Death and its Inhibition by T-Cell Receptor/CD3 Complex-Mediated Stimulation *Makoto Iwata, Rieko Iseki and Yoshihisa Kudo*	31
Chapter 4	Protein Modification in Apoptosis *Martin F. Lavin, Glenn D. Baxter, Qizhong Song, Duygu Findik and Eva Kovacs*	45
Chapter 5	Relationship Between Apoptosis and the Cell Cycle *P. Roy Walker, Joanna Kwast-Welfeld and Marianna Sikorska*	59
Chapter 6	Intracellular Zinc and the Regulation of Apoptosis *Peter D. Zalewski and Ian J. Forbes*	73

Chapter 7	Induction of Apoptosis by the Immunomodulating Agent Gliotoxin	87
	Paul Waring	

DNA FRAGMENTATION

Chapter 8	Glucocorticoid-Induced Programmed Cell Death (Apoptosis) in Leukemia and Pre-B Cells	99
	Emad S. Alnemri and Gerald Litwack	
Chapter 9	Evidence for the Role of an Endo-Exonuclease in the Chromatin DNA Fragmentation which Accompanies Apoptosis	111
	Murray J. Fraser, Christine M. Ireland, Stephen J. Tynan and Arthur Papaioannou	
Chapter 10	Critical Changes in Apoptosis in Thymocytes Precede Endonuclease Activation	123
	Gerald M. Cohen, Xiao-Ming Sun, Roger T. Snowden, Michael G. Ormerod and David Dinsdale	
Chapter 11	Structural Analysis of DNA Related to Apoptosis: Evidence for a Specific Lesion Associated with DNA Replication	133
	Bernard W. Stewart, Robert J. Sleiman, Daniel R. Catchpoole and Sally M. Pittman	

GENETIC REGULATION

Chapter 12	Regulation of Programmed Cell Death	145
	J. John Cohen	
Chapter 13	Macromolecular Synthesis, c-*myc*, and Apoptosis	153
	Douglas R. Green and Thomas G. Cotter	
Chapter 14	The *bcl*-2 Oncogene Regulates Lymphocyte Survival and Potentiates Lymphomagenesis	167
	Andreas Strasser, Alan W. Harris and Suzanne Cory	

| Chapter 15 | Bcl-2 Expression in the Prostate and its Association with Androgen-Independent Prostate Cancer | 179 |

Timothy J. McDonnell, Patricia Troncoso, Shawn Brisbay, Leland Chung, Christapher Logothetis, Jer-Tsong Hsieh, Martin Cambell and Shi-Ming Tu

| Chapter 16 | The Role of Tumour Suppressor Genes in Apoptosis | 187 |

Elisheva Yonish-Rouach, Doron Ginsberg and Moshe Oren

| Chapter 17 | Altered Gene Expression During Apoptosis | 203 |

Sherilyn D. Goldstone and Martin F. Lavin

APOPTOSIS IN THE IMMUNE SYSTEM

| Chapter 18 | The Role of Apoptosis in the Regulation of the Immune Response | 219 |

J. J. T. Owen, I. M. Allan, N. C. Moore and E. J. Jenkinson

| Chapter 19 | The APO-1 System: At the Crossroad of Lymphoproliferation, Autoimmunity and Apoptosis | 229 |

Peter H. Krammer

| Chapter 20 | Apoptosis in the Target Organ of an Autoimmune Disease | 235 |

Michael P. Pender

| Chapter 21 | Peptide Epitope-Induced Apoptosis of Human Cytotoxic T Lymphocytes | 245 |

A. Suhrbier and S. R. Burrows

APOPTOSIS IN CANCER

| Chapter 22 | Apoptosis in Irradiated Murine Tumours and Human Tumour Xenografts | 259 |

Raymond E. Meyn, L. Clifton Stephens, K. Kian Ang, Luka Milas, Nancy R. Hunter and Lester J. Peters

| Chapter 23 | Anti-Cancer Drugs and Apoptosis | 269 |

L. I. Huschtscha, C. E. Andersson, W. A. Bartier and M. H. N. Tattersall

Chapter 24	**Apoptosis Resulting from Anti-Cancer Agent Activity *in vivo* is Enhanced by Biochemical Modulation of Tumour Cell Energy**	**279**
	Daniel S. Martin, Robert L. Stolfi, Joseph R. Colofiore, Jason A. Koutcher, Alan Alfieri, Stephen Sternberg and L. Dee Nord	
Chapter 25	**Hyperthermia-Induced Apoptosis**	**297**
	Brian V. Harmon, Trevor H. Forster and Russell J. Collins	
Chapter 26	**Tubulin in Apoptotic Cells**	**315**
	Sally M. Pittman, Melissa Geyp, Steven J. Tynan, Cilinia M. Gramacho, Deborah H. Strickland, Murray J. Fraser and Christine M. Ireland	
Index		**325**

PREFACE

The study of apoptosis, programmed cell death or physiological cell death, has come of age in recent years after lingering on the edge of acceptance for more than two decades. It is ironic that the rather general features that have identified this process for more than twenty years, morphological characteristics, are still the only universal and consistent changes associated with this process. However, an enormous amount has recently been achieved in studying the cellular and molecular characteristics of apoptosis. The complexity of the process is evident from the variety of changes reported in gene expression. The realisation of the importance of apoptosis in the immune response during T-cell development, particularly in preventing autoimmune attack and in the process of cytotoxic T-cell killing, represented the first milestone in popularising this process. It became quickly evident that apoptosis mirrors cell proliferation and involves a complex network of signal transduction pathways related to those controlling cell growth. It seems likely that both newly synthesised proteins and pre-existing proteins contribute to the multiple steps involved.

This volume results from the second Bancroft Symposium held at the Queensland Institute of Medical Research in September 1992, which we believe was the first open meeting entirely dedicated to the subject of apoptosis. The intention was to discuss the biochemical basis of apoptosis, to consider the involvement of various genes in apoptosis, notably proto-oncogenes and tumour suppressor genes, and to outline the importance of apoptosis in the immune system and in cancer.

We hope that this volume will provide a landmark in the progress of research on apoptosis from the cellular to the molecular basis of this process.

Martin F. Lavin
Dianne Watters

CONTRIBUTORS

Alan Alfieri, Department of Medical Physics and Radiology, MSKCC, 1275 York Avenue, New York, NY 10021, USA

I. M. Allan, Department of Anatomy, Medical School, University of Birmingham, Birmingham, UK

Emad S. Alnemri, Department of Pharmacology and the Jefferson Cancer Institute, Jefferson Medical College of Thomas Jefferson University, Philadelphia, PA 19107, USA

C. E. Andersson, Department of Cancer Medicine, Blackburn Building, University of Sydney, Sydney, NSW 2006, Australia

K. Kian Ang, Department of Clinical Radiotherapy, The University of Texas, M. D. Anderson Cancer Center, Houston, Texas 77030, USA

W. A. Bartier, Department of Cancer Medicine, Blackburn Building, University of Sydney, Sydney, NSW 2006, Australia

Glenn D. Baxter, Queensland Cancer Fund Research Unit, Queensland Institute of Medical Research, The Bancroft Centre, 300 Herston Road, Brisbane, Queensland 4029, Australia

Shawn Brisbay, Department of Molecular Pathology, The University of Texas, M.D. Anderson Cancer Center, Houston, Texas 77030, USA

S. R. Burrows, Queensland Institute of Medical Research, The Bancroft Centre, 300 Herston Road, Brisbane, Queensland 4029, Australia

Martin Cambell, Department of Molecular Pathology, The University of Texas, M.D. Anderson Cancer Center, Houston, Texas 77030, USA

Daniel R. Catchpoole, Children's Leukaemia and Cancer Research Centre, University of New South Wales, Prince of Wales Children's Hospital, Sydney, NSW 2031, Australia

Leland Chung, Department of Urology, The University of Texas, M.D. Anderson Cancer Center, Houston, Texas 77030, USA

Gerald M. Cohen, MRC Toxicology Unit, Medical Research Council Laboratories, Woodmansterne Road, Carshalton, Surrey SM5 4EF, UK

J. John Cohen, Department of Microbiology and Immunology, University of Colorado School of Medicine, Denver, CO 80262, USA

Russell J. Collins, Department of Pathology, Royal Brisbane Hospital, Queensland 4029, Australia

Joseph R. Colofiore, Cancer Research, Catholic Medical Center, 89–15 Woodhaven Blvd, Woodhaven, NY 11421, USA

Suzanne Cory, The Walter and Eliza Hall Institute of Medical Research, PO Royal Melbourne Hospital, Victoria 3050, Australia

Thomas G. Cotter, The Department of Biology, St Patrick's College, Maynooth, Co. Kildare, Ireland

David Dinsdale, MRC Toxicology Unit, Medical Research Council Laboratories, Woodmansterne Road, Carshalton, Surrey SM5 4EF, UK

Duygu Findik, Queensland Cancer Fund Research Unit, Queensland Institute of Medical Research, The Bancroft Centre, 300 Herston Road, Brisbane, Queensland 4029, Australia

Ian J. Forbes, Department of Medicine, University of Adelaide, The Queen Elizabeth Hospital, Woodville, South Australia 5011, Australia

Trevor H. Forster, School of Life Science, Queensland University of Technology, Brisbane, Queensland 4000, Australia

Murray J. Fraser, Children's Leukaemia and Cancer Research Centre, University of New South Wales, Prince of Wales Children's Hospital, Sydney, NSW 2031, Australia

Melissa Geyp, Children's Leukaemia and Cancer Research Centre, University of New South Wales, Prince of Wales Children's Hospital, Sydney, NSW 2031, Australia

Doron Ginsberg, Dana-Farber Cancer Institute, Boston, MA 02115, USA

Sherilyn D. Goldstone, Department of Pathology, University of Sydney, NSW 2006, Australia

Cilinia M. Gramacho, Children's Leukaemia and Cancer Research Centre, University of New South Wales, Prince of Wales Children's Hospital, Sydney, NSW 2031, Australia

Douglas R. Green, Division of Cellular Immunology, La Jolla Institute for Allergy and Immunology, La Jolla, CA 92037, USA

Brian V. Harmon, School of Life Science, Queensland University of Technology, Brisbane, Queensland 4000, Australia

Alan W. Harris, The Walter and Eliza Hall Institute of Medical Research, PO Royal Melbourne Hospital, Victoria 3050, Australia

Jer-Tsong Hsieh, Department of Urology, The University of Texas, M.D. Anderson Cancer Center, Houston, Texas 77030, USA

Nancy R. Hunter, Department of Experimental Radiotherapy, The University of Texas, M. D. Anderson Cancer Center, Houston, Texas 77030, USA

L. I. Huschtscha, Department of Cancer Medicine, Blackburn Building, University of Sydney, Sydney, NSW 2006, Australia

Christine M. Ireland, Children's Leukaemia and Cancer Research Centre, University of New South Wales, Prince of Wales Children's Hospital, Sydney, NSW 2031, Australia

Rieko Iseki, Department of Cellular Recognition Research, Mitsubishi Kasei Institute of Life Sciences, 11, Minamiooya, Machida-shi, Tokyo 194, Japan

Makoto Iwata, Department of Cellular Recognition Research, Mitsubishi Kasei Institute of Life Sciences, 11, Minamiooya, Machida-shi, Tokyo 194, Japan

E. J. Jenkinson, Department of Anatomy, Medical School, University of Birmingham, Birmingham, UK

Mikael Jondal, Department of Immunology, Karolinska Institutet, Box 60400, S-104 01 Stockholm, Sweden

John F. R. Kerr, Department of Pathology, University of Queensland Medical School, Herston, Brisbane, Queensland 4006, Australia

Jason A. Koutcher, Department of Medical Physics and Radiology, MSKCC, 1275 York Avenue, New York, NY 10021, USA

Eva Kovacs, Queensland Cancer Fund Research Unit, Queensland Institute of Medical Research, The Bancroft Centre, 300 Herston Road, Brisbane, Queensland 4029, Australia

Peter H. Krammer, German Cancer Research Center, Tumourimmunology Program, Im Neuenheimer Feld 280, D-6900 Heidelberg, Germany

Yoshihisa Kudo, Department of Neuroscience, Mitsubishi Kasei Institute of Life Sciences, 11, Minamiooya, Machida-shi, Tokyo 194, Japan

Joanna Kwast-Welfeld, Apoptosis Research Group, Institute for Biological Sciences, National Research Council of Canada, Ottawa, K1A 0R6, Canada

Martin F. Lavin, Queensland Cancer Fund Research Unit, Queensland Institute of Medical Research, The Bancroft Centre, 300 Herston Road, Brisbane, Queensland 4029, Australia

Gerald Litwack, Department of Pharmacology and the Jefferson Cancer Institute, Jefferson Medical College of Thomas Jefferson University, Philadelphia, PA 19107, USA

Christopher Logothetis, Department of Medical Oncology, The University of Texas, M.D. Anderson Cancer Center, Houston, Texas 77030, USA

Daniel S. Martin, Developmental Chemotherapy, Memorial Sloan Kettering Cancer Center (MSKCC), 1275 York Avenue, New York, NY 10021 and Cancer Research, Catholic Medical Center, 89–15 Woodhaven Blvd, Woodhaven, NY 11421, USA

David J. McConkey, Laboratory of Immunobiology, Dana-Farber Cancer Institute and Department of Pathology, Harvard Medical School, Boston, MA 02115, USA

Timothy J. McDonnell, Department of Molecular Pathology, The University of Texas, M.D. Anderson Cancer Center, Houston, Texas 77030, USA

Raymond E. Meyn, Department of Experimental Radiotherapy, The University of Texas, M.D. Anderson Cancer Center, Houston, Texas 77030, USA

Luka Milas, Department of Experimental Radiotherapy, The University of Texas, M.D. Anderson Cancer Center, Houston, Texas 77030, USA

N. C. Moore, Department of Anatomy, Medical School, University of Birmingham, Birmingham, UK

L. Dee Nord, Cancer Research, Catholic Medical Center, 89–15 Woodhaven Blvd, Woodhaven, NY 11421, USA

Moshe Oren, Department of Chemical Immunology, The Weizmann Institute of Science, Rehovot 76100, Israel

Michael G. Ormerod, MRC Toxicology Unit, Medical Research Council Laboratories, Woodmansterne Road, Carshalton, Surrey SM5 4EF, UK

Sten Orrenius, Department of Toxicology, Karolinska Institutet, Box 60400, S-104 01 Stockholm, Sweden

J. J. T. Owen, Department of Anatomy, Medical School, University of Birmingham, Birmingham, UK

Arthur Papaioannou, Children's Leukaemia and Cancer Research Centre, Prince of Wales Children's Hospital, Randwick, NSW 2031, Australia

Michael P. Pender, Neuroimmunology Research Unit, Department of Medicine, Clinical Sciences Building, Royal Brisbane Hospital, Queensland 4029, Australia

Lester J. Peters, Department of Clinical Radiotherapy, The University of Texas, M.D. Anderson Cancer Center, Houston, Texas 77030, USA

Sally M. Pittman, Children's Leukaemia and Cancer Research Centre, University of New South Wales, Prince of Wales Children's Hospital, Sydney, NSW 2031, Australia

Marianna Sikorska, Apoptosis Research Group, Institute for Biological Sciences, National Research Council of Canada, Ottawa K1A 0R6, Canada

Robert J. Sleiman, Children's Leukaemia and Cancer Research Centre, University of New South Wales, Prince of Wales Children's Hospital, Sydney, NSW 2031, Australia

Roger T. Snowden, MRC Toxicology Unit, Medical Research Council Laboratories, Woodmansterne Road, Carshalton, Surrey SM5 4EF, UK

Qizhong Song, Queensland Cancer Fund Research Unit, Queensland Institute of Medical Research, The Bancroft Centre, 300 Herston Road, Brisbane, Queensland 4029, Australia

L. Clifton Stephens, Department of Veterinary Medicine and Surgery, The University of Texas, M.D. Anderson Cancer Center, Houston, Texas 77030, USA

Stephen Sternberg, Department of Pathology, MSKCC, 1275 York Avenue, New York, NY 10021, USA

Bernard W. Stewart, Children's Leukaemia and Cancer Research Centre, University of New South Wales, Prince of Wales Children's Hospital, Sydney, NSW 2031, Australia

Robert L. Stolfi, Cancer Research, Catholic Medical Center, 89–15 Woodhaven Blvd, Woodhaven, NY 11421, USA

Andreas Strasser, The Walter and Eliza Hall Institute of Medical Research, PO Royal Melbourne Hospital, Victoria 3050, Australia

Deborah H. Strickland, Children's Leukaemia and Cancer Research Centre, University of New South Wales, Prince of Wales Children's Hospital, Sydney, NSW 2031, Australia

A. Suhrbier, Queensland Institute of Medical Research, The Bancroft Centre, 300 Herston Road, Brisbane, Queensland 4029, Australia

Xiao-Ming Sun, MRC Toxicology Unit, Medical Research Council Laboratories, Woodmansterne Road, Carshalton, Surrey SM5 4EF, UK

M. H. N. Tattersall, Department of Cancer Medicine, Blackburn Building, University of Sydney, NSW 2006, Australia

Patricia Troncoso, Department of Molecular Pathology, The University of Texas, M.D. Anderson Cancer Center, Houston, Texas 77030, USA

Shi-Ming Tu, Department of Medical Oncology, The University of Texas, M.D. Anderson Cancer Center, Houston, Texas 77030, USA

Stephen J. Tynan, Children's Leukaemia and Cancer Research Centre, University of New South Wales, Prince of Wales Children's Hospital, Sydney, NSW 2031, Australia

P. Roy Walker, Apoptosis Research Group, Institute for Biological Sciences, National Research Council of Canada, Ottawa K1A 0R6, Canada

Paul Waring, Division of Cell Biology, John Curtin School of Medical Research, Australian National University, PO Box 334, Canberra City 2601, Australia

Elisheva Yonish-Rouach, Department of Chemical Immunology, The Weizmann Institute of Science, Rehovot 76100, Israel and Institut de Recherches Scientifiques sur le Cancer, 94891 Villejuif Cedex, France

Peter D. Zalewski, Department of Medicine, University of Adelaide, The Queen Elizabeth Hospital, Woodville, South Australia 5011, Australia

Chapter 1: Introduction

DEFINITION OF APOPTOSIS AND OVERVIEW OF ITS INCIDENCE

John F. R. Kerr
Department of Pathology, University of Queensland Medical School,
Herston, Brisbane, Queensland 4006, Australia

Apoptosis was originally defined and distinguished from necrosis on the basis of its ultrastructure (Kerr, 1969, 1971; Kerr *et al.*, 1972). Subsequently, the two processes were found to be accompanied by different patterns of DNA degradation (Wyllie, 1980; Afanas'ev *et al.*, 1986; Duvall and Wyllie, 1986). The circumstances of their occurrence have now been studied extensively (for reviews, see Wyllie *et al.*, 1980; Wyllie, 1981; Searle *et al.*, 1982; Wyllie, 1987; Walker *et al.*, 1988; Cohen, 1991; Kerr and Harmon, 1991; Cohen and Duke, 1992; Raff, 1992). On the whole, they have turned out to be disparate. Thus, whilst necrosis is always an outcome of severe injury to the cell, apoptosis occurs in normal tissues and, where it is observed pathologically, a homeostatic function can often be attributed to it.

In this chapter, the defining characteristics of apoptosis will be outlined and contrasted with those of necrosis, and its incidence will be briefly surveyed. Such an overview should provide a background for the later chapters, which mostly deal with recent advances in knowledge of the biochemical mechanisms and regulation of apoptosis. The ultrastructural features of cell death will be illustrated only diagrammatically and photographs of electrophoretic gels showing patterns of DNA degradation will not be reproduced; the reader interested in the detailed appearances and the justification of the descriptions given is referred to copiously illustrated reviews, which have extensive bibliographies (Wyllie *et al.*, 1980; Kerr *et al.*, 1987; Arends and Wyllie, 1991; Kerr and Harmon, 1991).

DEFINING CHARACTERISTICS OF APOPTOSIS AND COMPARISON WITH NECROSIS

Morphology

The main ultrastructural changes occurring in apoptosis and necrosis respectively are shown in Figure 1.1.

Early manifestations of apoptosis detectable with the electron microscope include compaction and segregation of the nuclear chromatin to produce sharply delineated, uniformly finely granular masses that lie against the nuclear envelope and condensation of the cytoplasm. These changes are often associated with convolution of the cell and nuclear outlines, and translucent cytoplasmic vacuoles may be evident at this stage. Continuing cellular condensation is then accompanied in many cell types by exuberant cell surface protrusion and breaking-up of the nucleus to form multiple fragments in which the compacted chromatin is either localised in peripheral crescents or distributed throughout most of their cross-sectional area. Finally, the cell surface protuberances separate to produce membrane-enclosed apoptotic bodies of varying size in which the closely packed cytoplasmic organelles remain well preserved. Some cells such as thymocytes, however, undergo only restricted cellular and nuclear budding during apoptosis. In tissues, apoptotic bodies are quickly taken-up by nearby cells and digested within lysosomes. Epithelial cells as well as professional mononuclear phagocytes participate in this mopping-up process. In cell cultures, on the other hand, apoptotic bodies usually escape phagocytosis and such bodies fairly rapidly undergo a process of degeneration that resembles necrosis.

By light microscopy, apoptotic bodies are evident as round or oval masses of cytoplasm, which are invariably smaller than the cell of origin and which often, but not always, contain one or more fragments of compacted chromatin. The formation of apoptotic bodies is completed in several minutes and their subsequent rapid digestion within neighbouring cells results in their becoming undetectable histologically in only a few hours. This latter fact, taken in conjunction with the small size of many apoptotic bodies and the characteristic absence of inflammation, makes the process relatively inconspicuous. Uninvolved cells close ranks, and cell deletion by this means is accomplished without destruction of basic tissue architecture.

In contrast to the changes described above, in the irreversibly injured cell destined to undergo necrosis, breakdown of fluid homeostasis leads to swelling of all cytoplasmic compartments (Figure 1.1). Subsequent rupture of membranes, release of lysosomal enzymes and dissolution of organelles result in conversion of the cell into a mass of debris. Clumping of nuclear chromatin is present at an early stage, but the compaction is less uniform than in apoptosis and the aggregates have less sharply defined edges. Later, the chromatin disappears and the necrotic cell is evident by light microscopy as a structureless eosinophilic ghost. Necrosis occurring *in vivo* is usually accompanied by the development of inflammation. The cellular debris is removed by mononuclear phagocytes, many of which are derived from blood monocytes that have

1: DEFINITION OF APOPTOSIS AND OVERVIEW OF ITS INCIDENCE

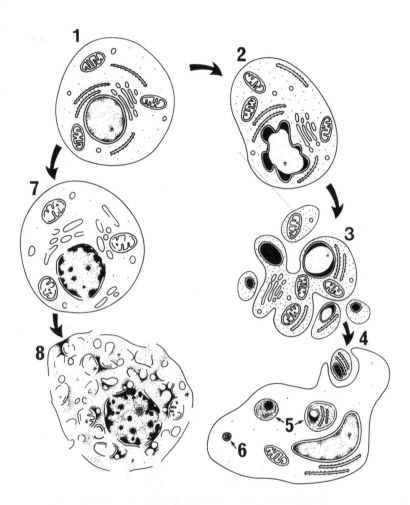

Figure 1.1 Sequence of ultrastructural changes in apoptosis (right) and necrosis (left). 1, normal cell. The earliest recognisable stage of apoptosis (2) is characterised by compaction and margination of nuclear chromatin, condensation of cytoplasm and variable convolution of nuclear and cell outlines. These changes are followed (3) by nuclear fragmentation and budding of the cell as a whole to produce membrane-bounded apoptotic bodies, which are phagocytosed (4) by nearby cells and degraded within lysosomes (5) to produce lysosomal residual bodies (6). In the irreversibly injured cell, signs of early necrosis (7) include irregular clumping of chromatin, swelling of all cytoplasmic compartments, the appearance of densities in the matrix of mitochondria and focal disruption of membranes. At a later stage (8), the cell degenerates to form a mass of debris.

migrated from small vessels during the inflammatory response. Large areas of confluent necrosis are replaced by granulation tissue with the eventual formation of a scar.

DNA Degradation Patterns

In 1980, Wyllie showed that the morphological changes of apoptosis are associated with distinctive, double-strand cleavage of nuclear DNA at the linker regions between nucleosomes. The resulting oligonucleosomal fragments can be demonstrated by agarose gel electrophoresis, where a so-called ladder develops (Wyllie, 1980; Cohen and Duke, 1984; Wyllie et al., 1984; Arends et al., 1990). Necrosis, in contrast, is accompanied by random DNA degradation and digestion of histone (Afanas'ev et al., 1986; Duvall and Wyllie, 1986), a diffuse smear being seen in gels (Kerr and Harmon, 1991). Moreover, the kinetics of the DNA changes differ in the two processes: cleavage is an early event in apoptosis, but is only extensive some hours after the onset of morphological degeneration in cells undergoing necrosis (Duvall and Wyllie, 1986; Collins et al., 1992).

The discovery of the distinctive pattern of DNA cleavage in apoptosis was of major importance for two reasons. Firstly, it provided a method for recognizing the process biochemically, which has since been used extensively. Secondly, it afforded a starting point for study of the mechanism of apoptosis.

The correlation between morphology and DNA degradation pattern in cell death outlined above has been shown to hold good in a great variety of circumstances. It may, however, not be invariable (Cohen and Duke, 1992; Collins et al., 1992; Ucker et al., 1992).

INCIDENCE OF APOPTOSIS

Cell Death in Normal Development

Programmed cell death occurring during the normal development of mammals displays the typical ultrastructural features of apoptosis. Well studied examples include the cell deletion that takes place in regressing interdigital webs (Kerr et al., 1987) and that involved in development of the gut mucosa (Harmon et al., 1984) and the retina (Penfold and Provis, 1986). In amphibia, the cell death responsible for involution of larval organs during metamorphosis is also morphologically apoptotic in type (Kerr et al., 1974; Ishizuya-Oka and Shimozawa, 1992). There is, however, uncertainty at present as to whether this developmental apoptosis in vertebrates is accompanied by the usual pattern of internucleosomal DNA cleavage (Lockshin and Zakeri, 1991). Developmental death in invertebrates undoubtedly exhibits some morphological and biochemical differences from classical apoptosis observed in higher animals (Robertson and Thomson, 1982; Lockshin and Zakeri, 1991).

Cell Death in Healthy Adult Animals

Apoptosis occurs at a low rate in most healthy adult mammalian tissues, where it complements mitosis in steady state kinetics (Kerr, 1969, 1971; Wyllie et al., 1973; Potten, 1977; Benedetti et al., 1988; Allan et al., 1992). There is growing evidence that the two processes are regulated by both circulating trophic factors and factors produced locally at the tissue level (Duke and Cohen, 1986; Lynch et al., 1986; Williams et al., 1990; Crompton, 1991; Her et al., 1991; Mangan and Wahl, 1991; Oberhammer et al., 1991, 1992; Rotello et al., 1991; Tai et al., 1991; Kelley et al., 1992; Rodriguez-Tarduchy et al., 1992), and the traditional emphasis placed on cell proliferation as the dominant parameter in cell population control is being tempered by a realization that cell deletion is also of cardinal importance (Raff, 1992).

Normal involution of endocrine-dependent tissues induced by changes in blood levels of trophic hormones is mediated by apoptosis (Sandow et al., 1979; Ferguson and Anderson, 1981; Walker et al., 1989), as are other normal involutional processes such as ovarian follicular atresia (O'Shea et al., 1978; Hughes and Gorospe, 1991; Tilly et al., 1991; Hurwitz and Adashi, 1992) and catagen regression of hair follicles (Weedon and Strutton, 1981), where the triggers to involution are poorly understood.

In the immune system, apoptosis of lymphocytes subserves a variety of essential regulatory functions (for reviews, see Cohen, 1991; Golstein et al., 1991). For example, there is evidence that it is involved in the selection of B-cells in lymphoid germinal centres (Liu et al., 1989) and in the deletion of autoreactive T cells in the thymus during the development of immune tolerance (Smith et al., 1989); the latter topic will be covered in detail in Chapter 18.

Megakaryocyte remnants undergo apoptosis after the release of platelets (Radley and Haller, 1983) and neutrophil leukocytes, which are terminally differentiated cells with a short life-span, are also deleted by this means (Savill et al., 1989). Terminal differentiation of epidermal keratinocytes to form squames is associated with morphological and biochemical changes in their nuclei that are similar to those found in apoptosis (McCall and Cohen, 1991). Here the apoptotic pathway has apparently been modified during evolution, with the cell flattening-out prior to being shed rather than rounding-up and budding to form discrete bodies.

Cell Death Occurring During Pathological Atrophy and Regression of Hyperplasia

Some types of pathological atrophy merely represent an exaggeration of the processes involved in physiological involution of tissues and here, as would be expected, apoptosis is involved in the tissue shrinkage. Thus atrophy of endocrine-dependent organs following pathological withdrawal of trophic hormonal stimulation is accompanied by a massive wave of apoptosis, an occurrence that has, for example, been observed in the prostate after castration (Kerr and Searle, 1973; Kyprianou and Isaacs, 1988) and in the adrenal cortex after suppression of ACTH secretion by glucocorticoid administration (Wyllie et al., 1973). Induction of extensive thymocyte apoptosis by artificial exposure to high levels of glucocorticoid (Wyllie, 1980) is another example of exaggeration of a phenomenon that occurs under physiological conditions. Apoptosis is, however, also responsible for certain other types of atrophy that have no physiological counterpart; these include atrophy of hepatic (Kerr, 1971) and renal (Gobé et al., 1990) tissues subjected to mild ischaemia, of the pancreas after duct ligation (Walker, 1987) and of the kidney following obstruction of the ureter (Gobé and Axelsen, 1987). The way in which apoptosis is initiated in these latter situations is unknown. Examples of regression of hyperplasia in which apoptosis is implicated are restoration of increased liver size to normal after cessation of administration of cyproterone acetate (Bursch et al., 1985) or lead (Columbano et al., 1985) and disappearance of abnormally proliferated small bile ducts after relief of biliary obstruction (Bhathal and Gall, 1985).

Apoptosis Occurring in Certain Pathological Contexts Viewed as Altruistic Cell Suicide

The apparently paradoxical induction of "normal cell death" by injurious stimuli may, in some cases, be explained teleologically by regarding the death as a homeostatic mechanism that effects the selective deletion of cells whose survival would prejudice the welfare of the host. Thus it can be argued that apoptosis of cells with significant, unrepaired genetic damage and of cells infected by a variety of viruses represents altruistic cell suicide.

Moderate doses of radiation induce extensive apoptosis in rapidly proliferating cell populations (Hugon and Borgers, 1966; Kerr and Searle, 1980; Gobé et al., 1988; Stephens et al., 1991) and in lymphoid organs (Trowell, 1966; Sellins and Cohen, 1987; Yamada and Ohyama, 1988). Such massive cell death is, of course, not beneficial to an animal, and may indeed prove fatal. However, during evolution, animals would rarely have encountered radiation in doses sufficient to produce extensive apoptosis. The selective elimination of the occasional cell with unrepaired radiation-induced DNA damage, on the other hand, would clearly have been advantageous (Searle et al., 1975; Potten, 1977). In rapidly renewing tissues, the genetic abnormality might otherwise become fixed and magnified. The peculiar sensitivity of non-proliferating lymphocytes to the induction of apoptosis by radiation might be explained teleologically by the potential danger to the host of clonal expansion of cells with mutations in the genes coding for their antigen receptors, which might render them autoreactive. As Cohen and Duke (1992) have pointed out, "better dead than wrong" may be a compelling maxim governing the behaviour of lymphocytes. The molecular processes implicated in the induction of apoptosis by radiation are, however, still unknown.

A variety of cancer-chemotherapeutic agents also enhance apoptosis in spontaneously rapidly proliferating populations of cells (Searle et al., 1975; Eastman, 1990; Anilkumar et al., 1992). Whilst many of these agents are known to affect DNA in one way or another, to attempt an all-embracing teleological explanation of their apoptosis-inducing capacity along the lines outlined above for radiation would be simplistic. One of the current challenges confronting research on apoptosis is elucidation of the way in which these drugs trigger the process (Eastman, 1990; Dive and Hickman, 1991). The dramatically different sensitivities of cells at different locations in structured cellular hierarchies, such as exist in intestinal crypts, to various anticancer drugs (Ijiri and Potten, 1987) might provide clues to the mechanisms involved, and studies at the molecular level may prove rewarding.

Apoptosis induced by T cells (Battersby et al., 1974; Sanderson, 1976; Don et al., 1977; Sanderson and Glauert, 1977; Matter, 1979; Russell and Dobos, 1980; Duke et al., 1983; Ucker, 1987), K-cells (Sanderson and Thomas, 1977; Stacey et al., 1985) and NK-cells (Bishop and Whiting, 1983) can also be argued to fall within the category of altruistic cell suicide. Lymphocyte-induced cell death plays a critical role in the containment of many viral infections, the benefits to the host of such death being well recognized (Martz and Howell, 1989). Lymphocytes may have acquired the ability to initiate apoptosis during evolution, thus exploiting the existence of a mechanism for selectively deleting cells with minimal tissue disruption. It is interesting in this regard that cytotoxic T-cell induced apoptosis is not blocked by protein synthesis inhibition (Duke et al., 1983) or *bcl-2* expression (Vaux et al., 1992; for evidence that *bcl-2* is involved in regulating apoptosis, see Chapters 14 and 15). This suggests that T cells may plug-in down-stream in the apoptosis activation pathway. In some cases, however, apoptosis may be initiated intrinsically following viral infection, without involvement of the cellular immune system (Walker et al., 1988). At least one insect virus appears to have developed a means of overcoming this intrinsic activation by producing an apoptosis-inhibitory protein, which greatly increases virus yield (Clem et al., 1991).

The occurrence of apoptosis in viral infection may not, nevertheless, always represent a beneficial, protective response. There is evidence that depletion of T4 lymphocytes in human beings with HIV infection (Laurent-Crawford et al., 1991; Terai et al., 1991; Groux et al., 1992; Meyaard et al., 1992) and in cats with feline retrovirus infection (Rojko et al., 1992) is mediated by apoptosis. These may represent special cases in which the viruses have developed ways of initiating apoptosis of a specific cell type that is essential to cellular immune functions, thus enabling them to cripple the cellular immune defences of the body.

Apoptosis Produced by Other Forms of Injury

Apoptosis is induced by certain injurious agents that are not known to have a major effect on DNA and that are also capable of producing necrosis. Here the severity of the injury appears to determine the type of cell death that results, mild damage being followed by enhanced apoptosis and severe damage by confluent necrosis. Injury due to ischaemia (Kerr, 1971; Gobé et al., 1990), to hyperthermia (Harmon et al., 1990; Sellins and Cohen, 1991; Van Bruggen et al., 1991; Mosser and Martin, 1992; see also Chapter 25) and to a variety of toxins (Kerr, 1969; Reynolds et al., 1984; Ledda-Columbano et al., 1991) satisfies these criteria. A homeostatic role for apoptosis occurring under such circumstances is not clearly apparent and its biological significance is at present uncertain.

Apoptosis Occurring Spontaneously in Tumours

Apoptosis occurs in all malignant tumours and is often of major kinetic importance (Kerr and Searle, 1972; Sarraf and Bowen, 1988). Some of this spontaneous apoptosis is likely to be due to mild ischaemia (Moore, 1987), to cellular immune attack (Curson and Weedon, 1979) or to release of cytokines such as TNF (Bellomo et al., 1992; Wright et al., 1992) by infiltrating mononuclear cells. However, the amount of apoptosis in a tumour often appears to be just as much a characteristic of that tumour as the rate of mitosis. There is evidence that certain proto-oncogenes (Hockenbery et al., 1990, 1991; Askew et al., 1991; Strasser et al., 1991; Alnemri et al., 1992; Evan et al., 1992; Shi et al., 1992) and tumour suppressor genes (Yonish-Rouach et al., 1991; Shaw et al., 1992) are involved in regulating apoptosis, and it is possible that much of the enhanced apoptosis seen in tumours is dependent on genetic abnormalities intrinsic to the tumour cells themselves (Wyllie et al., 1987). Research in this field is described in several chapters.

PERSPECTIVES

Recent advances in knowledge of the biochemistry of apoptosis fall into two main categories. One type of study has been directed towards understanding the biochemical basis of the events observed morphologically such as the dramatic alteration in cell shape (Cotter et al., 1992; see also Chapter 13) and the recognition and phagocytosis of apoptotic bodies by adjacent normal cells (Duvall et al., 1985; Savill et al., 1990; Dini et al., 1992). Other studies have been concerned with its initiation and regulation. In both areas, much still remains to be done. For example, little is known about the processes responsible for the rapid condensation of the cell. As far as initiation is concerned, one of the most intriguing unanswered questions relates to the way in which extremely circumscribed DNA damage induced by radiation can trigger a cascade that results in cell deletion. Moreover, as was pointed out earlier, the molecular mechanisms involved in the induction of apoptosis by anti-cancer drugs should be a rewarding topic of investigation.

ACKNOWLEDGEMENTS

I am most grateful to Clay Winterford for drawing the diagram. Some of the research on which this review is based was supported by the Queensland Cancer Fund, the National Health and Medical Research Council of Australia and the University of Queensland.

REFERENCES

Afanas'ev, V. N., Korol', B. A., Mantsygin, Y. A., Nelipovich, P. A., Pechatnikov, V. A. and Umansky, S. R. (1986) Flow cytometry and biochemical analysis of DNA degradation characteristic of two types of cell death. *FEBS Letters*, **194**, 347–350.

Allan, D. J., Harmon, B. V. and Roberts, S. A. (1992) Spermatogonial apoptosis has three morphologically recognizable phases and shows no circadian rhythm during normal spermatogenesis in the rat. *Cell Proliferation*, **25**, 241–250.

Alnemri, E. S., Fernandes, T. F., Haldar, S., Croce, C. M. and Litwack, G. (1992) Involvement of bcl-2 in glucocorticoid-induced apoptosis of human pre-B leukaemias. *Cancer Research*, **52**, 491–495.

Anilkumar, T. V., Sarraf, C. E., Hunt, T. and Alison, M. R. (1992) The nature of cytotoxic drug-induced cell death in murine intestinal crypts. *British Journal of Cancer*, **65**, 552–558.

Arends, M. J., Morris, R. G. and Wyllie, A. H. (1990) Apoptosis: the role of the endonuclease. *American Journal of Pathology*, **136**, 593–608.

Arends, M. J. and Wyllie, A. H. (1991) Apoptosis: mechanisms and roles in pathology. *International Review of Experimental Pathology*,, **32**, 223–254.

Askew, D. S., Ashmun, R. A., Simmons, B. C. and Cleveland, J. L. (1991) Constitutive c-*myc* expression in an IL-3-dependent myeloid cell line suppresses cell cycle arrest and accelerates apoptosis. *Oncogene*, **6**, 1915–1922.

Battersby, C., Egerton, W. S., Balderson, G., Kerr, J.F. and Burnett, W. (1974) Another look at rejection in pig liver homografts. *Surgery*, **76**, 617–623.

Bellomo, G., Perotti, M., Taddei, F., Mirabelli, F., Finardi, G., Nicotera, P. and Orrenius, S. (1992) Tumor necrosis factor α induces apoptosis in mammary adenocarcinoma cells by an increase in intracellular free Ca^{2+} concentration and DNA fragmentation. *Cancer Research*, **52**, 1342–1346.

Benedetti, A., Jézéquel, A. M. and Orlandi, F. (1988) Preferential distribution of apoptotic bodies in acinar zone 3 of normal human and rat liver. *Journal of Hepatology*, **7**, 319–324.

Bhathal, P. S., and Gall, J. A. M. (1985) Deletion of hyperplastic biliary epithelial cells by apoptosis following removal of the proliferative stimulus. *Liver*, **5**, 311–325.

Bishop, C. J. and Whiting, V. A. (1983) The role of natural killer cells in the intravascular death of intravenously injected murine tumour cells. *British Journal of Cancer*, **48**, 441–444.

Bursch, W., Taper, H. S., Lauer, B. and Schulte-Hermann, R. (1985) Quantitative histological and histochemical studies on the occurrence and stages of controlled cell death (apoptosis) during regression of rat liver hyperplasia. *Virchows Archive of B. Cell Pathology*, **50**, 153–166.

Clem, R. J., Fechheimer, M. and Miller, L. K. (1991) Prevention of apoptosis by a baculovirus gene during infection of insect cells. *Science*, **254**, 1388–1390.

Cohen, J. J. (1991) Programmed cell death in the immune system. *Advances in Immunology*, **50**, 55–85.

Cohen, J. J. and Duke, R. C. (1984) Glucocorticoid activation of a calcium-dependent endonuclease in thymocyte nuclei leads to cell death. *Journal of Immunology*, **132**, 38–42.

Cohen, J. J. and Duke, R. C. (1992) Apoptosis and programmed cell death in immunity. *Annual Review of Immunology*, **10**, 267–293.

Collins, R. J., Harmon, B. V., Gobé, G. C. and Kerr, J.F.R. (1992) Internucleosomal DNA cleavage should not be the sole criterion for identifying apoptosis. *International Journal of Radiation Biology*, **61**, 451–453.

Columbano, A., Ledda-Columbano, G. M., Coni, P. P., Faa, G., Liguori, C., Santa Cruz, G. and Pani, P. (1985) Occurrence of cell death (apoptosis) during the involution of liver hyperplasia. *Laboratory Investigation*, **52**, 670–675.

Cotter, T. G., Lennon, S. V., Glynn, J. M. and Green, D. R. (1992) Microfilament-disrupting agents prevent the formation of apoptotic bodies in tumor cells undergoing apoptosis. *Cancer Research*, **52**, 997–1005.

Crompton, T. (1991) IL3-dependent cells die by apoptosis on removal of their growth factor. *Growth Factors*, **4**, 109–116.

Curson, C. and Weedon, D. (1979) Spontaneous regression in basal cell carcinomas. *Journal of Cutaneous Pathology*, **6**, 432–437.

Dini, L., Autuori, F., Lentini, A., Oliverio, S. and Piacentini, M. (1992) The clearance of apoptotic cells in the liver is mediated by the asialoglycoprotein receptor. *FEBS Letters*, **296**, 174–178.

Dive, C. and Hickman, J. A. (1991) Drug-target interactions: only the first step in the commitment to a programmed cell death. *British Journal of Cancer*, **64**, 192–196.

Don, M. M., Ablett, G., Bishop, C. J., Bundesen, P. G., Donald, K. J., Searle, J. and Kerr, J.F.R. (1977) Death of cells by apoptosis following attachment of specifically allergized lymphocytes *in vitro*. *Australian Journal of Experimental Biology and Medical Science*, **55**, 407–417.

Duke, R. C., Chernevak, R. and Cohen, J. J. (1983) Endogenous endonuclease-induced DNA fragmentation: an early event in cell-mediated cytolysis. *Proceedings of the National Academy of Science, USA*, **80**, 6361–6365.

Duke, R. C. and Cohen, J. J. (1986) IL-2 addiction: withdrawal of growth factor activates a suicide program in dependent T cells. *Lymphokine Research*, **5**, 289–299.

Duvall, E. and Wyllie, A. H. (1986) Death and the cell. *Immunology Today*, **7**, 115–119.

Duvall, E., Wyllie, A. H. and Morris, R.G. (1985) Macrophage recognition of cells undergoing programmed cell death (apoptosis). *Immunology*, **56**, 351–358.

Eastman, A. (1990) Activation of programmed cell death by anticancer agents: cisplatin as a model system. *Cancer Cells*, **2**, 275–280.

Evan, G. I., Wyllie, A. H., Gilbert, C. S., Littlewood, T. D., Land, H., Brooks, M., Waters, C. M., Penn, L. Z. and Hancock, D. (1992) Induction of apoptosis in fibroblasts by c-*myc* protein. *Cell*, **69**, 119–128.

Ferguson, D. J. P. and Anderson, T. J. (1981) Morphological evaluation of cell turnover in relation to the menstrual cycle in the "resting" human breast. *British Journal of Cancer*, **44**, 177–181.

Gobé, G. C. and Axelsen, R. A. (1987) Genesis of renal tubular atrophy in experimental hydronephrosis in the rat. Role of apoptosis. *Laboratory Investigation*, **56**, 273–281.

Gobé, G. C., Axelsen, R. A., Harmon, B. V. and Allan, D. J. (1988) Cell death by apoptosis following X-irradiation of the foetal and neonatal kidney. *International Journal of Radiation Biology*, **54**, 567–576.

Gobé, G. C., Axelsen, R. A. and Searle, J. W. (1990) Cellular events in experimental unilateral ischemic renal atrophy and in regeneration after contralateral nephrectomy. *Laboratory Investigation*, **63**, 770–779.

Golstein, P., Ojcius, D. M. and Young, D.-E. (1991) Cell death mechanisms and the immune system. *Immunological Reviews*, **121**, 29–65.

Groux, H., Torpier, G., Monté, D., Mouton, Y., Capron, A. and Ameisen, J. C. (1992) Activation-induced death by apoptosis in CD4$^+$ T cells from human immunodeficiency virus-infected asymptomatic individuals. *Journal of Experimental Medicine*, **175**, 331–340.

Harmon, B., Bell, L. and Williams, L. (1984) An ultrastructural study on the "meconium corpuscles" in rat foetal intestinal epithelium with particular reference to apoptosis. *Anatomy and Embryology*, **169**, 119–124.

Harmon, B. V., Corder, A. M., Collins, R. J., Gobé, G. C., Allen, J., Allan, D. J. and Kerr, J. F. R. (1990) Cell death induced in a murine mastocytoma by 42–47°C heating *in vitro*: evidence that the form of death changes from apoptosis to necrosis above a critical heat load. *International Journal of Radiation Biology*, **58**, 845–858.

Her, E., Frazer, J., Austen, K. F. and Owen, W. F. (1991) Eosinophil hematopoietins antagonize the programmed cell death of eosinophils. *Journal of Clinical Investigation*, **88**, 1982–1987.

Hockenbery, D., Nuñez, G., Milliman, C., Schreiber, R. D. and Korsmeyer, S. J. (1990) Bcl-2 is an inner mitochondrial membrane protein that blocks programmed cell death. *Nature*, **348**, 334–336.

Hockenbery, D. M., Zutter, M., Hickey, W., Nahm, M., and Korsmeyer, S. J. (1991) Bcl-2 protein is topographically restricted in tissues characterized by apoptotic cell death. *Proceedings of the National Academy of Science, USA*, **88**, 6961–6965.

Hughes, F. M. and Gorospe, W. C. (1991) Biochemical identification of apoptosis (programmed cell death) in granulosa cells: evidence for a potential mechanism underlying follicular atresia. *Endocrinology*, **129**, 2415–2422.

Hugon, J. and Borgers, M. (1966) Ultrastructural and cytochemical studies on karyolytic bodies in the epithelium of the duodenal crypts of whole body X-irradiated mice. *Laboratory Investigation*, **15**, 1528–1543.

Hurwitz, A. and Adashi, E. Y. (1992) Ovarian follicular atresia as an apoptotic process: a paradigm for programmed cell death in endocrine tissues. *Molecular and Cellular Endocrinology*, **84**, C19–C23.

Ijiri, K. and Potten, C. S. (1987) Cell death in cell hierarchies in adult mammalian tissues. In *Perspectives on Mammalian Cell Death*, edited by C. S. Potten, pp. 326–356. Oxford: Oxford University Press.

Ishizuya-Oka, A. and Shimozawa, A. (1992) Programmed cell death and heterolysis of larval epithelial cells by macrophage-like cells in the anuran small intestine *in vivo* and *in vitro*. *Journal of Morphology*, **213**, 185–195.

Kelley, L. L., Koury, M. J. and Bondurant, M. C. (1992) Regulation of programmed death in erythroid progenitor cells by erythropoietin: effects of calcium and of protein and RNA syntheses. *Journal of Cellular Physiology*, **151**, 487–496.

Kerr, J. F. R. (1969) An electron-microscope study of liver cell necrosis due to heliotrine. *Journal of Pathology*, **97**, 557–562.

Kerr, J. F. R. (1971) Shrinkage necrosis: a distinct mode of cellular death. *Journal of Pathology*, **105**, 13–20.

Kerr, J. F. R. and Harmon, B. V. (1991) Definition and incidence of apoptosis: an historical perspective. In *Apoptosis: The Molecular Basis of Cell Death*, edited by L. D. Tomei and F. O. Cope, pp. 5–29. New York: Cold Spring Harbor Laboratory Press.

Kerr, J. F. R., Harmon, B. and Searle, J. (1974) An electron-microscope study of cell deletion in the anuran tadpole tail during spontaneous metamorphosis with special reference to apoptosis of striated muscle fibres. *Journal of Cell Science*, **14**, 571–585.

Kerr, J. F. R. and Searle, J. (1972) A suggested explanation for the paradoxically slow growth rate of basal-cell carcinomas that contain numerous mitotic figures. *Journal of Pathology*, **107**, 41–44.

Kerr, J. F. R. and Searle, J. (1973) Deletion of cells by apoptosis during castration-induced involution of the rat prostate. *Virchows Archiv. B. Cell Pathology*, **13**, 87–102.

Kerr, J. F. R. and Searle, J. (1980) Apoptosis: its nature and kinetic role. In *Radiation Biology in Cancer Research*, edited by R. E. Meyn and H. R. Withers, pp. 367–384. New York: Raven Press.

Kerr, J. F. R., Searle, J., Harmon, B. V. and Bishop, C. J. (1987) Apoptosis. In *Perspectives on Mammalian Cell Death*, edited by C. S. Potten, pp. 93–128. Oxford: Oxford University Press.

Kerr, J. F. R., Wyllie, A. H. and Currie, A. R. (1972) Apoptosis: a basic biological phenomenon with wide-ranging implications in tissue kinetics. *British Journal of Cancer*, 26, 239–257.

Kyprianou, N. and Isaacs, J. T. (1988) Activation of programmed cell death in the rat ventral prostate after castration. *Endocrinology*, 122, 552–562.

Laurent-Crawford, A. G., Krust, B., Muller, S., Rivière, Y., Rey-Cuillé, M.-A., Béchet, J.-M., Montagnier, L. and Hovanessian, A. G. (1991) The cytopathic effect of HIV is associated with apoptosis. *Virology*, 185, 829–839.

Ledda-Columbano, G. M., Coni, P., Curto, M., Giacomini, L., Faa, G., Oliverio, S., Piacentini, M. and Columbano, A. (1991) Induction of two different modes of cell death, apoptosis and necrosis, in rat liver after a single dose of thioacetamide. *American Journal of Pathology*, 139, 1099–1109.

Liu, Y.-J., Joshua, D. E., Williams, G. T., Smith, C. A., Gordon, J. and MacLennan, I.C.M. (1989) Mechanism of antigen-driven selection in germinal centres. *Nature*, 342, 929–931.

Lockshin, R. A. and Zakeri, Z. (1991) Programmed cell death and apoptosis. In *Apoptosis: The Molecular Basis of Cell Death*, edited by L. D. Tomei and F. O. Cope, pp. 47–60. New York: Cold Spring Harbor Laboratory Press.

Lynch, M. P., Nawaz, S. and Gerschenson, L. E. (1986) Evidence for soluble factors regulating cell death and cell proliferation in primary cultures of rabbit endometrial cells grown on collagen. *Proceedings of the National Academy of Science, USA*, 83, 4784–4788.

Mangan, D. F. and Wahl, S. M. (1991) Differential regulation of human monocyte programmed cell death (apoptosis) by chemotactic factors and pro-inflammatory cytokines. *Journal of Immunology*, 147, 3408–3412.

Martz, E. and Howell, D. M. (1989) CTL: virus control cells first and cytolytic cells second? DNA fragmentation, apoptosis and the prelytic halt hypothesis. *Immunology Today*, 10, 79–86.

Matter, A. (1979) Microcinematographic and electron microscopic analysis of target cell lysis induced by cytotoxic T lymphocytes. *Immunology*, 36, 179–190.

McCall, C. A. and Cohen, J. J. (1991) Programmed cell death in terminally differentiating keratinocytes: role of endogenous endonuclease. *Journal of Investigative Dermatology*, 97, 111–114.

Meyaard, L., Otto, S. A., Jonker, R. R., Mijnster, M. J., Keet, R. P. M. and Miedema, F. (1992) Programmed death of T cells in HIV-1 infection. *Science*, 257, 217–219.

Moore, J. V. (1987) Death of cells and necrosis of tumours. In *Perspectives on Mammalian Cell Death*, edited by C. S. Potten, pp. 295–325. Oxford: Oxford University Press.

Mosser, D. D. and Martin, L. H. (1992) Induced thermotolerance to apoptosis in a human T lymphocyte cell line. *Journal of Cellular Physiology*, 151, 561–570.

Oberhammer, F., Bursch, W., Parzefall, W., Breit, P., Erber, E., Stadler, M. and Schulte-Hermann, R. (1991) Effect of transforming growth factor β on cell death of cultured rat hepatocytes. *Cancer Research*, 51, 2478–2485.

Oberhammer, F. A., Pavelka, M., Sharma, S., Tiefenbacher, R., Purchio, A. F., Bursch, W. and Schulte-Hermann, R. (1992) Induction of apoptosis in cultured hepatocytes and in regressing liver by transforming growth factor β 1. *Proceedings of the National Academy of Science, USA*, 89, 5408–5412.

O'Shea, J. D., Hay, M. F. and Cran, D. G. (1978) Ultrastructural changes in the theca interna during follicular atresia in sheep. *Journal of Reproductive Fertility*, 54, 183–187.

Penfold, P. L. and Provis, J. M. (1986) Cell death in the development of the human retina: phagocytosis of pyknotic and apoptotic bodies by retinal cells. *Graefe's Archive for Clinical and Experimental Ophthalmology*, 224, 549–553.

Potten, C. S. (1977) Extreme sensitivity of some intestinal crypt cells to X and γ irradiation. *Nature*, 269, 518–521.

Radley, J. M. and Haller, C. J. (1983) Fate of senescent megakaryocytes in the bone marrow. *British Journal of Haematology*, **53**, 277–287.

Raff, M. C. (1992) Social controls on cell survival and cell death. *Nature*, **356**, 397–400.

Reynolds, E. S., Kanz, M. F., Chieco, P. and Moslen, M.T. (1984) 1,1-dichloroethylene: an apoptotic hepatotoxin? *Environmental Health Perspectives*, **57**, 313–320.

Robertson, A. M. G. and Thomson, J. N. (1982) Morphology of programmed cell death in the ventral nerve cord of *Caenorhabditis elegans* larvae. *Journal of Embryology and Experimental Morphology*, **67**, 89–100.

Rodriguez-Tarduchy, G., Collins, M. K. L., García, I. and López-Rivas, A. (1992) Insulin-like growth factor-I inhibits apoptosis in IL-3-dependent hemopoietic cells. *Journal of Immunology*, **149**, 535–540.

Rojko, J. L., Fulton, R. M., Rezanka, L. J., Williams, L. L., Copelan, E., Cheney, C. M., Reichel, G. S., Neil, J. C., Mathes, L. E., Fisher, T. G. and Cloyd, M. W. (1992) Lymphocytotoxic strains of feline leukemia virus induce apoptosis in feline T4-thymic lymphoma cells. *Laboratory Investigation*, **66**, 418–426.

Rotello, R. J., Lieberman, R. C., Purchio, A. F. and Gerschenson, L. E. (1991) Coordinated regulation of apoptosis and cell proliferation by transforming growth factor β 1 in cultured uterine epithelial cells. *Proceedings of the National Academy of Science, USA*, **88**, 3412–3415.

Russell, J. H. and Dobos, C. B. (1980) Mechanisms of immune lysis II. CTL-induced nuclear disintegration of the target begins within minutes of cell contact. *Journal of Immunology*, **125**, 1256–1261.

Sanderson, C. J. (1976) The mechanism of T cell mediated cytotoxicity. II. Morphological studies of cell death by time-lapse micro-cinematography. *Proceedings of the Royal Society of London. Series B: Biological Sciences*, **192**, 241–255.

Sanderson, C. J. and Glauert, A. M. (1977) The mechanism of T cell mediated cytotoxicity V. Morphological studies by electron microscopy. *Proceedings of the Royal Society of London. Series B: Biological Sciences*, **198**, 315–323.

Sanderson, C. J. and Thomas, J. A. (1977) The mechanism of K cell (antibody-dependent) cell mediated cytotoxicity. II. Characteristics of the effector cell and morphological changes in the target cell. *Proceedings of the Royal Society of London. Series B: Biological Sciences*, 197, 417–424.

Sandow, B. A., West, N. B., Norman, R. L. and Brenner, R. M. (1979) Hormonal control of apoptosis in hamster uterine luminal epithelium. *American Journal of Anatomy*, **156**, 15–36.

Sarraf, C. E. and Bowen, I. D. (1988) Proportions of mitotic and apoptotic cells in a range of untreated experimental tumours. *Cell and Tissue Kinetics*, **21**, 45–49.

Savill, J., Dransfield, I., Hogg, N. and Haslett, C. (1990) Vitronectin receptor-mediated phagocytosis of cells undergoing apoptosis. *Nature*, **343**, 170–173.

Savill, J. S., Wyllie, A. H., Henson, J. E., Walport, M. J., Henson, P. M. and Haslett, C. (1989) Macrophage phagocytosis of aging neutrophils in inflammation. Programmed cell death in the neutrophil leads to its recognition by macrophages. *Journal of Clinical Investigation*, **83**, 865–875.

Searle, J., Kerr, J. F. R. and Bishop, C. J. (1982) Necrosis and apoptosis: distinct modes of cell death with fundamentally different significance. *Pathology Annual*, **17**(2), 229–259.

Searle, J., Lawson, T. A., Abbott, P. J., Harmon, B. and Kerr, J. F. R. (1975) An electron-microscope study of the mode of cell death induced by cancer-chemotherapeutic agents in populations of proliferating normal and neoplastic cells. *Journal of Pathology*, **116**, 129–138.

Sellins, K. S. and Cohen, J. J. (1987) Gene induction by γ-irradiation leads to DNA fragmentation in lymphocytes. *Journal of Immunology*, **139**, 3199–3206.

Sellins, K. S. and Cohen, J. J. (1991) Hyperthermia induces apoptosis in thymocytes. *Radiation Research*, **126**, 88–95.

Shaw, P., Bovey, R., Tardy, S., Sahli, R., Sordat, B. and Costa, J. (1992) Induction of apoptosis by wild-type p53 in a human colon tumor-derived cell line. *Proceedings of the National Academy of Science, USA*, **89**, 4495–4499.

Shi, Y., Glynn, J. M., Guilbert, L. J., Cotter, T. G., Bissonnette, R. P. and Green, D. R. (1992) Role for c-*myc* in activation-induced apoptotic cell death in T cell hybridomas. *Science*, **257**, 212–214.

Smith, C. A., Williams, G. T., Kingston, R., Jenkinson, E. J. and Owen, J. J. T. (1989) Antibodies to CD3/T-cell receptor complex induce death by apoptosis in immature T cells in thymic cultures. *Nature*, **337**, 181–184.

Stacey, N. H., Bishop, C. J., Halliday, J. W., Halliday, W. J., Cooksley, W. G. E., Powell, L. W. and Kerr, J. F. R. (1985) Apoptosis as the mode of cell death in antibody-dependent lymphocytotoxicity. *Journal of Cell Science*, **74**, 169–179.

Stephens, L. C., Ang, K. K., Schultheiss, T. E., Milas, L. and Meyn, R. E. (1991) Apoptosis in irradiated murine tumours. *Radiation Research*, **127**, 308–316.

Strasser, A., Harris, A. W. and Cory, S. (1991) *bcl-2* transgene inhibits T cell death and perturbs thymic self-censorship. *Cell*, **67**, 889–899.

Tai, P.-C., Sun, L. and Spry, C. J. F. (1991) Effects of IL-5, granulocyte/macrophage colony-stimulating factor (GM-CSF) and IL-3 on the survival of human blood eosinophils *in vitro*. *Clinical and Experimental Immunology*, **85**, 312–316.

Terai, C., Kornbluth, R. S., Pauza, C. D., Richman, D. D. and Carson, D. A. (1991) Apoptosis as a mechanism of cell death in cultured T lymphoblasts acutely infected with HIV-1. *Journal of Clinical Investigation*, **87**, 1710–1715.

Tilly, J. L., Kowalski, K. I., Johnson, A. L. and Hsueh, A. J .W. (1991) Involvement of apoptosis in ovarian follicular atresia and postovulatory regression. *Endocrinology*, **129**, 2799–2801.

Trowell, O. A. (1966) Ultrastructural changes in lymphocytes exposed to noxious agents *in vitro*. *Quarterly Journal of Experimental Physiology*, **51**, 207–220.

Ucker, D. S. (1987) Cytotoxic T lymphocytes and glucocorticoids activate an endogenous suicide process in target cells. *Nature*, **327**, 62–64.

Ucker, D. S., Obermiller, P. S., Eckhart, W., Apgar, J. R., Berger, N. A. and Meyers, J. (1992) Genome digestion is a dispensable consequence of physiological cell death mediated by cytotoxic T lymphocytes. *Molecular Cell Biology*, **12**, 3060–3069.

Van Bruggen, I., Robertson, T. A. and Papadimitriou, J. M. (1991) The effect of mild hyperthermia on the morphology and function of murine resident peritoneal macrophages. *Experimental and Molecular Pathology*, **55**, 119–134.

Vaux, D. L., Aguila, H. L. and Weissman, I. L. (1992) *bcl-2* prevents death of factor-deprived cells but fails to prevent apoptosis in targets of cell mediated killing. *International Immunology*, **4**, 821–824.

Walker, N. I. (1987) Ultrastructure of the rat pancreas after experimental duct ligation I. The role of apoptosis and intraepithelial macrophages in acinar cell deletion. *American Journal of Pathology*, **126**, 439–451.

Walker, N. I., Bennett, R. E. and Kerr, J. F. R. (1989) Cell death by apoptosis during involution of the lactating breast in mice and rats. *American Journal of Anatomy*, **185**, 19–32.

Walker, N. I., Harmon, B. V., Gobé, G. C. and Kerr, J. F. R. (1988) Patterns of cell death. *Methods and Achievements in Experimental Pathology*, **13**, 18–54.

Weedon, D. and Strutton, G. (1981) Apoptosis as the mechanism of the involution of hair follicles in catagen transformation. *Acta Dermato-venereologica*, **61**, 335–339.

Williams, G. T., Smith, C. A., Spooncer, E., Dexter, T. M., and Taylor, D. R. (1990) Haemopoietic colony stimulating factors promote cell survival by suppressing apoptosis. *Nature*, **343**, 76–79.

Wright, S. C., Kumar, P., Tam, A. W., Shen, N., Varma, M. and Larrick, J. W. (1992) Apoptosis and DNA fragmentation precede TNF-induced cytolysis in U937 cells. *Journal of Cellular Biochemistry*, **48**, 344–355.

Wyllie, A. H. (1980) Glucocorticoid-induced thymocyte apoptosis is associated with endogenous endonuclease activation. *Nature*, **284**, 555–556.

Wyllie, A. H. (1981) Cell death: a new classification separating apoptosis from necrosis. In *Cell Death in Biology and Pathology*, edited by I.D. Bowen and R.A. Lockshin, pp. 9–34. London: Chapman and Hall.

Wyllie, A. H. (1987) Cell death. *International Review of Cytology, Supplement* **17**, 755–785.

Wyllie, A. H., Kerr, J. F. R. and Currie, A. R. (1980) Cell death: the significance of apoptosis. *International Review of Cytology*, **68**, 251–306.

Wyllie, A. H., Kerr, J. F. R., Macaskill, I. A. M. and Currie, A. R. (1973) Adrenocortical cell deletion: the role of ACTH. *Journal of Pathology*, **111**, 85–94.

Wyllie, A. H., Morris, R. G., Smith, A. L. and Dunlop, D. (1984) Chromatin cleavage in apoptosis: association with condensed chromatin morphology and dependence on macromolecular synthesis. *Journal of Pathology*, **142**, 67–77.

Wyllie, A. H., Rose, K. A., Morris, R. G., Steel, C. M., Foster, E. and Spandidos, D. A. (1987) Rodent fibroblast tumours expressing human *myc* and *ras* genes: growth, metastasis and endogenous oncogene expression. *British Journal of Cancer*, **56**, 251–259.

Yamada, T. and Ohyama, H. (1988) Radiation-induced interphase death of rat thymocytes is internally programmed (apoptosis). *International Journal of Radiation Biology*, **53**, 65–75.

Yonish-Rouach, E., Resnitzky, D., Lotem, J., Sachs, L., Kimchi, A. and Oren, M. (1991) Wild-type p53 induces apoptosis of myeloid leukaemic cells that is inhibited by interleukin-6. *Nature*, **352**, 345–347.

BIOCHEMICAL MECHANISMS

Chapter 2

SIGNAL TRANSDUCTION IN THYMOCYTE APOPTOSIS

David J. McConkey[1], Sten Orrenius[2] and Mikael Jondal[3]
[1] *Laboratory of Immunobiology, Dana-Farber Cancer Institute and Department of Pathology, Harvard Medical School, Boston, MA 02115, USA*
Department of [2]*Toxicology and* [3]*Immunology, Karolinska Institutet, Box 60400, S-104 01 Stockholm, Sweden*

INTRODUCTION

One of the key features of apoptosis that distinguishes it from accidental cell death (necrosis) is that it is extrinsically and intrinsically regulated. An intense area of investigation is therefore concerned with the identification and characterization of receptor-mediated signalling mechanisms that control the process in various tissues. Because immature thymic lymphocytes can be readily induced to undergo apoptosis in response to diverse stimuli both *in vitro* and *in vivo*, they remain one of the most useful model systems for studying the biochemical mechanisms underlying apoptosis. In addition, recent work has shown that apoptosis mediates cell death during the critical process of negative selection of autoreactive progenitor T cells within the thymus, and understanding how signal transduction controls this procsss is therefore of relevance to elucidating the mechanisms controlling T-cell tolerance and the prevention of autoimmunity. Several signalling pathways have now been identified that positively or negatively regulate thymocyte apoptosis. These mechanisms will be discussed, particularly with regard to how they may relate to other processes such as proliferation and proliferative nonresponsiveness (anergy).

SIGNAL TRANSDUCTION IN THYMOCYTE APOPTOSIS

Surface receptors function by conveying the information of extracellular ligand binding to the intracellular millieu. This is accomplished by the activation of signalling cascades that involve activation of protein kinases and phosphatases and production of bioactive small signalling molecules known as "second messengers". The earliest of these mechanisms to be described in detail involve production of cAMP and stimulation of polyphosphoinositide hydrolysis, mediated by the G protein-coupled activation of adenylate cyclase and phospholipase C, respectively. Although the focus of these earlier efforts was concerned with how these events mediated activation events such as proliferation and metabolic modulation, evidence supporting their involvement in control of apoptosis is now emerging. In particular, thymocyte apoptosis is regulated by these signal transduction systems in a fashion that suggests tight coupling to regulation of cellular differentiation and proliferation.

Role of Ca^{2+}

The involvement of Ca^{2+} in the control of thymocyte apoptosis was first suggested directly by results obtained by Kaiser and Edelman, who showed that glucocorticoid-induced cytolysis was associated with enhanced Ca^{2+} influx (Kaiser and Edelman, 1977) and could be mimicked by incubating the cells with Ca^{2+} ionophores (Kaiser and Edelman, 1978). Following the observation by Wyllie (1980) that thymocyte apoptosis is associated with endogenous endonuclease activation, further evidence for a role for Ca^{2+} was derived from characterization of the candidate enzymatic activity. An endogenous Ca^{2+}/Mg^{2+} dependent endonuclease found originally by Hewish and Burgoyne (1973) and Vanderbilt and colleagues (Vanderbilt et al., 1982) to be constitutive in nuclei obtained from thymocytes, liver, and various other tissues was implicated in glucocorticoid-induced DNA fragmentation in thymocytes by Cohen and Duke (1984), and Wyllie and colleagues (Wyllie et al., 1984) showed that it could be directly activated by Ca^{2+} ionophore treatment. Several groups are now in the process of purifying candidate endonuclease(s). Most notably, Gaido and Cidlowski (1990) have reported the purification and biochemical characterization of an 18 kDa endonuclease from thymocyte nuclei that exhibits the Ca^{2+} dependence and inhibitors sensitivity characteristic of the endonuclease responsible for DNA fragmentation in intact apoptotic cells.

Elevations of the cytosolic Ca^{2+} level are involved in the effects of a large number of hormonal and other stimuli (Berridge and Irvine, 1984) and commonly mediate cytotoxicity (Orrenius et al., 1989; Trump and Berezsky, 1992). We found that cytosolic Ca^{2+} increases precede endonuclease activation in thymocytes exposed to Ca^{2+} ionophores, glucocorticoid hormones, antibodies to the T-cell receptor, and the environmental contaminant dioxin; blocking these increases with extracellular Ca^{2+} chelation or intracellular Ca^{2+} buffering prevented both DNA fragmentation and cell death (reviewed in McConkey et al., 1990c). Interestingly, the level of the Ca^{2+}

increase induced by Ca^{2+} ionophore treatment appeared to be critical in determining whether cell death involves DNA fragmentation characteristic of apoptosis. Concentrations of the agent that induced physiological-range increases stimulated endonuclease activity, whereas higher levels actually inhibited the response (McConkey et al., 1989b). Together these results indicate that moderate elevations of the cytosolic Ca^{2+} concentration represent an important stimulatory mechanism for thymocyte apoptosis.

The induction of apoptosis by T-cell receptor stimulation is perhaps the most interesting model system of programmed cell death in thymocytes, since the response likely mimics the process of TCR-mediated negative selection. The observation that antibodies to the T-cell receptor induce apoptosis in thymocytes was first reported by Smith and colleagues (1989), and more recent work by Murphy and coworkers (1990) with TCR transgenic mice has confirmed that specific antigenic peptide induces the response. A role for cytosolic Ca^{2+} increases in antigen-induced thymocyte apoptosis in a similar transgenic mouse model has been suggested by Nakayama et al. (1992), who reported elevated cytosolic Ca^{2+} levels in transgenic thymocytes exposed to a negatively selecting environment *in vivo*; the elevated levels were restricted to the subpopulation of cells known to be the target of negative selection.

Role of Protein Kinase C

The breakdown of phosphatidylinositol trisphosphate by phospholipase C results in the production of diacylglycerol, a second messenger cofactor that mediates activation of the serine/threonine protein kinase, protein kinase C (Nishizuka, 1984). Among its important substrates are the protein components of the AP-1 transcriptional regulatory complex, *fos* and *jun*, and phosphorylation leads to alterations in their activities by a number of different mechanisms (Hunter and Karin, 1992). Activation of PKC appears to be critically involved in the stimulation of resting human T lymphocytes, in that TCR triggering does not lead to activation resulting in production of the autocrine growth factor interleukin-2 (IL-2) and subsequent proliferation in the absence of accessory cell-derived co-stimuli, and actually leads to a state of prolonged nonresponsiveness known as anergy (Schwartz, 1990). This co-stimulus can be substituted with phorbol esters (Crabtree, 1989), potent and specific activators of PKC. Indeed, T-cell activation can be achieved by combined treatment with calcium ionophores and phorbol esters, suggesting that elevated calcium plus PKC activation are sufficient to induce all of the subsequent biochemical and molecular events involved in TCR-mediated T-cell activation. This simple interpretation is supported by the molecular requirements for activation of the IL-2 promoter leading to gene induction (Crabtree, 1989), and that inefficient PKC activation is involved in anergy is suggested by recent work showing that anergy is due to a defect in activation of AP-1 (Kang et al., 1992).

The observation that calcium ionophores induce apoptosis in thymocytes is particularly intriguing in light of what is known about the parallel regulation of proliferation and anergy in mature T cells. We and others have found that phorbol esters are capable of blocking the effects of various apoptosis stimuli in thymocytes via a mechanism that involves PKC activation (McConkey et al., 1989a; Kizaki et al., 1989). In addition, we have shown that interleukin 1, a well known mitogenic agent for thymocytes, is capable of blocking apoptosis via a mechanism that involves enhanced diacylglycerol production and subsequent PKC activation (McConkey et al., 1990d). Thus, the signalling requirements for apoptosis in thymocytes appear similar to the mechanisms underlying anergy in mature T cells. Intriguingly, incubation of thymocytes with phorbol esters alone is capable of inducing apoptosis via a mechanism that also appears to involve PKC activation (Kizaki et al., 1989). The latter is characterized by a prolonged lag period between treatment and endonuclease activation (6 h) during which the process is reversible. We are currently investigating whether or not the known capacity of phorbol esters to downregulate PKC protein levels is involved in this response.

Role of cAMP

Early work suggested that there might be a link between elevations of cAMP and programmed cell death. Pharmacologic agents that elevate cAMP were reported to induce cell death in tumour cell lines *in vitro* (Basile et al., 1973; Bourgeous, 1986), and Pratt and Martin (1975) noted that increases in cAMP accompany cell death in the secondary palatal epithelium during fusion of the palate. We became interested in the possibility that cAMP regulates apoptosis because of work suggesting that elevations of cAMP induce anergy in mature T cells. It was subsequently found that agents that elevate cAMP levels stimulate endogenous endonuclease activation and cell death characteristic of apoptosis in thymocytes (McConkey et al., 1990b; Kizaki et al., 1990). Included among these agents were E-series prostaglandins, which had been shown previously to elevate cAMP in thymocytes and represent a physiologically relevant triggering mechanism. More recently we have shown that the adenosine analog NECA (5'-(N-ethyl)-carboxamidoadenosine), which also elevates cAMP in thymocytes, is capable of inducing deletion of thymocytes *in vivo* (manuscript in preparation).

Role of Tyrosine Kinase Activation

It has been well over a decade since the original recognition that protein tyrosine phosphorylation plays an important role in oncogenesis, and it is now clear that tyrosine kinases are overrepresented relative to other proteins among known oncogenes. The development of monoclonal anti-phosphotyrosine antibodies have now allowed investigators to more easily investigate how tyrosine kinase activation is involved in signal transduction, and it is now clear that the earliest intracellular effects of the majority of growth regulatory factors involve increased substrate tyrosine phosphorylation. In T cells the critical role tyrosine kinase activation plays in signal transduction has been demonstrated in recent work showing that tyrosine kinase inhibitors block TCR-mediated T-cell activation (Mustelin et al., 1990).

Preliminary evidence supports a role for tyrosine kinase regulation of apoptosis. Julius and coworkers (1990) have shown that stimulation via CD4 sensitizes splenic T cells to TCR-mediated apoptosis. We have recently found that antibodies to either CD4 or CD8 potentiate endonuclease activation in response to TCR triggering in thymocytes, both *in vitro* and *in vivo* (Figure 2.1). The mechanism involved prolongation and potentiation of the cytosolic Ca^{2+} increase observed *in vitro*, and low concentrations of the tyrosine kinase inhibitors genistein and tyrphostin partially blocked the effect. Because both CD4 and CD8 are physically associated with the protein tyrosine kinase $p56^{lck}$, we investigated whether changes in intracellular substrate tyrosine phosphorylation were involved. Anti-phosphotyrosine Western blotting revealed potentiated tyrosine phosphorylation of substrates at 120, 56, and 40 kDa in cells treated with anti-CD3 plus anti-CD4 or anti-CD8 antibodies. Further characterization of these proteins indicated that the 56 kDa phosphoprotein was the tyrosine kinase $p56^{lck}$, and preliminary results indicate that the 120 kDa phosphoprotein is *ras*-GTPase activating protein, otherwise known as *ras*-GAP. Because the latter is widely expressed and has been implicated in signal transduction in diverse tissues, it is possible that GAP is involved in regulation of apoptosis in other cell types. In addition, we observed rapid dephosphorylation of the 21 kDa tyrosine phosphorylated form of CD3 zeta following anti-CD3 antibody treatment. This effect had not been observed previously, although the observation that CD3 zeta is constitutively tyrosine phosphorylated in thymocytes has been previously documented (Nakayama et al., 1989). Although it appears that tyrosine kinase activation can potentiate TCR-mediated thymocyte apoptosis, it is not apparent that it is absolutely required for the response. In a recent study by Nakayama and Loh (1992) experiments were conducted to determine the effects of genistein and another specfic tyrosine kinase inhibitor, herbimycin A, on antigen-induced apoptosis in TCR transgenic thymocytes. Although both inhibitors blocked thymidine incorporation in a dose-dependent fashion, and optimal doses blocked *in vitro* activation of $p56^{lck}$, no significant inhibition of TCR-induced cell loss or DNA fragmentation was observed. Therefore, it is conceivable that the primary signalling pathway responsible for TCR-mediated thymocyte apoptosis does not require trigger-induced tyrosine kinase activation, although the pathway may be potentiated by activation of $p56^{lck}$. It is possible that the constitutive phosphorylation of CD3 zeta is symptomatic of *in situ* tyrosine kinase activation that may obviate the need for additional tyrosine phosphorylation events typically induced following TCR triggering and required for T-cell activation of mature cells. However, a potential caveat in the interpretation of these findings is that moderate doses of tyrosine kinase inhibitors can by themselves induce DNA fragmentation in thymocytes (D. J. McConkey, unpublished observations).

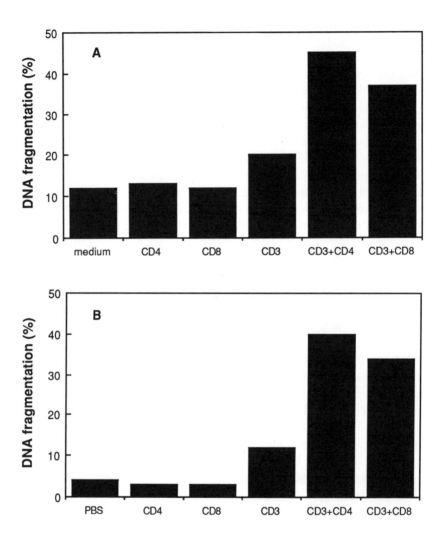

Figure 2.1 Potentiation of anti-CD3-induced thymocyte apoptosis by anti-CD4 or anti-CD8 antibodies. A. Thymocyte suspensions were treated with 10 μg anti-CD3 antibody (2C11) in the absence or presence of 10 μg Gk1.5 (anti-CD4) or 53.6 (anti-CD8), and DNA fragmentation was determined after 5 h as described previously (Wyllie, 1980). Results are mean values from 3 separate experiments. B. Balb/c mice were injected i.p. with PBS or 100 μg 2C11 in the absence or presence of 100 μg Gk1.5 or 53.6. Thymi were harvested 6 h later and DNA fragmentation was determined. Results are from one experiment that were typical of 3 replicates.

SIGNALLING CROSSTALK AND CONTROL OF APOPTOSIS

The concept that signal transduction pathways rarely act in isolation from one another is becoming more apparent, and the interaction of signalling pathways with each other now referred to as "crosstalk". The inhibition of Ca^{2+}-mediated apoptosis by PKC activation represents one example of how crosstalk regulates the process. In our most recent work we have investigated the possibility that crosstalk is a general theme in the regulation of thymocyte apoptosis.

cAMP and the Glucocorticoid Response

Previous work has shown that elevations of cAMP potentiate steroid receptor-mediated functional responses. Groul, Bourgeois and coworkers (1986, 1989) demonstrated that cAMP potentiates glucocorticoid-induced cell death in lymphoid cells via a mechanism that involves upregulation of glucocorticoid receptor (GR) function. The laboratory has also shown that cAMP can potentiate glucocorticoid responses at the level of transcription (Harrigan et al., 1989). Even more striking are the observations of O'Malley and colleagues suggesting that elevations of cAMP in the absence of specific ligand can stimulate steroid receptor-mediated transcription (Power et al., 1991). The biochemical mechanisms underlying these interactions are unknown at present.

With these observations in mind, we investigated the possibility that cAMP could amplify the glucocorticoid response in thymocytes. When cell suspensions were treated with cAMP plus glucocorticoid in vitro, dramatic potentiation of glucocorticoid-induced Ca^{2+} increases, DNA fragmentation and cell death were observed (Figure 2.2). The effects were linked to increased cellular binding of hormone but not to increases in glucocorticoid receptor protein levels; in fact, combined treatment actually led to a decrease in GR protein levels. Because recent work has shown that the GR is a phosphoprotein, we investigated the possibility that cAMP-mediated changes in GR phosphorylation were involved in the response. Treatment with either cAMP or glucocorticoid alone resulted in early (by 1 h) GR phosphorylation that was substantially increased in the cells incubated with both cAMP and glucocorticoid simultaneously. Future efforts will be aimed at determining how phosphorylation affects GR function.

We have obtained additional evidence for crosstalk between cAMP and glucocorticoids in promoting thymocyte apoptosis in vivo using the nonhydrolyzable adenosine analog NECA. The work of Kizaki and colleagues has shown that adenosine analogs stimulate DNA fragmentation in isolated thymocytes via a mechanism that involves production of cAMP (Kizaki et al., 1990). Intraperitoneal injection of NECA resulted in depletion of steroid-sensitive $CD4^+CD8^+$ thymocytes, suggesting that cAMP is sufficient to induce apoptosis by itself in vivo. However, suboptimal doses of NECA that failed to induce cell loss were capable of potentiating the effects of low doses of

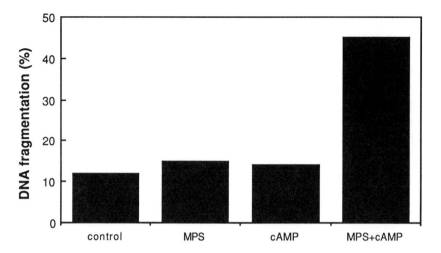

Figure 2.2 cAMP potentiates glucocorticoid-induced DNA fragmentation in thymocytes. Cell suspensions were incubated for 5 h with 10 nM methylprednisolone, 250 μM dibutyryl cAMP, or both, and DNA fragmentation was determined as described previously (Wyllie, 1980). Results are mean values from 5 separate experiments.

glucocorticoids. Thus, cAMP may regulate steroid receptor function *in vivo*, although indirect systemic effects of NECA may also be involved.

Effect of Steroid Receptor Antagonists on Thymocyte Apoptosis *In Vivo*

The observation that certain steroid receptors can be activated in the absence of ligand by cAMP opens the question of whether they are more generally involved in the responses to surface receptor signalling than has been thought previously. To investigate this possibility in terms of thymocyte apoptosis, we have designed experiments to test the effects of the steroid receptor antagonist, RU-486, on NECA and anti-CD3-induced apoptosis in murine thymocytes *in vivo*. Preliminary experiments confirmed that RU-486 blocks glucocorticoid-induced loss of thymocytes. Short-term (less than two days) treatment with the inhibitor alone did not appear to adversely affect thymocyte subpopulation phenotype as measured by fluorescence-activated cell sorting (FACS) or overall cell numbers; if anything, a modest increase in overall cell numbers was observed. As reported previously, injection of microgram quantities of the anti-CD3 antibody 2C11 led to a dramatic reduction in the number of $CD4^+CD8^+$ immature thymocytes after overnight exposure. In contrast, coinjection of RU-486 with 2C11 led to a total inhibition of cell death. Moreover, RU-486 completely blocked cell loss due to injection of the adenosine compound NECA (Table 2.1). At least two models could be used to explain these effects. In one, RU-486 blocks TCR-or cAMP-mediated potentiation of the effects of subthreshold levels of glucocorticoid hormones on thymocytes (a ligand-dependent response). In the other,

Table 2.1 Inhibition of thymocyte cell loss by RU-486 *in vivo*.[a]

Treatment	Mean surviving cells ($\times 10^6$)	
	Control	+RU-486
Control	200	201
anti-CD3	46	136
NECA	85	242

a. RU-486 in combination with 50 µg anti-CD3 monoclonal antibody 2C11 or 100 µl of a 200 µM solution of NECA in PBS. Thymus glands were harvested 24 h later and cell numbers determined. Results represent mean values from 3–5 separate experiments.

ligand-independent activation of the glucocorticoid receptor occurs in response to TCR or adenosine receptor triggering, perhaps via a mechanism that involves GR phosphorylation. We are currently analyzing the effects of RU-486 *in vitro* to provide evidence for one or the other scenario. At any rate, the data indicate that any physiological model for thymocyte apoptosis *in vivo* must account for crosstalk between several receptor-mediated signalling pathways.

FUTURE DIRECTIONS

Despite recent advances, there is still much to be learned about the manner in which signal transduction pathways influence apoptosis in thymocytes and other cell types. Perhaps the largest unresolved issue within this context is the basis for how similar signalling systems can mediate different responses. We have presented an argument for a role for crosstalk in these responses, although other mechanisms are likely to be involved as well. It also seems likely that particular signals (Ca^{2+}, cAMP, PKC activation, etc.) will have different effects on apoptosis in different tissues, perhaps due to tissue-specific programming.

Another important issue is how inherent sensitivity to apoptosis at different stages of the cell cycle and/or development is regulated. For example, in the thymus sensitivity to diverse stimuli that trigger apoptosis is confined to one particular ($CD4^+CD8^+$) subpopulation of cells. This subpopulation is also the target of T-cell receptor-mediated clonal deletion (Murphy *et al.*, 1990). Moreover, although Ca^{2+}-dependent endonuclease activity is found constitutively in nuclei isolated from resting thymocytes, the activity is not observed in nuclei prepared from mitogenically activated cells (Cohen and Duke, 1984; McConkey *et al.*, 1990a), the basis for which is unclear at present but may involve signalling processes associated with cellular activation. Recent work by Lenardo (1990) has identified one mechanism that regulates inherent sensitivity to apoptosis in mature T cells. He found that peripheral T cells previously exposed to IL-2 underwent apoptosis upon subsequent exposure to antigen in an IL-2 dose-dependent fashion. Intriguingly, thymocyte IL-2 receptor levels increase and subsequently decline just prior to acquisition of CD4 and CD8 that signals sensitization to apoptosis (Shimonkevitz et al., 1987). Whether IL-2 primes thymocytes for apoptosis during development also remains a topic for further investigation. Answers to this and other questions concerning the role of developmentally regulated signalling processes in the regulation of thymocyte apoptosis will no doubt provide information relevant to understanding other developmental processes (i.e. synapse formation in the developing brain) that involve programmed cell death.

REFERENCES

Basile, D. V., Wood, H. N. and Braun, A. C. (1973) Programming cells for death under defined experimental conditions: Relevance to the tumor problem. *Proceedings of the National Academy of Sciences, USA*, **70**, 3055–3059.

Berridge, M. J. and Irvine, R. F. (1984) Inositol trisphosphate, a novel second messenger in cellular signal transduction. *Nature*, **312**, 315–321.

Cohen, J. J. and Duke, R. C. (1984) Glucocorticoid activation of a calcium-dependent endonuclease in thymocyte nuclei leads to cell death. *Journal of Immunology*, **132**, 38–42.

Crabtree, G. R. (1989) Contingent genetic regulatory events in T lymphocyte activation. *Science*, **243**, 355–361.

Gaido, M. L. and Cidlowski, J. A. (1990) Identification, purification, and characterization of a calcium-dependent endonuclease (NUC 18) from apoptotic rat thymocytes. *Journal of Biological Chemistry*, **266**, 18 580–18 585.

Groul, D. J., Campbell, N. F. and Bourgeois, S. (1986) Cyclic AMP-dependent protein kinase promotes glucocorticoid receptor function. *Journal of Biological Chemistry*, **261**, 4909–4914.

Groul, D. J., Rajah, F. M. and Bourgeois, S. (1989) Cyclic AMP-dependent kinase modulation of the glucocorticoid-induced cytolytic response in murine T-lymphoma cells. *Molecular Endocrinology*, **3**, 2119–2127.

Harrigan, M. T., Baughman, G., Campbell, F. and Bourgeois, S. (1989) Isolation and characterization of glucocorticoid- and cyclic AMP-induced genes in T lymphocytes. *Molecular and Cellular Biology*, **9**, 3438–3446.

Hewish, D. R. and Burgoyne, L. A. (1973) The calcium dependent endonuclease activity of isolated nuclear preparations. Relationships between its occurrence and the occurrence of other classes of enzymes found in nuclear preparations. *Biochemical and Biophysical Research Communications*, **52**, 475–482.

Hunter, T. and Karin, M. (1992) The regulation of transcription by phosphorylation. *Cell*, **70**, 375–387.

Kaiser, N. and Edelman, I. S. (1977) Calcium dependence of glucocorticoid-induced lymphocytolysis. *Proceedings of the National Academy of Sciences, USA*, **74**, 638–642.

Kaiser, N. and Edelman, I. S. (1978) Further studies on the role of calcium in glucocorticoid-induced lymphocytolysis. *Endocrinology*, **103**, 936–942.

Kang, S.-M., Beverly, B., Tran, A.-C., Brorson, K., Schwartz, R. H. and Lenardo, M. J. (1992) Transactivation by AP-1 is a molecular target of T-cell clonal anergy. *Science*, **257**, 1134–1138.

Kizaki, H., Tadakuma, T., Odaka, C., Muramatsu, J. and Ishimura, Y. (1989) Activation of a suicide process of thymocytes through DNA fragmentation by calcium ionophores and phorbol esters. *Journal of Immunology*, **143**, 1790–1794.

Kizaki, H., Suzuki, K., Tadakuma, T. and Ishimura, Y. (1990) Adenosine receptor-mediated accumulation of cyclic AMP-induced T-lymphocyte death through internucleosomal DNA cleavage. *Journal of Biological Chemistry*, **265**, 5280–5284.

Lenardo, M. J. (1991) Interleukin-2 programs mouse alpha/beta T lymphocytes for apoptosis. *Nature*, **353**, 858–860.

McConkey, D. J., Hartzell, P., Chow, S. C., Orrenius, S. and Jondal, M. (1990d) Interleukin 1 inhibits T-cell receptor-mediated apoptosis in immature thymocytes. *Journal of Biological Chemistry*, **265**, 3009–3011.

McConkey, D. J., Hartzell, P., Jondal, M. and Orrenius, S. (1989a) Inhibition of DNA fragmentation in thymocyte and isolated thymocyte nuclei by agents that stimulate protein kinase C. *Journal of Biological Chemistry*, **264**, 13 399–13 402.

McConkey, D. J., Hartzell, P., Nicotera, P. and Orrenius, S. (1989b) Calcium-activated DNA fragmentation kills immature thymocytes. *FASEB Journal*, **3**, 1843–1849.

McConkey, D. J., Hartzell, P. and Orrenius, S. (1990a) Rapid turnover of endogenous endonuclease activity in thymocytes: Effects of inhibitors of macromolecular synthesis. *Archives of Biochemistry and Biophysics*, **278**, 284–287.

McConkey, D. J., Orrenius, S. and Jondal, M. (1990b) Agents that elevate cAMP stimulate DNA fragmentation in thymocytes. *Journal of Immunology*, **145**, 1227–1230.

McConkey, D. J., Orrenius, S. and Jondal, M. (1990c) Cellular signalling in programmed cell death (apoptosis). *Immunology Today*, **11**, 120–121.

Murphy, K. M., Heimberger, K. B. and Loh, D. Y. (1990) Induction by antigen of intrathymic apoptosis of $CD4^+CD8^+TCR^{lo}$ thymocytes *in vivo*. *Science*, **250**, 1720–1723.

Mustelin, T., Coggeshall, K. M., Isakov, N. and Altman, A. (1990) T-cell antigen receptor-mediated activation of phospholipase C requires tyrosine phosphorylation. *Science*, **247**, 1584–1587.

Nakayama, K. and Loh, D. Y. (1992) No requirement for $p56^{lck}$ in the antigen-stimulated clonal deletion of thymocytes. *Science*, **257**, 94–96.

Nakayama, T., Singer, A., Hsi, E. D. and Samelson, L. E. (1989) Intrathymic signalling in immature $CD4^+CD8^+$ thymocytes results in tyrosine phosphorylation of the T-cell receptor zeta chain. *Nature*, **341**, 651–654.

Nakayama, T., Ueda, Y., Yamada, H., Shores, E. W., Singer, A. and June, C. H. (1992) In vivo calcium elevations in thymocytes with T cell receptors that are specific for self ligands. *Science*, **257**, 96–99.

Newell, M. K., Haughn, L. J., Maroun, C. R. and Julius, M. H. (1990) Death of mature T cells by separate ligation of CD4 and the T-cell receptor for antigen. *Nature*, **347**, 286–289.

Nishizuka, Y. (1984) The role of protein kinase C in cell surface signal transduction and tumour promotion. *Nature*, **308**, 693–698.

Orrenius, S., McConkey, D. J., Bellomo, G. and Nicotera, P. (1989) Role of Ca^{2+} in toxic cell killing. *Trends in Pharmacological Science*, **10**, 281–285.

Power, R. F., Mani, S. K., Codina, J., Conneely, O. M. and O'Malley, B. W. (1991) Dopaminergic and ligand-independent actvation of steroid hormone receptors. *Science*, **254**, 1636–1639.

Pratt, R. M. and Martin, G. R. (1975) Epithelial cell death and cyclic AMP increase during palatal development. *Proceedings of the National Academy of Science, USA*, **72**, 874–877.

Schwartz, R. H. (1990) A cell culture model for T lymphocyte clonal anergy. *Science*, **248**, 1349–1356.

Shimonkevitz, R. P., Husmann, L. A., Bevan, M. J. and Crispe, I. N. (1987) Transient expression of IL-2 receptor precedes differentiation of immature thymocytes. *Nature*, **329**, 157–159.

Smith, C. A., Williams, G. T., Kingston, R., Jenkinson, E. J. and Owen, J. J. T. (1989) Antibodies to CD3/T-cell receptor complex induce death by apoptosis in immature T cells in thymic cultures. *Nature*, **337**, 181–184.

Trump, B. F. and Berezesky, I. K. (1992) The role of cytosolic Ca^{2+} in cell injury, necrosis, and apoptosis. *Current Opinions in Cell Biology*, **4**, 227–232.

Vanderbilt, J. N., Bloom, K. S. and Anderson, J. N. (1982) Endogenous nuclease: properties and effects on transcribed genes in chromatin. *Journal of Biological Chemistry*, **257**, 13 009–13 017.

Wyllie, A. H. (1980) Glucocorticoid-induced thymocyte apoptosis is associated with endogenous endonuclease activation. *Nature*, **284**, 555–556.

Wyllie, A. H., Morris, R. G., Smith, A. L. and Dunlop, D. (1984) Chromatin cleavage in apoptosis: Association with condensed chromatin morphology and dependence on macromolecular synthesis. *Journal of Pathology*, **142**, 67–77.

Chapter 3

REGULATION OF THYMOCYTE APOPTOSIS
Glucocorticoid-Induced Death and its Inhibition by T-Cell Receptor/CD3 Complex-Mediated Stimulation

Makoto Iwata,[1] Rieko Iseki[1] and Yoshihisa Kudo[2]

Departments of [1]Cellular Recognition Research and [2]Neuroscience, Mitsubishi Kasei Institute of Life Sciences, 11, Minamiooya, Machida-shi, Tokyo194, Japan

INTRODUCTION

The immune systems of vertebrates have evolved for the efficient defense against the invasion of pathogenic microorganisms and certain forms of cancer by acquiring the mechanism to generate specific responses to these threats. The specific responses have been attained by producing the diversity of specific antigen receptors, T-cell receptors (TCR) and immunoglobulins. The diversity of these receptors is created by rearrangement of the receptor genes. However, all the created receptors are not necessarily expressed on the mature and functional lymphocytes. The mechanism of clonal selection, especially in T cells, has become one of the biggest issues in the field of immunology.

APOPTOSIS AND CLONAL SELECTION IN THE THYMUS

The precursors of T cells are mainly produced in the bone marrow, immigrate into the thymus, and begin to express CD4 and CD8 molecules on their surfaces. After some cycles of proliferation, they express T-cell receptors (TCR) (Figure 3.1). Each T-cell

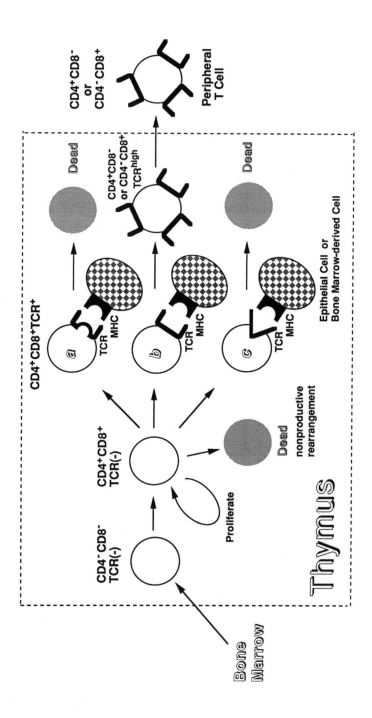

Figure 3.1 T-cell differentiation and maturation in the thymus. *a*) A T-cell clone with strong affinity for self MHC. *b*) A T-cell clone with proper affinity for self MHC. *c*) A T-cell clone with little affinity for self MHC.

clone usually bears only one species of TCR, while the clones that cannot perform productive gene rearrangement are considered to undergo apoptosis. It is known that the vast majority of immature TCR$^+$ clones also undergo apoptosis in the thymus. The T-cell repertoire is largely influenced by selection events in the thymus. Two types of the selection events are generally considered (Blackman et al., 1990; Sprent et al., 1990; von Boehmer and Kisielow, 1990); positive selection of clones that bear TCR biased for recognition of antigen in association with self MHC (major histocompatibility complex)-encoded proteins, and negative selection of clones that bear TCR with strong affinity for self MHC molecules (± self peptide). As for the negative selection, it has been suggested that the self-reactive immature clones undergo apoptotic cell death upon stimulation by binding of the TCR to self antigens (Smith et al., 1989), a process leading to self-tolerance. On the other hand, little is known about the mechanism of the positive selection. It is still not clear if positive selection precedes negative selection, or if there are specialized thymic stroma cells to present self MHC (± self antigen) only for positive selection or for negative selection. Nonetheless, the affinity of the TCR for self MHC (± self peptide) might be the key for the selection events. The clones bearing TCR with proper affinity (not too strong and not too weak) for self MHC may be somehow positively selected. The clones bearing TCR with little or very low affinity for self MHC should be useless clones, and are considered to undergo apoptosis in the thymus.

The majority of normal murine thymocytes are immature CD4$^+$CD8$^+$ (double-positive) cells that bear intermediate levels of TCR. The cells that are involved in the selection events in the thymus belong to this population. CD4$^+$CD8$^-$ or CD4$^-$CD8$^+$ (single-positive) cells that bear high levels of TCR are mature type cells, and are ready to emigrate from the thymus. The CD3 complex is associated with TCR, and transduces signals from TCR to the intracellular machinery. If an anti-TCR antibody or an anti-CD3 antibody is injected into a normal mouse, mainly CD4$^+$CD8$^+$ double positive cells are going to die and the relative proportion of single-positive cells increases (Shi et al., 1991). This phenomenon is considered to reflect the mechanism of negative selection. A similar change can be observed when glucocorticoids are injected into a normal mouse (Hugo et al., 1991). The major population of glucocorticoid-sensitive thymocytes appears to be also immature, double-positive cells. It is known that adrenalectomy induces a marked increase in the thymus size and the number of thymocytes (Shortman and Jackson, 1974), suggesting that physiological concentrations of glucocorticoid hormones play a role in controlling thymocyte apoptosis.

CIRCADIAN RHYTHM OF PLASMA GLUCOCORTICOID HORMONE LEVELS

The concentration of glucocorticoid hormones in the plasma is regulated by circadian rhythm. In nocturnal animals, such as mice and rats, it reaches peak levels in the evening, and comes back to minimum levels in the morning. In the peak period, a normal animal can attain 10^{-7} to 10^{-6} M glucocorticoid in its plasma, and that concentration can induce apoptosis in their thymocytes *in vitro* (Cohen and Duke, 1984; Wyllie *et al.*, 1984; Iwata *et al.*, 1991).

REGULATION OF GLUCOCORTICOID-INDUCED APOPTOSIS IN THYMOCYTES

Corticosterone (CS) is the major glucocorticoid in rats and mice. As shown in Figure 3.2a, 10^{-7} to 10^{-6} M CS induced apoptosis as determined by DNA fragmentation. Thus, the clones that are positively selected in the thymus have to be rescued from glucocorticoid-induced death by some mechanism. Because the primary differences among $CD4^+CD8^+$ double-positive T-cell clones should be their TCR, we considered that signals via the TCR might contribute to the rescue. We wondered what would happen if both glucocorticoids and signals through TCR/CD3 complex were given to thymocytes. It has been generally considered, however, that both glucocorticoid-induced death and the death induced by TCR-mediated stimulation depend on similar mechanisms that involve an increase in intracellular Ca^{2+} level (Kaiser and Edelman, 1977, 1978; McConkey *et al.*, 1989a, 1989b). We challenged this concept, and performed the following experiments.

We coated culture plates with various concentrations of a monoclonal anti-CD3 antibody, 145-2C11, and cultured normal murine thymocytes in the plates with various concentration of CS. As shown in Figure 3.2b, without adding CS, a small but significant population of the cells underwent apoptosis upon incubation with anti-CD3 in a dose-dependent fashion. In the presence of 10^{-7} M CS that can be observed in normal plasma, a V-shape curve was obtained; apoptosis was induced in the presence of either a very low concentration of anti-CD3 or high concentrations of anti-CD3. However, in the presence of a limited concentration range of anti-CD3, glucocorticoid-induced death was inhibited. Thus, too weak or too strong stimulation via the TCR/CD3 complex may not be able to rescue thymocytes from glucocorticoid-induced death, but a proper stimulation via the complex may be able to do so. From these results, we speculated what may be happening *in vivo*. Glucocorticoid at its peak concentration can induce death in immature thymocytes that receive only a weak signal via the TCR. These TCR may have little affinity for self MHC. Self-reactive clones also die by the TCR-mediated signals, and glucocorticoids may enhance the death. Only the clones that can receive proper signals via the TCR/CD3 complex can survive, and the nature of the signals might be regulated by the affinity of TCR to self MHC (± self peptide). In the presence of 10^{-6} M CS, a small inhibition was observed in the presence of a higher concentration of anti-CD3. From these results, we propose the following hypothesis.

"Positive selection of thymocytes for the formation of the T-cell repertoire is based on the inhibition of glucocorticoid-induced apoptosis in immature thymocytes

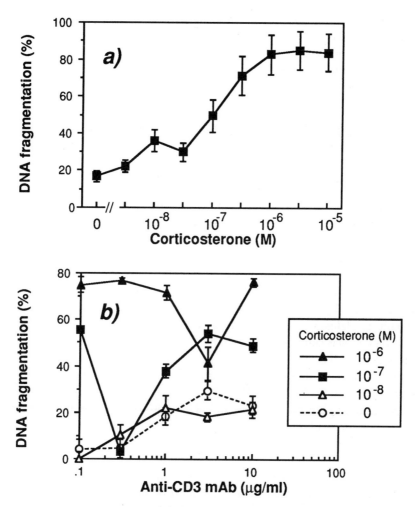

Figure 3.2 Glucocorticoid-induced apoptosis in murine thymocytes and its inhibition. *a)* Induction of DNA fragmentation by CS *in vitro*. *b)* Inhibition of CS-induced DNA fragmentation by an appropriate concentration of anti-CD3 antibody. The culture plates were coated with graded concentrations (abscissa) of the antibody. The cultures contained 10% FCS. The FCS we used endogenously contained 1.0×10^{-8} M hydrocortisone, 2.6×10^{-9} M progesterone, and less than 1×10^{-9} M CS (Iwata *et al.*, 1991). (Iwata *et al.*, 1991, reprinted with permission from the *European Journal of Immunology*.)

bearing TCR with proper affinity for self by the TCR-mediated signals, and the selection may be influenced by the glucocorticoid level."

It is becoming evident that not only TCR engagement but also other molecular interactions especially LFA-1 and ICAM-1 or -2 are required for the process of thymic selection. We found that an anti-LFA-1 antibody (M17/4.2) alone coated on a plate does not induce apoptosis in murine thymocytes. However, the dose response curve of anti-CD3-induced DNA fragmentation in thymocytes shifted to the left (lower concentrations of anti-CD3) in the presence of anti-LFA-1. Glucocorticoid-induced death was also inhibited by the combination of anti-CD3 and anti-LFA-1, and the extent of the inhibition was much more stable than that by anti-CD3 alone (Iwata and Iseki, unpublished data).

APOPTOSIS IN T-CELL HYBRIDOMAS AS A MODEL FOR STUDYING THYMOCYTE APOPTOSIS

Antagonism Between Glucocorticoid- and Activation-Induced Apoptosis

Biochemical studies are necessary to analyze what the proper signal for the rescue is. The heterogeneity of normal thymocytes makes this difficult. On the other hand, monoclonal T-cell hybridomas can be easily obtained by the fusion of a thymoma, BW5147, and mature T cells, and they often retain some properties of immature thymocytes. Thus, we first employed cloned T-cell hybridomas to analyze the signals via the TCR/CD3 complex. DO-11.10 and BOG8 are among the hybridomas that undergo apoptosis by TCR/CD3-mediated stimulation and by glucocorticoids.

The culture plates were coated with various concentrations of anti-CD3 (Figure 3.3a). DO-11.10 was cultured for 16 h in the plates in the presence of various concentrations of dexamethasone (DEX), a potent synthetic glucocorticoid. In the absence of DEX, the cells died in a dose-dependent fashion by anti-CD3. In the presence of a low concentration of DEX, a V-shape curve was obtained as we observed in the experiment with thymocytes. In our system, DEX was always 50 to 100 times more potent than CS. In the presence of higher concentrations of DEX, higher concentrations of anti-CD3 were required for the inhibition. However, unlike the results with thymocytes, T-cell hybridomas did not undergo apoptosis in the presence of high concentrations of both anti-CD3 and the glucocorticoid. We found that glucocorticoids and TCR/CD3-mediated stimulation are mutually antagonistic in the induction of apoptosis in T-cell hybridomas. Thus, it is clear that apoptosis can be induced by different pathways. The antagonism in a T-cell hybridoma was also observed by Zacharchuk *et al.* (1990), independently.

The growth of T-cell hybridomas is inhibited by either glucocorticoids or TCR/CD3-mediated stimulation. The growth was, however, not inhibited when both glucocorticoids and TCR/CD3-mediated stimulation were given together. T-cell hybridomas such as DO-11.10 and BOG8 produce IL-2 upon stimulation with antigen plus antigen-presenting cells, anti-CD3, or the combination of ionomycin and PMA. At the same time, however, they stop growing and eventually die. In the presence of one of these stimuli and glucocorticoids together, the cells not only produce IL-2 but also survive and grow. Thus, glucocorticoids do not shut off all the signals through TCR, and the signals required for the induction of apoptosis and the production of IL-2 are not necessarily the same (Iwata *et al.*, 1991).

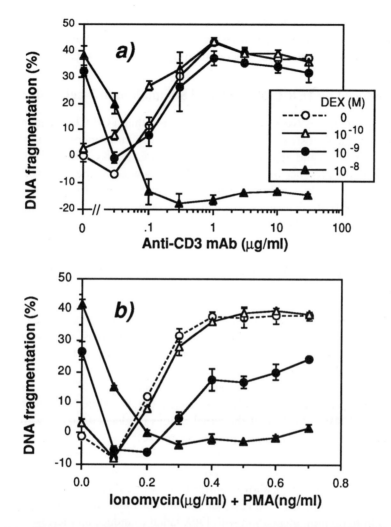

Figure 3.3 Glucocorticoid-induced apoptosis in a T-cell hybridoma (DO-11.10) and its inhibition. DNA fragmentation was determined by the recovery of radioactivity in the detergent soluble fraction of 5-[^{125}I]iodo-UdR-labelled cells. *a)* Antagonism between DEX-induced death and anti-CD3-induced death. *b)* The combination of ionomycin and PMA could substitute anti-CD3 for the regulation of DNA fragmentation in the T-cell hybridoma. (Iseki *et al.*, 1991, reprinted with permission, Copyright ©1991, *The Journal of Immunology*.)

Glucocorticoids induce apoptosis in T-cell hybridomas, while they inhibit TCR/CD3-mediated apoptosis. The potency of various steroids for the two kinds of effect was in the order of DEX > hydrocortisone > CS >> β-estradiol, testosterone. It was parallel to the order of their anti-inflammatory potency. Cortisone did not have any activity in regulating apoptosis in T-cell hybridomas *in vitro*. It is reasonable as cortisone is known to exert its anti-inflammatory activity after it is converted to other forms of steroid, such as hydrocortisone *in vivo*. Sex hormones had little activity, except that high doses of progesterone exerted glucocorticoid-like activity that might be related to its weak affinity for glucocorticoid receptor (Iwata *et al.*, 1991).

Signals Induced by TCR/CD3-Mediated Stimulation in T-cell Hybridomas

Our subsequent studies (Iseki *et al.*, 1991) indicated that TCR/CD3-mediated stimulation regulates apoptosis in T-cell hybridomas through an increase in intracellular Ca^{2+} levels ($[Ca^{2+}]_i$) and activation of protein kinase C (PKC). Indeed, as shown in Figure 3.3b, in the presence of various concentrations of DEX, the effects of anti-CD3 can be mimicked by the combination of ionomycin, a calcium ionophore, and PMA, a PKC activator. Note that the V-shape was also mimicked with this combination in the presence of 10^{-9} M DEX.

THE ROLE OF $[Ca^{2+}]_i$ IN GLUCOCORTICOID-INDUCED APOPTOSIS IN THYMOCYTES

Before we analyzed TCR-mediated signals in thymocytes, we re-examined Ca^{2+} mobilization in glucocorticoid-induced death. In T-cell hybridomas, we have previously shown that an increase in $[Ca^{2+}]_i$ is neither induced by glucocorticoids nor required for glucocorticoid-induced death (Iseki *et al.*, 1991). In rat thymocytes, however, it has been suggested by some groups (Kaiser and Edelman, 1977, 1978; McConkey *et al.*, 1989a) that the process was dependent on calcium uptake, while Nicolson and Young (1979) concluded it unlikely that the lethal actions of glucocorticoids were initiated by glucocorticoid-induced changes in calcium uptake.

We analyzed the effect of glucocorticoids on $[Ca^{2+}]_i$ in murine thymocytes by using Fura-2/AM and fluorescence microscopy. As shown in Figure 3.4, DEX failed to induce an increase in $[Ca^{2+}]_i$ above control level, while thapsigargin, an inhibitor of endoplasmic reticulum Ca^{2+}-ATPase, or ionomycin induced a rapid and dramatic increase. Up to 60 min, incubation with DEX failed to induce an increase (Iseki *et al.*, unpublished data).

Thus, we re-examined the role of Ca^{2+} in glucocorticoid-induced apoptosis. In rat thymocytes, it has been reported that EGTA, Quin 2/AM, or calmodulin inhibitors inhibited glucocorticoid-induced apoptosis or DNA fragmentation (McConkey *et al.*, 1989a). In murine thymocytes, however, we found that EGTA failed to inhibit glucocorticoid-induced DNA fragmentation. To deplete intracellular Ca^{2+}, the cells were incubated with various concentrations of Quin 2/AM in the presence or absence of DEX. At a concentration of 25 μM or higher, DNA fragmentation was inhibited. But the same concentrations induced cytolysis as assessed by trypan blue dye exclusion. Similarly, calmodulin inhibitors, trifluoperazine and calmidazolium, inhibited DNA fragmentation only at toxic doses. Thus, we could not obtain any results that proved a requirement for an increase in $[Ca^{2+}]_i$ in this process (Iseki *et al.*, unpublished data).

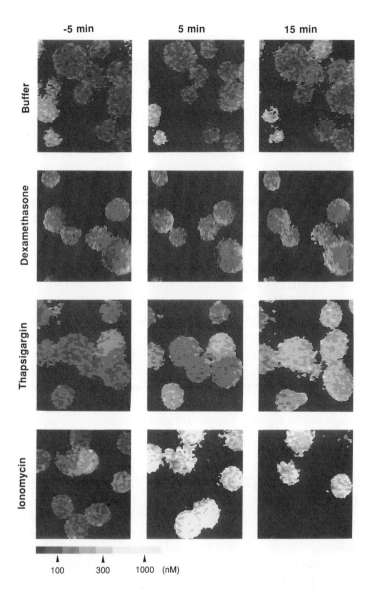

Figure 3.4 DEX fails to increase $[Ca^{2+}]_i$ in murine thymocytes. The buffer alone, DEX (10^{-6} M), thapsigargin (10^{-7} M), or ionomycin (10^{-6} M) was administered at time 0. The images of Fura-2-labelled cells taken by flourescence microscopy at times 0.5 min, 5 min, and 15 min are shown. Upon ionomycin administration, some cells were lost after the vigorous increase in $[Ca^{2+}]_i$.

We do not have an explanation at this stage for the discrepancy between our results and those of previous studies in rat thymocytes. The experiments with rat thymocytes need to be repeated in our hands for further comparison.

SIGNALS INDUCED BY TCR/CD3-MEDIATED STIMULATION IN THYMOCYTES

Crosslinking of CD3 molecules induced a small but significant increase in $[Ca^{2+}]_i$. It was, however, difficult to prove that the increase in $[Ca^{2+}]_i$ is essential for anti-CD3-induced apoptosis in murine thymocytes, because calcium chelating agents by themselves induced death in the cells during the time required to detect anti-CD3-induced apoptosis. Anti-CD3-induced apoptosis was inhibited by inhibitors of PKC but not by those of cyclic nucleotide-dependent or other kinases. Thus, it appears that activation of PKC is essential for anti-CD3-induced apoptosis (Iwata and Iseki, unpublished data).

Ionomycin or PMA alone induced apoptosis in murine thymocytes, but failed to inhibit glucocorticoid-induced death. An appropriate combination of ionomycin and PMA, however, inhibited glucocorticoid-induced apoptosis. In the absence of CS, the combination induced DNA fragmentation at higher doses (Figure 3.5). CS-induced DNA fragmentation was inhibited by a limited dose range of the combination. Similar results were obtained in cytolysis. The requirement for an increase in $[Ca^{2+}]_i$ with ionomycin for the inhibition of glucocorticoid-induced apoptosis supports our conclusion that an increase in $[Ca^{2+}]_i$ is not required for the induction of apoptosis by glucocorticoids. The inhibition pattern was similar to that observed with the anti-CD3 antibody in Figure 3.2b. From these results, as in T-cell hybridomas, TCR/CD3-mediated signals might regulate thymocyte apoptosis by an increase in $[Ca^{2+}]_i$ and activation of PKC.

REGULATION OF APOPTOSIS BY GLUCOCORTICOIDS AND CLEFT PALATE

Do glucocorticoids disturb negative selection of thymocytes? We have shown that a proper stimulation via the TCR/CD3 complex inhibits glucocorticoid-induced apoptosis in thymocytes, but have not obtained any direct evidence that glucocorticoids inhibit the TCR/CD3-mediated induction of apoptosis in thymocytes. The heterogeneity of thymocytes prevents us from drawing a conclusion. In Figure 3.2b, a small population of thymocytes was rescued from CS (10^{-6} M)-induced death upon stimulation with anti-CD3. It is possible that this population would have been dead upon stimulation with anti-CD3 in the absence of 10^{-6} M CS. Madar and Chesnokova (1992) reported that CS treatment markedly inhibited neonatal induction of

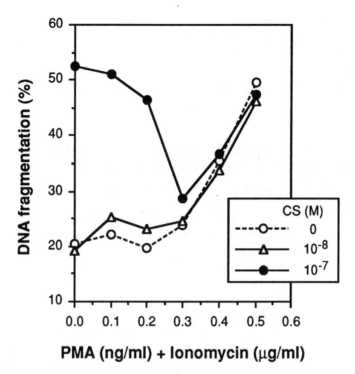

Figure 3.5 Combination of ionomycin and PMA over a limited concentration range inhibits CS-induced DNA fragmentation in murine thymocytes. BALB/c thymocytes were cultured with graded concentrations of ionomycin and PMA with their ratio kept at 1000:1 for 16 h in the presence (10^{-8} M, *open triangle*; 10^{-7} M, *closed circle*) or absence (*open circle*) of CS.

immunological tolerance to hen egg lysozyme in mice. Interference with negative selection of thymocytes might explain their findings.

The inhibition of apoptotic cell death by glucocorticoids has been shown not only in T-cell hybridomas but also in glucocorticoid-induced cleft palate. The injection of glucocorticoids into a pregnant animal often induces cleft palate in the foetus. The clefting action of glucocorticoids involves a delay of the elevation of the embryonic palatal shelves from a vertical position to a horizontal position and inhibition of programmed cell death in epithelial cells of embryonic palatal shelves. Gupta *et al.* (1984) reported that DEX itself or phospholipase A_2-inhibitory proteins obtained from DEX-treated thymocytes or embryonic palates inhibited the programmed cell death in the medial-edge epithelium of single mouse embryonic palatal shelves in culture. The degree of cleft palate produced by glucocorticoids is in direct proportion to their anti-inflammatory activity. It has been postulated that glucocorticoids induce their anti-inflammatory action by producing a protein that inhibits phospholipase A_2, cyclooxygenase, or other enzymes. The results suggest that similar proteins may be produced by thymocytes under the influence of glucocorticoids and may affect thymic selection. It is known that A/J mice are highly susceptible to glucocorticoid-induced cleft palate and about 8% of their progeny have cleft palate without the glucocorticoid treatment. Autoimmunity often develops in this strain of mice (Teague, 1968). Apoptosis in self-reactive thymocytes of these mice might be spontaneously inhibited in the presence of physiologic concentration of glucocorticoids, and that may result in autoimmunity.

REGULATION OF THYMOCYTE APOPTOSIS BY LIGANDS OF OTHER MEMBERS OF THE STEROID HORMONE RECEPTOR SUPERFAMILY

Feeding a mouse a vitamin A-supplemented ration causes thymus enlargement and an increase in the number of thymocytes. It also increases the susceptibility of mice to glucocorticoid-induced cleft palate (Tyan, 1987). Retinoic acid, the active metabolite of vitamin A, at near-physiologic concentrations inhibited activation-induced apoptosis in both thymocytes and T-cell hybridomas, while it enhanced glucocorticoid-induced apoptosis in these cells (Iwata and Iseki, 1992). Retinoic acid by itself did not induce apoptosis. It has been shown that retinoic acid influences the proliferation and differentiation of a variety of cell types, and that it plays an important role in vertebrate development. The receptors of glucocorticoids, sex hormones, and retinoic acid belong to the *erb*A oncogene-related steroid hormone receptor superfamily. The receptors of thyroid hormones and $1\alpha,25$-dihydroxy-vitamin D_3 also belong to the superfamily. Thyroid hormones are known to play a pivotal role in frog-tadpole metamorphosis that is accompanied by massive programmed cell death (apoptosis) and cell transformation. $1\alpha,25$-dihydroxy-vitamin D_3 is known to play a critical role in mineral and skeletal homeostasis, but is also known to be involved in the regulation of immune responses. Both thyroid hormones and $1\alpha,25$-dihydroxy-vitamin D_3, however, failed to modulate apoptosis in both T-cell hybridomas and thymocytes (Iwata and Iseki, 1992).

The mechanism of the regulation of apoptosis by the combination of TCR/CD3-mediated stimulation and glucocorticoids or retinoic acid is not yet clear. Modulation of PKC-induced AP-1 activity by activated glucocorticoid receptors or retinoic acid receptors might be involved in the mechanism.

SUMMARY AND CONCLUSIONS

The peak levels of glucocorticoid hormones in normal plasma can induce apoptosis in murine thymocytes *in vitro*. Proper stimulation via the TCR/CD3 complex rescues thymocytes from glucocorticoid-induced apoptosis. Thus, we proposed a hypothesis that the positive selection of thymocytes for the formation of the T-cell repertoire is based on the inhibition of glucocorticoid-induced apoptosis in immature T-cell clones bearing TCR with proper affinity for self by the TCR-mediated signals *in situ*. The role of glucocorticoids in the formation of the T-cell repertoire have to be directly examined *in vivo* for the physiological foundations of the hypothesis using TCR transgenic mice or super antigens. Retinoic acid inhibits activation-induced apoptosis in thymocytes at near-physiological concentrations, while it enhances glucocorticoid-induced apoptosis. It remains to be determined how retinoids can be involved in thymic selection *in vivo*. An increase in $[Ca^{2+}]_i$ appears to be neither induced by glucocorticoids nor responsible for glucocorticoid-induced apoptosis in murine thymocytes, although the basal level of $[Ca^{2+}]_i$ may be required for the activation of an endonuclease. It is suggested that TCR/CD3-mediated stimulation rescues thymocytes from glucocorticoid-induced apoptosis via an appropriate increase in $[Ca^{2+}]_i$ and proper activation of PKC. The analysis of the possible involvement of calcineurin and c-*myc* as well as *bcl*-2 and APO-1/Fas antigen in the regulation of thymocyte apoptosis may provide a further insight into the mechanism of thymic selection.

REFERENCES

Blackman, M., Kappler, J. and Marrack, P. (1990) The role of the T-cell receptor in positive selection of developing T cells. *Science*, **248**, 1335–1341.

Cohen, J. J. and Duke, R. C. (1984) Glucocorticoid activation of a calcium-dependent endonuclease in thymocyte nuclei leads to cell death. *Journal of Immunology*, **132**, 38–42.

Gupta, C., Katsumata, M., Goldman, A. S. and Herold, R. (1984) Glucocorticoid-induced phospholipase A$_2$-inhibitory proteins mediate glucocorticoid teratogenicity *in vitro*. *Proceedings of the National Academy of Sciences, USA*, **81**, 1140–1143.

Hugo, P., Boyd, R. L., Waanders, G. A. and Scollay, R. (1991) CD4$^+$CD8$^+$CD3^{high} thymocytes appear transiently during ontogeny: evidence from phenotypic and functional studies. *European Journal of Immunology*, **21**, 2655–2660.

Iseki, R., Mukai, M. and Iwata, M. (1991) Regulation of T lymphocyte apoptosis: Signals for the antagonism between activation- and glucocorticoid-induced death. *Journal of Immunology*, **147**, 4286–4292.

Iwata, M., Hanaoka, S. and Sato, K. (1991) Rescue of thymocytes and T-cell hybridomas from glucocorticoid-induced apoptosis by stimulation via the T-cell receptor/CD3 complex: a possible *in vitro* model for selection of the T-cell repertoire. *European Journal of Immunology*, **21**, 643–648.

Iwata, M., Mukai, M., Nakai, Y. and Iseki, R. (In press) Retinoic acid inhibit activation-induced apoptosis in T-cell hybridomas and thymocytes. *Journal of Immunology*, **159**.

Kaiser, N and Edelman, I. S. (1977) Calcium dependence of glucocorticoid-induced lymphocytolysis. *Proceedings of the National Academy of Sciences, USA*, **74**, 638–642.

Kaiser, N. and Edelman, I. S. (1978) Further studies on the role of calcium in glucocorticoid-induced lymphocytolysis. *Endocrinology*, **103**, 936–942.

Madar, J. and Chesnokova, V. M. (1992) Corticosterone treatment affects neonatal induction of immunological tolerance to hen egg lysozyme in mice. *Folia Biologica (Praha)*, **38**, 103–112.

McConkey, D. J., Hartzell, P., Amador-Perez, J. F., Orrenius, S. and Jondal, M. (1989b) Calcium-dependent killing of immature thymocytes by stimulation via the CD3/T-cell receptor complex. *Journal of Immunology*, **143**, 1801–1806.

McConkey, D. J., Nicotera, P., Hartzell, P., Bellomo, G., Wyllie, A. H. and Orrenius, S. (1989a) Glucocorticoids activate a suicide process in thymocytes through an elevation of cytosolic Ca^{2+} concentration. *Archives of Biochemistry and Biophysics*, 269, 365–370.

Nicholson, M. L. and Young, D. A. (1979) Independence of the lethal actions of glucocorticoids on lymphoid cells from possible hormone effects on calcium uptake. *Journal of Supramolecular Structure*, 10, 165–174.

Shi, Y., Bissonnette, R. P., Parfrey, N., Szalay, M., Kubo, R. T., and Green, D. R. (1991) In vivo administration of monoclonal antibodies to the CD3 T-cell receptor complex induces cell death (apoptosis) in immature thymocytes. *Journal of Immunology*, 146, 3340–3346.

Shortman, K. and Jackson, H. (1974) The differentiation of T lymphocytes. I. Proliferation kinetics and interrelationships of subpopulations of mouse thymus cells. *Cellular Immunology*, 12, 230–246.

Smith, C. A., Williams, G. T., Kingston, R., Jenkinson, E. J. and Owen, J. J. T. (1989) Antibodies to CD3/T-cell receptor complex induce death by apoptosis in immature T cells in thymic cultures. *Nature*, 337, 181–184.

Sprent, J., Gao, E. K. and Webb, S. R. (1990) T-cell reactivity to MHC molecules: Immunity versus tolerance. *Science*, 248, 1357–1363.

Teague, P. O., Friou, G. J., Myers, L. L. (1968) Anti-nuclear antibodies in mice. I. Influence of age and possible genetic factors on spontaneous and induced responses. *Journal of Immunology*, 101, 791–798.

Tyan, M. L. (1987) Vitamin A-enhanced cleft palate susceptibility associated with H-2. *Journal of Immunogenetics*, 14, 239–245.

von Boehmer, H. and Kisielow, P. (1990) Self-nonself discrimination by T cells. *Science*, 248, 1369–1373.

Wyllie, A. H. (1980) Glucocorticoid-induced thymocyte apoptosis is associated with endogenous endonuclease activation. *Nature*, 284, 555–556.

Wyllie, A. H., Morris, R. G., Smith, A. L. and Dunlop, D. (1984) Chromatin cleavage in apoptosis: Association with condensed chromatin morphology and dependence on macromolecular synthesis. *Journal of Pathology*, 142, 67–77.

Zacharchuk, C. M., Mercep, M., Chakraborti, P. K., Simons, S. S., Jr. and Ashwell, J. D. (1990) Programmed T lymphocyte death: cell activation- and steroid-induced pathways are mutually antagonistic. *Journal of Immunology*, 145, 4037–4045.

Chapter 4

PROTEIN MODIFICATION IN APOPTOSIS

Martin F. Lavin, Glenn D. Baxter, Qizhong Song,
Duygu Findik and Eva Kovacs

Queensland Cancer Fund Research Unit,
Queensland Institute of Medical Research,
The Bancroft Centre, 300 Herston Road, Brisbane, Queensland 4029, Australia

INTRODUCTION

In recent years it has become evident that cell proliferation, differentiation and cell death (apoptosis) are controlled by an elaborate network of signalling pathways. These pathways function through receptors, protein kinases, second messengers, phosphorylated protein intermediates and factors that bind to and regulate the expression of individual genes or groups of genes. Phosphorylation is a key element in the equation, leading to the regulation of metabolism, control of passage of cells through the cycle, change in the distribution of cellular proteins and transcriptional control. Different signals are amplified through a series of pathways involving several protein kinases, and "cross-talk" occurs between the pathways. Perhaps the best described are those utilizing cAMP to activate protein kinase A; diacylglycerol to activate protein kinase C and the tyrosine kinases which in many cases are the receptors for growth factors.

PROTEIN PHOSPHATASES

Phosphorylation state of proteins depends not only on protein kinases but also on protein phosphatases. While the latter have received less attention than the kinases it is now clear that they also play a critical role in the regulation of cellular processes (Cohen, 1992). The serine/threonine phosphatases have been classified into four groups, 1, 2A, 2B and 2C and can be distinguished by their substrate specificities, catalytic and regulatory subunits, and response to inhibitors (Cohen and Cohen, 1989). Phosphatase type 1 (PP1) has a well described role in the regulation of glycogen metabolism (Ballou and Fischer, 1986) and is required for mitotic progression in *Drosophila* (Axton et al., 1990). Mammalian cells are blocked in metaphase by microinjection of anti-PP1 antibodies while introduction of the enzyme into one pole of an anaphase cell accelerated completion of telophase (Fernandez et al., 1992). PP1 also appears to be a major regulator of cAMP-responsive element binding protein (CREB) activity in cAMP-responsive cells (Hagiwara et al., 1992).

Since PP1 and PP2A are serine/threonine phosphatases they would be expected to have the opposite effect to protein kinase C. This kinase phosphorylates a series of proteins which play a role in cell proliferation (Nishizuka, 1988), and in tumour promotion (Delclos et al., 1980). Thus it is not unreasonable to propose that these protein phosphatases would act as tumour suppressors similar to the retinoblastoma protein and p53. Evidence for this comes from the observation that okadaic acid, a 38 carbon polyether isolated from the marine sponge *Halichondria okadai* (Tachibana et al., 1981), a potent inhibitor of PP1 and PP2A, is a tumour promoter not functioning through PKC (Suganuma et al., 1988). In this context these phosphatases when active would be anti-proliferative and perhaps assist in steering cells into death pathways.

CELL DEATH

Two major processes for cell death have been described, necrosis and apoptosis or programmed cell death (Wyllie et al., 1980). The latter is observed during development (Hammar and Mottet, 1971), morphogenesis (Wyllie et al., 1980), in the immune system (Duke et al., 1983; Moss et al., 1991) and after exposure of cells to steroids, toxins and other damaging agents (Cohen and Duke, 1984; McConkey et al., 1988; Baxter and Lavin, 1992). Apoptosis is an active process showing an absolute requirement for *de novo* protein synthesis in some cases (Lieberman et al., 1970; Cohen and Duke, 1984), and in others, while this requirement does not appear to exist, expression of specific genes is associated with apoptosis (Owens et al., 1991; Goldstone and Lavin, 1991; Shi et al., 1992). We have attempted to investigate the molecular basis of apoptosis by considering the importance of protein modification as well as *de novo* protein synthesis in this process.

IMPORTANCE OF PROTEIN SYNTHESIS

It is well established that a requirement exists for RNA and protein synthesis in murine thymocytes undergoing apoptosis (Wyllie et al., 1980; Duke and Cohen, 1986; McConkey et al., 1988). One exception to this in the murine system is cytotoxic T cell (CTL) killing of targets (Cohen et al., 1985). The picture is much less clear with human cells, since the majority of reports do not support a role for RNA or protein synthesis

(Green and Cotter, Chapter 13, pp. 153–156). Compounds such as cycloheximide that are used to inhibit protein synthesis actually exacerbate the process (Heine et al., 1966; Ijiri, 1987).

Protein synthesis alone is not sufficient for diphtheria toxin-induced cytolysis of U937 cells (Chang et al., 1989). Cycloheximide prevents DNA degradation in erythroid progenitor cells deprived of erythropoietin but puromycin, another protein synthesis inhibitor, failed to prevent degradation (Kelley et al., 1992). In addition glucocorticoid receptor, from which a number of key functional domains had been deleted, was capable of causing cell death apparently based only on its ability to bind to DNA (Nazareth et al., 1991). The results outlined in Figure 4.1 demonstrate that cycloheximide is incapable of inhibiting DNA fragmentation in the Burkitt's lymphoma cell line BM13674 exposed to heat or ionizing radiation, both of which have been shown to cause apoptosis in these cells (Baxter and Lavin, 1992). Previous studies have revealed that not only does cycloheximide not inhibit apoptosis in acute leukaemia blasts deprived of IL-2, but it enhances the process (Baxter et al., 1989).

ROLE OF PROTEIN MODIFICATION

If protein synthesis is not the only or the major mechanism responsible for establishing the apoptotic machinery in the cell, protein modification by phosphorylation, acetylation or some other post-translational change could contribute to the process. As outlined above phosphorylation/dephosphorylation plays a central role in signal transduction controlling cell growth and proliferation (Cohen, 1992). Changes in the phosphorylation state of proteins might also be significant in the process of cell death. Exposure of lymphocytes to calcium ionophore and phorbol ester leads to a proliferative stimulus similar to that induced by mitogens (Truneh et al., 1985). This is achieved by phorbol ester activation of protein kinase C (PKC) followed by phosphorylation of a number of cellular factors (Nishizuka et al., 1984). It is of interest that phorbol esters prevent DNA fragmentation and cell death in thymocytes exposed to Ca^{2+} ionophore and glucocorticoid hormone (Kanter et al., 1984; McConkey et al., 1989). Keeping a phosphate on appears to be important in avoiding cell death. We have provided further evidence in support of this with phosphatase inhibitors which have the same nett effect as stimulating protein kinase activity (Baxter and Lavin, 1992; Song et al., 1992). Okadaic acid an inhibitor of serine/threonine phosphatases completely prevented DNA fragmentation in BM13674 cells exposed to heat or ionizing radiation (Figure 4.1). This compound also inhibited fragmentation induced by several cytotoxic compounds (Song et al., 1992). As expected, okadaic acid also prevented apoptosis in these cells (Table 4.1). TPA (12-O-tetradecanoyl-phorbol 13-acetate), a stimulator of PKC, failed to arrest apoptosis induced in these cells (Table 4.1). In preventing apoptosis it was evident that okadaic acid had caused premature chromatin condensation (Figure 4.2B) which was readily distinguished from the peripheral crescents of compacted chromatin and the fragmentation present in nuclei of cells undergoing apoptosis (Figure 4.2C). Recent results from this laboratory (Song

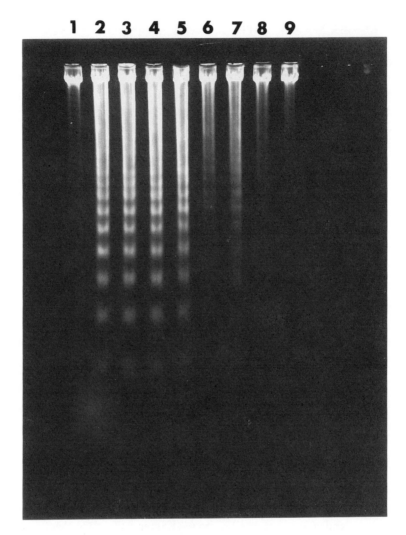

Figure 4.1 DNA fragmentation in BM13674 cells exposed to ionizing radiation or heat. Cells were either exposed to 20 Gy of γ-rays or heat shocked. After 4 h DNA was extracted and electrophoresed in 1.8% agarose gels. Lane 1, Control; lane 2, 20 Gy of ionizing radiation; lane 3, heat shock; lane 4, 20 Gy of ionizing radiation with 1 µg/ml cycloheximide; lane 5, heat shock with 1 µg/ml cycloheximide; lane 6, 1 µg/ml cycloheximide; lane 7, 50 µg/ml cycloheximide; lane 8, 20 Gy of ionizing radiation with 500 nM okadaic acid; lane 9, heat shock with 500 nM okadaic acid. (Baxter and Lavin, 1992, reprinted with permission, Copyright © 1992, *The Journal of Immunology*.)

and Lavin, 1993), employing the potent phosphatase (PP1 and PP2A) inhibitor calyculin A again demonstrated inhibition of DNA fragmentation and the morphological features of apoptosis, but this time in the absence of chromatin condensation (Figure 4.2D). The latter results would seem to rule out the formation of a physical

Table 4.1 Effect of inhibitor/activators on induction of apoptosis in human cell lines by heat and ionizing radiation.

Treatment	% Apoptosis Cell Type	
	BM 13674	CEM C7
Untreated	4.6 ± 1.0	4.2 ± 1.0
Heat Treated	41.0 ± 2.6	35.0 ± 3.2
Heat treated + cycloheximide	63.0 ± 3.5	55.0 ± 3.6
Heat treated + TPA	42.0 ± 2.5	38.0 ± 2.8
Heat treated + okadaic acid	6.0 ± 0.5	7.0 ± 1.5
Irradiation	45.0 ± 5.9	48.0 ± 2.9
Irradiation + cycloheximide	73.0 ± 9.1	60.0 ± 2.5
Irradiation + TPA	46.0 ± 5.4	48.0 ± 2.8
Irradiation + okadaic acid	6.0 ± 1.1	6.3 ± 1.4

barrier to endogenous endonuclease created by chromatin condensation. More likely these phosphatase inhibitors are preventing the dephosphorylation of proteins necessary to initiate apoptosis rather than through dephosphorylation of $p34^{cdc2}$ which would cause premature mitosis.

PROTEIN DEPHOSPHORYLATION

Two dimensional gel electrophoresis demonstrates that specific proteins are dephosphorylated during apoptosis and these changes are prevented in the presence of okadaic acid (Figure 4.3). One such protein of size ~40 kDa and isoelectric point (pI = 6.5) appears to undergo dephosphorylation in response to different agents that cause apoptosis, and in different cell lines undergoing apoptosis (Baxter and Lavin, 1992; Song et al., 1992). Inhibition of apoptosis was only observed at 500 nM okadaic acid or higher concentrations while the effective concentration was 15 nM and above for calyculin A. Based on IC_{50} values of 0.5 nM–1 nM for both of these compounds for inhibition of PP2A, and IC_{50} values of 2 nM (calyculin A) and 60–500 nM (okadaic acid) for inhibition of PP1 (Ishihara et al., 1989), it would suggest that PP1 inhibition is important in preventing apoptosis. Alternatively it is possible that phosphatases PP1 and PP2A both need to be inhibited. As discussed above these are critical enzymes for a number of important cellular processes and it is not unexpected that continued inhibition would lead to cell death. The inhibition of apoptosis by both okadaic acid and

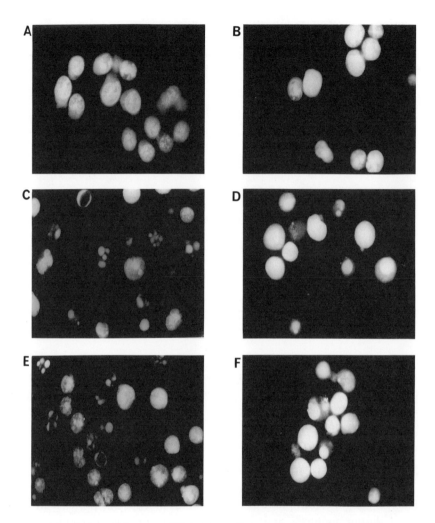

Figure 4.2 Morphological features of BM13674 cells undergoing apoptosis. Cells were treated to induce apoptosis with various agents in the presence and absence of 1 μM okadaic acid. Cells were stained with acridine orange (5 μg/ml) after 8 h incubation and observed using a BioRad MRC600 fluorescence confocal microscope. A, control; B, control + 1 μM okadaic acid; C, 10 μg/ml tetrandrine (Chen *et al.*, 1992); D, 10 μg/ml tetrandrine + 1 μM okadaic acid; E, 0.3 μM bistratene A (Watters *et al.*, 1990); F, 0.3 μM bistratene A + 1 μM okadaic acid. Arrows point to cells undergoing nuclear condensation or apoptosis. (Song *et al.*, 1992, reprinted with permission from the *Journal of Cellular Physiology*, © Copyright 1992, John Wiley & Sons, Inc.)

calyculin A is evident up to 8 h post-treatment. By 24 h after exposure to radiation, heat or chemicals, even in the presence of phosphatase inhibitors, BM13674 and CEM cells undergo apoptosis. This question of timing is a critical point and it should be appreciated that inhibitors which are effective in a defined window may in the longer term exacerbate the process. In relation to this point Ishida *et al.* (1992) have recently shown that 8 nM okadaic acid arrested cells at G2/M and caused the appearance of

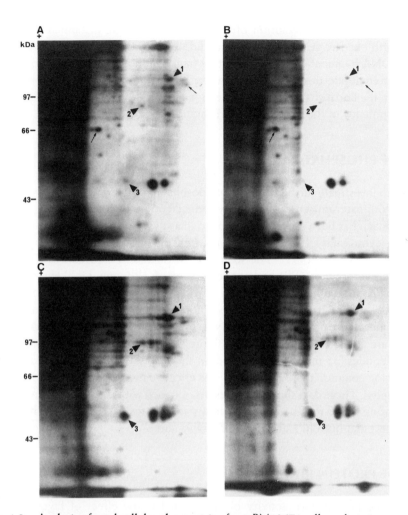

Figure 4.3 Analysis of total cellular phosproteins from BM13674 cells undergoing apoptosis. Cells were labelled with (^{32}P) phosphate for 2 h, treated with bistratene to induce apoptosis, and incubated for a further 1 h prior to lysis. A, untreated cells; B, cell treated with 0.3 μM bistratene; C, untreated cells incubated in the presence of okadaic acid (500 nM); D, bistratene-treated cells incubated in the presence of okadaic acid. Arrowheads point to proteins undergoing dephosphorylation during apoptosis, which are protected from dephosphorylation in the presence of okadaic acid. Arrows point to a protein whose phosphorylation status does not change after bistratene treatment. This was used as a standard for phosphorimager analysis. (Song *et al.*, 1992, reprinted with permission from the *Journal of Cellular Physiology*, © Copyright 1992, John Wiley & Sons, Inc.)

nuclear fragmentation in U937 cells only after 48 h incubation. Even at 500 nM okadaic acid, while premature chromosome condensation was evident by 3 h, apoptosis did not occur until later. Boe *et al.* (1991) have described cellular changes such as condensation of chromatin, shedding of cellular contents via surface bleb formation, and hyperconvolution of the nuclear membrane in human and rat cells treated with okadaic acid. These authors did not provide any data on okadaic acid-induced DNA fragmentation but stated that fragmentation was much less pronounced than with equitoxic doses of cAMP analog. The more pronounced morphological changes in the Boe *et al.* (1991) study after 3–6 h might be explained by a difference in purity or potency of the okadaic acid preparations and the cell types used.

ROLE OF PHOSPHATASES

In order to investigate further the role of phosphatases in apoptosis we have designed phosphorothioate antisense oligodeoxyribonucleotides specific for PP1 and PP2A. This approach has been successfully applied in a study of the role of perforin in CTL-mediated cytotoxicity (Acha-orbea *et al.*, 1990), a process that functions through apoptosis (Moss *et al.*, 1991). The sequence for human PP2A is known (Arino *et al.*, 1988) but that for human PP1 had not been reported, although it is likely to be closely related to PP1 from other species (Nitschke *et al.*, 1992). Accordingly we isolated a full-length cDNA clone from a human cDNA library (unpublished data), using a 455 bp PCR product as previously described (Nitschke *et al.*, 1992). The sequences of PP2A and PP1 are aligned in Figure 4.4 and the regions chosen for synthesis of antisense oligonucleotides (15 mers) are indicated. In short only the antisense oligonucleotide to the region common to PP2A and PP1 was effective in preventing apoptosis (unpublished results). These data support the results obtained with okadaic acid and calyculin A implicating a role for phosphatase activity in apoptosis. However, in this case the data suggest that both PP1 and PP2A are important.

ROLE OF PROTEIN KINASES

In inhibiting phosphatase activity okadaic acid has the effect of favouring kinase activity and maintaining proteins in hyperphosphorylated states (Haystead *et al.*, 1990; Cohen *et al.*, 1990). In shifting the balance to protein phosphorylation okadaic acid is having the same nett effect as TPA but perhaps differing with respect to the pattern of protein phosphorylation. Nevertheless, both compounds prevent the onset of apoptosis. In order to investigate the importance of protein kinase activity in preventing apoptosis we have employed inhibitors of two major kinases, PKC and cAMP-dependent PKA. Inhibition of protein kinase A (PKA) with the specific inhibitor H-89 (Chijiwa *et al.*, 1990) increased the level of ionizing radiation-induced apoptosis in BM13674 cells with time up to 8 h. On the other hand calphostin C, a PKC-specific inhibitor (Kobayashi *et al.*, 1989), failed to increase the extent of apoptosis in irradiated cells. We have previously shown that TPA did not protect against radiation or heat-induced apoptosis in these cells (Baxter and Lavin, 1992). These results are in agreement with the calphostin data suggesting that PKC activity is not always critical for protection. However, inhibition of PKA activity increased the

4: PROTEIN MODIFICATION IN APOPTOSIS

```
HUMAN PP1          M S D S E K L N L D S I I G R L L E V Q G S   22
HUMAN PP2A(HL-14)  M D . . V F T K E L D Q W R E Q L N E C       20

HUMAN PP1          R P G K N V Q L T E N E I R G L C L K S R E   44
HUMAN PP2A(HL-14)  K - - - - - . . S . S Q V K S . . E . A K .   37

HUMAN PP1          I F L S Q P I L L E L E A P L K I C G D I H   66
HUMAN PP2A(HL-14)  . L T K E S N V Q . V R C . V T V . . . V .   59

HUMAN PP1          G Q Y Y D L L R L F E Y G G F P P E S N Y L   88
HUMAN PP2A(HL-14)  . . F H . . M E . . R I . . K S . D T . . .   81

HUMAN PP1          F L G D Y V D R G K Q S L E T I C L L L A Y   110
HUMAN PP2A(HL-14)  . M . . . . . . . Y Y . V . . V T . V . L     103

HUMAN PP1          K I K Y P E N F F L L R G N H E C A S I N R   132
HUMAN PP2A(HL-14)  . V R . R . R I T I . . . . . S R Q . T Q     125

HUMAN PP1          I Y G F Y D E C K R R Y - N I K L W K T F T   153
HUMAN PP2A(HL-14)  V . . . . . . L . K . G . A N V . . Y . .     147

HUMAN PP1          D C F N C L P I A A I V D E K I F C C H G G   175
HUMAN PP2A(HL-14)  . L . D Y . . L T . L . . G Q . . . . L . .   169

HUMAN PP1          L S P D L Q S M E Q I R R I M R P T D V P D   197
HUMAN PP2A(HL-14)  . . . S I D T L D H . . A L D . L Q E . . . H 191

HUMAN PP1          Q G L L C D L L W S D P D K D V Q G W G E N   219
HUMAN PP2A(HL-14)  E . P M . . . . . . . . - . R G . . . I S     212

HUMAN PP1          D R G V S F T F G A E V V A K F L H K H D L   241
HUMAN PP2A(HL-14)  P . . A G Y . . . Q D I S E T . N . A N G .   234

HUMAN PP1          D L I C R A H Q V V E D G Y E F F A K R Q L   263
HUMAN PP2A(HL-14)  T . V S . . . . L . M E . . N W C H D . N V   256

HUMAN PP1          V T L F S A P N Y C G E F D N A G A M M S V   285
HUMAN PP2A(HL-14)  . . I . . . . . . Y R C G . Q A . I . E L     278

HUMAN PP1          D E T L M C S F Q I L K P A D K N K G K Y G   307
HUMAN PP2A(HL-14)  . D . . K Y . . L Q F D . . P R - - - - -     294

HUMAN PP1          Q L S G L N P G G R P I T P P R N S A K A K   329
HUMAN PP2A(HL-14)  - - - - - - R . E P H V . R R T P D Y F L     309

HUMAN PP1          K                                             330
```

Figure 4.4 Comparison of nucleotide sequence for human phosphatase 1 (PP1) and human phosphatase 2A (PP2A). Regions chosen for design of antisense oligonucleotides are boxed.

extent of radiation-induced apoptosis implying that substrates for this activity played a direct role in the process of apoptosis.

PROPOSED SCHEME

A simplified scheme to account for the various observations, involving both PKA and PP1 is depicted in Figure 4.5. It should be stressed that this does not accommodate a role for PP2A which may also play an important part in this pathway. A stimulus such as ionizing radiation or heat might activate adenylate kinase which in turn is capable of phosphorylating inhibitor 1, a specific inhibitor for PP1 in its phosphorylated form (Cohen, 1992). In the presence of the PKA inhibitor H-89 any soluble PP1 would be present in an active form thus favouring dephosphorylation. In muscle PP1 is associated with glycogen as a complex between the catalytic subunit and glycogen (G)-binding protein. Phosphorylation of the G-subunit by PKA leads to a dissociation of the complex releasing the soluble form of PP1 which would normally be rendered inactive by complexing with the phosphorylated form of inhibitor. In Figure 4.5 a

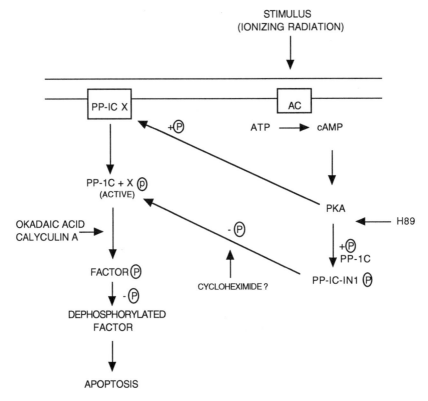

Figure 4.5 Proposed scheme for involvement of phosphatase in apoptosis. The series of events occuring in response to radiation damage, resulting in apoptosis, are described in the text. AC, adenylate kinase; PKA, protein kinase A; H-89, PKA inhibitor; PP-1C, phosphatase 1 catalytic subunit; PP-1C-IN1, complex between PP-1C and inhibitor 1; PP-1CX, complex between PP-1C and unknown protein sequestering the enzyme to the membrane or other subcellular structure; Factor P, phosphorylated factor undergoing dephosphorylation prior to apoptosis. Okadaic acid and calyculin A are inhibitors of PP-1C.

hypothetical subunit X substitutes for the G-subunit which would be responsible for the sequestration of PP1 to a membrane, chromatin or some other subcellular structure. There is good evidence that PP1 is localized to the nucleus at specific stages of the cell cycle (Fernandez *et al.*, 1992). In the presence of H-89 the PP1-X complex would be under-phosphorylated, remain in a particulate state and thus shift the equilibrium to an active state favouring dephosphorylation. Cycloheximide has been shown to enhance the degree of apoptosis in some systems (Ijiri, 1987). This compound also causes an increase in phosphatase activity in mammalian cells. This might be achieved by inhibiting *de novo* synthesis of PKA or inhibitor for PP1. Exposure of cells to agents inducing apoptosis in the presence of H-89 would lead to an increase in phosphatase activity for the reasons outlined above and as a consequence increased levels of apoptosis. This would be additive with any increased soluble activity due to an underphosphorylated ineffective inhibitor 1.

In summary it seems likely that protein modification as well as *de novo* protein synthesis is important in the process of apoptosis. Similar to the situation with cell proliferation a network of signal transduction pathways is likely to exist for apoptosis. The diversity of the stimuli giving rise to apoptosis together with the complexity of the changes in gene expression support this idea. As in the case of cell growth it is likely that phosphorylation/dephosphorylation plays a key role in cell death pathways. A greater understanding of the roles of PP1 and PP2A will assist in delineating the steps leading to apoptosis.

ACKNOWLEDGEMENTS

This work was supported by funds from the Queensland Cancer Fund and the University of Queensland Cancer Research Fund.

REFERENCES

Acha-Orbea, H., Scarpellino, L., Hertig, S., Dupuis, M. and Tschopp, J. (1990) Inhibition of lymphocyte mediated cytotoxicity by perforin. *The EMBO Journal*, 9, 3815–3819.

Arino. J., Woon, C. W., Brantigan, D. L., Miller, T. B. and Johnson, G. L. (1988) Human liver phosphatase 2A: cDNA and amino acid sequence of two catalytic subunit isotypes. *Proceedings of the National Academy of Science, USA*, 85, 4252–4256.

Axton, J. M., Dombradi, V., Cohen, P. T. W. and Glover, D. M. (1990) One of the protein phosphatases I isoenymes in Drosophila is essential for mitosis. *Cell*, 63, 33–46.

Ballou, L. M. and Fischer, E. H. (1986) Phosphoprotein phosphatases. *The Enzymes*, 3, 311–361.

Baxter, G. D. and Lavin, M. F. (1992) Specific protein dephosphorylation in apoptosis induced by ionizing radiation and heat shock in human lymphoid tumor lines. *Journal of Immunology*, 148, 1949–1954.

Boe, R., Gjertsen, B. T., Vitermyr, O. K., Houge, G., Lanotte, M. and Doskeland, S. O. (1991) The protein phosphatase inhibitor okadaic acid induces morphological changes typical to apoptosis in mammalian cells. *Experimental Cell Research*, 195, 237–246.

Chang, M., Bramhall, J., Graves, S., Bonavida, B. and Wisniesk, B. J. (1989) Internucleosomal DNA cleavage precedes diptheria toxin-induced cytolysis. *Journal of Biological Chemistry*, 264, 15 261–15 267.

Chen, P., Lavin, M. F., Teh, B. S., Seow, W. K. and Thong, Y. H. (1992) Induction of apoptosis by tetrandrine: comparison with other immunosuppressive agents. *International Journal of Immunotherapy*, VIII, 85–90.

Chijiwa, T., Mishima, A., Hagiwara, M., Sano, M., Hayashi, K., Inoue, T., Naito, K., Toshioka, T. and Hidaka, H. (1990) Inhibition of forskolin-induced neurite outgrowth and protein phosphorylation by a newly synthesized selective inhibitor of cyclic AMP-dependent protein kinase, N-[2-(p-Bromocinnanylamino)ethyl]-5-isoquinolinesulfonamide (H-89), of PC12D pheochromocytoma cells. *Journal of Biological Chemistry*, 265, 5267–5272.

Cohen, J. J. and Duke, R. C. (1984) Glucocorticoid activation of a calcium-dependent endonuclease in thymocyte nuclei leads to cell death. *Journal of Immunology*, 132, 38–42.

Cohen, J. J., Duke, R. C., Chervenak, R., Sellins, K. S. and Olson, L. K. (1985) DNA fragmentation in targets of CTL: an example of programmed cell death in the immune system. *Advances in Experimental Medicine and Biology*, **184**, 493–506.

Cohen, P. (1992) Signal integration at the level of protein kinases, protein phosphatases and their substrates. *Trends in Biochemical Sciences*, **17**, 408–413.

Cohen, P. and Cohen, P. T. W. (1989) Protein phosphatases come of age. *Journal of Biological Chemistry*, **264**, 21 435–21 438.

Cohen, P., Holmes, C. F. B and Tsukitani, Y. (1990) Okadaic acid: a new probe for the study of cellular regulation. *Trends in Biochemical Sciences*, **15**, 98–102.

Delclos, K. B., Nagle, D. S. and Blumberg, P. (1980) Specific binding of phorbol ester tumor promoters to mouse skin. *Cell*, **19**, 1025–1032.

Duke, R. C., Chervenak, R. and Cohen, J. J. (1983) Endogenous endonuclease induced DNA fragmentation: an early event in cell mediated cytolysis. *Proceedings of the National Academy of Sciences, USA*, **80**, 6361–6365.

Duke, R. C. and Cohen, J. J. (1986) IL-2 addiction: withdrawal of growth factor activates a suicide program in dependent T cells. *Lymphokine Research*, **5**, 289–299.

Fernandez, A., Brautigan, D. L. and Lamb, N. J. C. (1992) Protein phosphatase type 1 in mammalian cell mitosis: chromosomal localization and involvement in mitotic exit. *Journal of Cell Biology*, **116**, 1421–1430.

Goldstone, S. D. and Lavin, M. F. (1991) Isolation of a cDNA clone, encoding a human beta-galactoside binding protein, overexpressed during glucocorticoid-induced cell death. *Biochemical and Biophysical Research Communications*, **178**, 746–750.

Hagiwara, M., Alberts, A., Brindle, P., Meinkoth, J., Feramisco, J., Deng, T., Karin, M., Shenolikar, S. and Montminy, M. (1992) Transcriptional attenuation following cAMP induction requires PP-1-mediated dephosphorylation of CREB. *Cell*, **70**, 105–113.

Hammar, S. P. and Mottet, N. K. (1971) Tetrazolium salt and electron microscope studies of cellular degeneration and necrosis in the interdigital areas of the developing chick limb. *Journal of Cell Science*, **8**, 229–236.

Haystead, T. A. J., Weiel, J. E., Litchfield, D. W., Tsukatani, Y., Fisher, E. H. and Krebs, E. G. (1990) Okadaic acid mimics the action of insulin in stimulating protein kinase activity in isolated adipocytes. *Journal of Biological Chemistry*, **265**, 16 571–16 580.

Heine, U., Laglois, A. J. and Beard, J. W. (1966) Ultrastructural alterations in avian leukemic myeloblasts exposed to actinomycin D *in vitro*. *Cancer Research*, **26**, 1847–1858.

Ijiri, K. and Potten, C. S. (1987) Further studies on the response of intestinal crypt cells of different hierarchical status to eighteen different cytotoxic agents. *British Journal of Cancer*, **55**, 113–123.

Ishida, Y., Furukawa, Y., Decaprio. J. A., Saito, M. and Griffin, J. D. (1992) Treatment of myeloid leukemic cells with the phosphatase inhibitor okadaic acid induces cell cycle arrest at either G1/S or G2/M depending on dose. *Journal of Cellular Physiology*, **150**, 484–492.

Ishihara, H., Martin, B. L., Brautigan, D. L., Karaki, H., Ozaki, H., Kato, Y., Fusetani, N., Watabe, S., Hashimoto, K., Uemura, D. and Hartshorne, D. J. (1989) Calyculin A and okadaic acid: inhibitors of protein phosphatase activity. *Biochemical and Biophysical Research Communications*, **159**, 871–877.

Kanter, P., Leister, K. J., Tomei, L. D., Wenner, P. A. and Wenner, C. E. (1984) Epidermal growth factor and tumor promoters prevent DNA fragmentation by different mechanisms. *Biochemical and Biophysical Research Communications*, **118**, 392–399.

Kelley, L. L., Koury, M. J. and Bondurant, M. C. (1992) Regulation of programmed death in erythroid progenitor cells by erythroproietin: effects of calcium and of protein and RNA syntheses. *Journal of Cellular Physiology*, **151**, 487–496.

Kobayashi, E., Nakano, H., Morimoto, M. and Tamaoki, T. (1989) Calphostin C (UCN-1028C), a novel microbial compound, is a highly potent and specific inhibitor of protein kinase C. *Biochemical and Biophysical Research Communications*, **159**, 548–553.

Lieberman, M. W., Verbin, R. S., Landay, M., Liang, H., Farber, E., Lee, T. N. and Starr, R. (1970) A probable role for protein synthesis in intestinal epithelial cell damage induced *in vivo* by cytosine arabinoside, nitrogen mustard, or X-irradiation. *Cancer Research*, **30**, 942–951.

McConkey, D. J., Hartzell, P., Duddy, S. K., Hakansson, H. and Orrenius, S. (1988) 2,3,7,8-Tetrachloro-dibenzo-p-dioxin kills immature thymocytes by Ca^{2+}-mediated endonuclease activation. *Science*, **242**, 256–259.

McConkey, D. J., Hartzell, P., Jondal, M. and Orrenius, S. (1989) Inhibition of DNA fragmentation in thymocytes and isolated thymocyte nuclei by agents that stimulate protein kinase C. *Journal of Biological Chemistry*, **264**, 13 399–13 402.

Moss, D. J., Burrows, S. R., Baxter, G. D. and Lavin, M. F. (1991) T cell – T cell killing is induced by specific epitopes: evidence for an apoptotic mechanism. *Journal of Experimental Medicine*, **173**, 681–686.

Nazareth, L. V., Harbour, D. V. and Thompson, E. B. (1991) Mapping the human glucocorticoid receptor for leukemic cell death. *Journal of Biological Chemistry*, **266**, 12 976–12 980.

Nishizuka, Y. (1988) The molecular heterogenity of protein kinase C and its implications for cellular regulation. *Nature*, **334**, 661–665.

Nishizuka, Y. (1984) The role of protein kinase C in cell surface signal transduction and tumour promotion. *Nature*, **308**, 693–698.

Nitschke, K., Fleig, U., Schell, J. and Palme, K. (1992) Complementation of the cs *dis*2-11 cell cycle mutant of *Schizosaccharomyces pombe* by a protein phosphatase from *Arabidopsis thalianae*. *The EMBO Journal*, **11**, 1327–1333.

Owens, G. P., Hahn, W. E. and Cohen, J. J. (1991) Identification of mRNAs associated with programmed cell death in immature thymocytes. *Molecular and Cellular Biology*, **11**, 4177–4188.

Shi, Y. *et al.* (1992) Role of c-*myc* in activation induced apoptotic cell death in T-cell hybridomas. *Science*, **257**, 212–214.

Song, Q., Baxter, G. D., Kovacs, E. M., Findik, D. and Lavin, M. F. (1992) Inhibition of apoptosis in human tumour cells by okadaic acid. *Journal of Cellular Physiology*, **153**, 550–556.

Song, Q. and Lavin, M. F. (1993) Calyculin A, a potent inhibitor of phosphatases-1 and -2A, prevents apoptosis. *Biochemical and Biophysical Research Communications*, **190**, 47–55.

Tachibana, K., Scheuer, P. J., Tsukitani, Y., Kikuchi, H., Van Engen, D., Clardy, J., Gopichand, Y. and Schmitz, F. J. (1981) Okadaic acid, a cytotoxic polyether from two marine sponges of the genus Halichondria. *Journal of the American Chemical Society*, **103**, 2469–2471.

Truneh, A., Albert, F., Golstein, P. and Schmitt-Verhulst, A-M. (1985) Early stages of lymphocyte activation bypassed by synergy between calcium inophores and phorbol ester. *Nature*, **313**, 318–320.

Watters, D., Marshall, K., Hamilton, S., Michael, J., McArthur, M., Seymour, G., Hawkins, C. J., Gardiner, R. and Lavin, M. (1990) The bistratenes: new cytotoxic marine macrolides which induce some properties indicative of differentiation in HL-60 cells. *Biochemical Pharmacology*, **39**, 1609–1614.

Wyllie, A. H., Kerr, J. F. R. and Currie, A. R. (1980) Cell death: the significance of apoptosis. *International Review of Cytology*, **68**, 251–305.

Chapter 5

RELATIONSHIP BETWEEN APOPTOSIS AND THE CELL CYCLE

P. Roy Walker, Joanna Kwast-Welfeld and Marianna Sikorska

Apoptosis Research Group, Institute for Biological Sciences, National Research Council of Canada, Ottawa, K1A 0R6, Canada

Tissue formation during development, and maintenance or remodelling during adult life requires a fine balance between cell proliferation, differentiation and removal of damaged or unwanted cells by apoptosis or active cell death. That apoptosis plays a major role during development has been well established by the elegant studies in *C. elegans* (Ellis *et al.*, 1991), where the fate of every cell is known and at least 30% die by apoptosis. The role that apoptosis plays in the forming of the tissues and their maintenance in higher mammals is also becoming apparent. For example, apoptosis is now known to play a major role in the development of the nervous system (Clarke, 1990) and in the formation of digits (Lockshin and Zakeri, 1991) and in tissue regression (Tenniswood *et al.*, 1992). In addition, in the adult organism apoptosis plays a vital part in the functioning of the immune system.

Thymocytes and lymphocytes are susceptible to apoptosis at various times throughout their lifespan (Walker *et al.*, 1991; Roy *et al.*, 1992). Immature thymocytes that would become autoreactive are removed by negative selection. This process is termed activation induced cell death and occurs when the T-cell receptor (TCR) is triggered by a self antigen (Shi *et al.*, 1990). Instead of undergoing normal proliferative activation the cells are diverted into the cell death pathway and undergo apoptosis a few hours later. Cells that are positively selected differentiate into mature peripheral T cells and remain quiescent in G0 until they interact with a foreign antigen. In this quiescent state both the cell cycle and apoptosis are suspended until the cells are re-activated. Activation occurs following the interaction between the TCR and foreign antigen and the cells proliferate and differentiate into mature cytotoxic T cells. The proliferative expansion of such a clone of lymphocytes is strictly dependent upon the presence of the cytokine interleukin-2 (IL-2). When IL-2 levels decline these activated cells cannot become quiescent and are disposed of by apoptosis, with the exception of a few long-lived memory cells. Thus thymocytes and lymphocytes provide excellent model systems for studying the relationships between differentiation and cell death and the cell cycle and cell death.

THE CELL CYCLE AND CELL DEATH

At present virtually nothing is known about how cycling or differentiated cells enter the cell death pathway. Soloff *et al.* (1987) studying the induction of apoptosis in V79 fibroblasts following cold shock noticed that exponentially growing cells were much more resistant to the shock than confluent cells. The confluent cultures fragmented 85% of their DNA within 40 min after a 60 min cold shock. Similarly, Perotti *et al.* (1990) observed a cell cycle dependency in the response of McCoy's Human synovial cells to cold shock with exponential cultures being resistant and confluent cultures being very sensitive. Interestingly, phorbol esters partially protected the latter cells from apoptosis. These studies demonstrated a cell cycle dependency of cell death, although the apparent resistance to cold shock may have been a failure to wait until the cycling cells reach a particular stage of the cell cycle at which they would have died.

Most cells growing *in vitro* are dependent upon growth factors to stay in the cell cycle. These are usually provided by serum, but in some cases a specific growth factor must be added. In the absence of growth factor most cells become quiescent, i.e. they no longer proliferate, but remain viable until fresh growth factor is supplied. Quiescent cells are typically arrested in G0 after leaving the cell cycle in early G1. In contrast, IL-2-dependent CTL cells are different in that they cannot re-enter G0 or become quiescent. Instead, in the absence of the cytokine they arrest briefly in early G1 and then undergo apoptosis (Figure 5.1). We have been studying the mechanisms governing the switching between continued cell cycle progression and diversion into the cell death pathway in CTLL cells, an IL-2-dependent cytotoxic T-cell line (Gillis

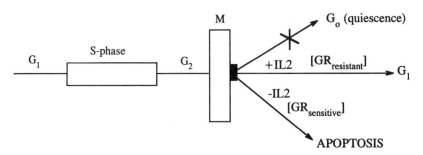

Figure 5.1 Alternative fates of IL-2-dependent CTLL cells immediately following mitosis. GR = glucocorticoid receptor.

and Smith, 1977). These cells require the continuous presence of IL-2 to progress through G1 and to pass the G1/S restriction point and initiate DNA replication. If the level of IL-2 is insufficient to support a further round of the cell cycle the cells complete mitosis and accumulate in early G1. After several hours without cytokine the cells begin to undergo apoptosis. Typical growth curves are shown in Figure 5.2. In the presence of IL-2 the cells usually undergo two rounds of cell division within 48 h and only a small fraction of cells undergo apoptosis. In contrast, in the absence of IL-2, more than 50% of the cells have undergone apoptosis by 48 h. CTLL cells have an added advantage for cell death studies since glucocorticoids can also induce apoptosis. Thus, in the absence of IL-2, glucocorticoids greatly increase the rate at which the cells undergo apoptosis with virtually the entire population being dead in less than 24 h (Figure 5.2). However, this only occurs in the absence of IL-2. In the presence of the cytokine the cells become resistant to the effects of glucocorticoids. Thus, IL-2 not only protects against apoptosis that occurs as a result of a failure to re-enter the cell cycle, but also makes them resistant to the effect of glucocorticoids. Interleukin-2 generates the signals necessary to both repress cell death and promote cell cycle progression as a result of its interaction with its surface receptor. The interleukin-2 receptor (IL-2R) consists of two subunits of p55 (IL-2R alpha) and p75 (IL-2R beta). The receptor is unique amongst growth factor receptors in that it has only a very short intracellular domain (13 residues for the alpha subunit and 286 for the beta subunit) and no enzymatic activity has been ascribed to this region (Smith, 1989). Two signal transduction pathways do, however, appear to respond to the interaction between IL-2 and IL-2R. The first, which is still controversial, is the PKC pathway and the second is the activation of tyrosine kinase(s).

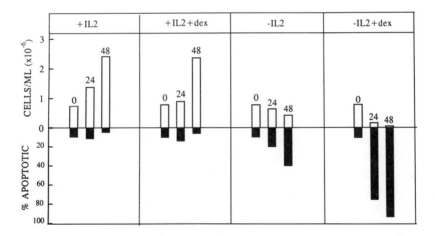

Figure 5.2 Effects of IL-2 and dexamethasone (dex) on the proliferation and viability of CTLL cells. Cells were plated at 0.8×10^6 cells/ml and the number of viable cells and the percentage of the cells that were apoptotic (estimated by flow cytometry, Walker et al., 1993) determined at 24 and 48 h. Dexamethasone was added at 0 h in each case.

THE ROLE OF PKC IN IL-2 SIGNAL TRANSDUCTION

The IL-2R does not induce phosphatidyl inositol hydrolysis (Smith, 1989), but does stimulate the hydrolysis of glycosylphosphatidyl inositol in CTLL cells (Merida et al., 1990). It is also not clear whether IL-2 can increase $[Ca^{2+}]_i$. Rossio et al. (1986) reported an elevation of intracellular calcium but others (LeGrue, 1987) have not been able to reproduce this. Furthermore, the whole relationship between PKC and IL-2 is controversial. Larsen et al. (1988) have shown that IL-2 does not cause the translocation of PKC to the membrane in activated human T cells. This is in complete contrast to the earlier results of Farrar and Anderson (1985) and Farrar and Taguchi (1985) who showed a rapid translocation of PKC to the membrane following IL-2 stimulation of mouse CT6 cells. The reasons for the discrepancy are not known, but it may relate to the stage of the cell cycle or species differences. Mills et al. (1988) have shown that PKC^- clones of IL-2-dependent mouse T cells still proliferate in the presence of IL-2 indicating that PKC is not essential for proliferation. Moreover, PKC down-regulated cells still proliferate in response to IL-2 indicating that there are PKC-dependent and independent pathways in these cells. Similar results were obtained by Valge et al. (1988) using a mouse T helper clone. However, these studies did not take into account the non-downregulable or inducible isoforms of PKC recently shown to exist (Friere-Moar et al., 1991) so they do not rule out PKC participation in the response to IL-2 in *normal* cells. Indeed, Redondo et al. (1988) have shown that phorbol esters stimulate proliferation of CTLL cells, but not as well as IL-2.

Further evidence for a functional role for PKC in CTLL cells came from studies which established that the phorbol ester, tetradecanoylphorbol 13-acetate (TPA), prevents the death of cells deprived of IL-2 for at least 10 hours (Rodriguez-Tarduchy and Lopez-Rivas, 1989). Thus PKC activation can prevent cells from moving along the cell death pathway, suggesting that IL-2 may switch cell death off by a PKC-mediated phosphorylation step. Similarly, phorbol esters have been found to inhibit ionophore and glucocorticoid-induced DNA fragmentation in immature rat thymocyte primary cultures (McConkey et al., 1989). Kizaki et al. (1989) and Ojeda et al. (1990), in contrast, provided evidence for PKC involvement in dexamethasone-induced apoptosis in mouse thymocytes by demonstrating that low concentrations of TPA appeared to induce DNA fragmentation and that H-7 (a PKC-specific inhibitor), but not HA-1004 (a cAMP-dependent protein kinase inhibitor), inhibited the process. Our initial work on rat thymocytes showed the same effect of H-7 (and HA-1004) but we had no stimulatory effect of TPA at 100 nM. In addition, Sato et al. (1988) have shown that whilst TPA had only a slight effect on human CEM-C7 cell viability it synergized strongly with low doses of glucocorticoids to induce apoptosis. In other studies, confluent C3H-10T1/2 cells eventually undergo apoptosis when incubated for up to 72 hours without serum. A population of cells becomes non-adherent and extensive DNA degradation typical of apoptosis (Wyllie, 1980) is found in these released cells (Kanter et al., 1984; Tomei et al., 1988). Epidermal growth factor or TPA were able to prevent this and induce proliferation. In these cells TPA also inhibited the apoptosis that occurs after ionizing radiation. Similarly, endothelial fibroblasts undergo apoptosis in the absence of serum (Araki *et al.*, 1990a, b) with the detached cells showing fragmented DNA. Fibroblast growth factor or TPA can prevent this. It is clear from these studies that PKC activation can replace several growth factors (IL-2, EGF, FGF) and inhibit the active cell death pathway in a number of different systems suggesting that it plays a fundamental role in the switching process between viability and cell death.

In our studies we have found that the phorbol ester, TPA, can protect CTLL cells from apoptosis for at least 48 hours after IL-2 withdrawal (Figure 5.3A). However, the cells do not proliferate so the activation of PKC by TPA, although it is sufficient to prevent apoptosis, cannot induce G1 progression. The cells remain arrested in early G1 in what might be considered a "pseudo G0". Interestingly, phorbol esters cannot prevent glucocorticoid-induced cell death and 80% of the cells are dead by 24–48 h in the presence of TPA and dexamethasone (Figure 5.3). It is clear, therefore, that IL-2 must generate additional signals which (a) allow the cells to undergo cell cycle progression and (b) simultaneously make them resistant to glucocorticoids.

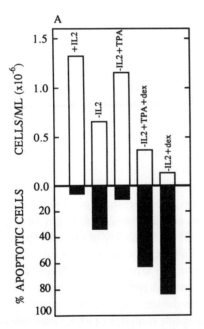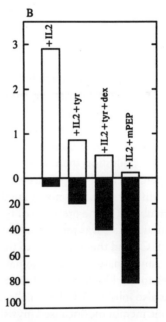

Figure 5.3 Effects of phorbol esters, PKC and tyrosine kinases on CTLL cell viability and proliferation. In A. cells were plated at 1.0×10^6 cells/ml and viability and percentage of apoptotic cells determined after 24 h. The cells were treated with either the phorbol ester, TPA, or dexamethasone in the presence or absence of IL-2. In B. cells were plated at 1.0×10^6 cells/ml and examined after 48 h. Cells received either tyrphostin (tyr), dexamethasone and/or the myristoylated peptide inhibitor of PKC (mPEP).

ROLE OF TYROSINE PHOSPHORYLATION IN IL-2 SIGNAL TRANSDUCTION

The interaction of IL-2 with the IL-2R has been shown to result in tyrosine phosphorylation of several substrates (p38, p57, p84, p100 and p120, Salzman et al., 1988). Similarly, Farrar et al. (1989) showed tyrosine phosphorylation of several proteins following IL-2 treatment of splenic lymphocytes. The latter included proteins, p42, p66, p70, p92, p180 that were phosphorylated within 5 minutes, whereas at 20 min the predominant proteins phosphorylated were p65–67, p70, p92, p116 and p180. The IL-2R beta subunit (p75) is now known to interact with $p56^{lck}$, the major lymphocyte membrane-associated tyrosine kinase, following IL-2 binding (Horak et al., 1991; Hatakeyama et al., 1991). The IL-2R beta chain is phosphorylated as a consequence of this interaction, along with several other proteins including the p72–74 raf-1 serine/threonine protein kinase (Saltzman et al., 1988; Asao et al., 1990; Merida and Gaulton, 1990; Turner et al., 1991). In human T cells tyrosine phosphorylation of these proteins is essential for IL-2 to drive proliferation since tyrphostin, a tyrosine kinase inhibitor, prevents cell cycle progression (Munoz et al., 1991). Only one major cytoplasmic protein was tyrosine phosphorylated in the latter study and is believed to be the 42 kDa microtubule-associated protein 2 kinase (MAP kinase) indicating that it plays a major role in IL-2 signal transduction. However, as described above additional substrates may also be phosphorylated. Tyrosine kinase(s) do, therefore, seem to be the principal transducers of information from the IL-2R, but an involvement of PKC cannot be ruled out.

Since a major signal transduction pathway used by IL-2 is mediated by the activation of tyrosine kinases we studied the effects of tyrphostins, which are specific inhibitors of tyrosine kinases, on cell death. In this experiment cell suspensions were seeded at 1×10^6 cells/ml and by 48 hours the cell number had increased approximately 3-fold in the presence of IL-2 (Figure 5.3B). When tyrphostin was added at the time of IL-2 addition there was no increase in cell number but only a modest increase in the percentage of apoptotic cells. Thus, inhibition of tyrosine kinases resulted in G1 arrest, but the cells did not immediately undergo apoptosis. Significantly however, cells incubated in the presence of tyrphostin re-acquired sensitivity to glucocorticoids even though IL-2 was present in the medium. Thus, the ability of the cells to progress through G1 and the resistance to the effects of glucocorticoids is, therefore, conferred by tyrosine kinase mediated signals. When tyrosine kinases are inhibited, cell cycle progression is blocked, but the cells do not immediately undergo apoptosis unless PKC is also inhibited. Indeed, cells incubated in the simultaneous presence of IL-2 and a specific membrane-localised myristoylated peptide inhibitor of PKC (Ioannides et al., 1990) rapidly undergo cell death confirming that PKC activity is required to disable the cell death pathway (Figure 5.3).

A ROLE FOR THE AP-1 TRANSCRIPTION FACTOR

Phorbol esters activate PKC in order to transduce signals to the nucleus and effect changes in gene expression. They have been shown to modify the expression of a number of genes through a common DNA regulatory element in their promoters, the TPA-responsive element, or TRE, which is the binding site for the transcription factor AP-1 (Angel *et al.*, 1988). The AP-1 transcription factor complex is composed mainly of *jun-jun* homodimers or *jun-fos* heterodimers, but other members of the *fos* and *jun* family of transcription factors are also be able to form complexes (Ryseck and Bravo, 1991). Thus, although collectively referred to as AP-1, several complexes may be involved in mediating changes in gene expression at the TRE. Since, the AP-1 transcription factor complex is required for cells to undergo G1 progression (Riabowol *et al.*, 1992) it may well play a role in switching between the cell death pathway and the cell cycle. Riabowol *et al.* (1992) have shown that fibroblasts that are unable to generate functional AP-1 are unable to proliferate and eventually become senescent. Senescence appears to result in part from a reduced ability to synthesise c-*fos* coupled with a failure to introduce the correct post-translational modifications that allow correct complex formation. The recent findings (reviewed by Diamond *et al.*, 1990; Miner *et al.*, 1991; Ponta *et al.*, 1992) that AP-1 can also modify the expression of glucocorticoid-inducible genes (and vice versa), through a complex series of interactions with the glucocorticoid receptor (GR) and the cognate DNA binding elements (TRE and GRE), are also relevant to studies on glucocorticoid-induced cell death. Because of these relationships between AP-1 and cell cycle progression as well as AP-1 and the GR we have investigated the role of AP-1 in switching between the cell cycle and apoptosis in CTLL cells in the presence and absence of IL-2 and glucocorticoids. In these experiments the ability of AP-1 to form DNA binding complexes was monitored using the gel retardation assay (Figure 5.4, Kwast-Welfeld *et al.*, 1991). In the presence of IL-2 the cells have high levels of AP-1 DNA binding activity which is compatible with its proposed essential role in cell cycle progression. Withdrawal of IL-2 leads to a decline in AP-1 DNA binding activity which precedes DNA fragmentation (Walker *et al.*, 1993). The glucocorticoid analogue, dexamethasone, dramatically increases the rate of loss of AP-1 DNA binding activity (Figure 5.4, lane 6) in IL-2-depleted cells and by 8 hours there is no detectable AP-1 DNA binding. Moreover, the appearance of DNA binding fragments of lower molecular weight suggest that the steroid is most likely triggering the proteolysis of the transcription factor. However, when IL-2 is present there is no loss of AP-1 DNA binding activity and, as described earlier, no cell death. Thus in the presence of IL-2 the AP-1 levels remain at the high level required for cell cycle progression (glucocorticoids have little or no effect on cell cycle progression in the presence of IL-2). Phorbol esters, as described above, are known to induce gene transcription through the TPA-responsive element which binds the AP-1 transcription factor. As shown in Figure 5.4 TPA can induce AP-1 DNA binding activity in IL-2-depleted cells that had lost their AP-1 DNA binding activity. However, unlike the AP-1 induced by IL-2, the AP-1 induced by TPA is not resistant

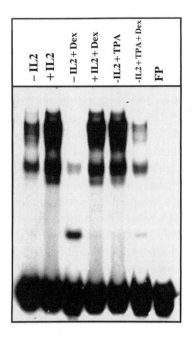

Figure 5.4 Levels of AP-1 DNA binding activity (determined by gel retardation assay, Kwast-Welfeld *et al.*, 1991). Cells were maintained with the additions indicated for 24 h. FP = free probe.

to glucocorticoids and AP-1 DNA binding activity is rapidly lost and the cells undergo apoptosis (Figure 5.2). There is, therefore, a positive correlation between the presence of AP-1 and the repression of apoptosis, but additional modifications to the protein components of AP-1 that do not affect its DNA binding activity are required to generate a form of the transcription factor complex that can promote G1 progression and confer resistance to glucocorticoids.

A tyrosine kinase-mediated step must confer these additional modifications. At present we do not know at what level this essential tyrosine phosphorylation takes place. It has been shown by Pulverer *et al.* (1991) that the transacting activity of the AP-1 component, c-*jun*, is stimulated by MAP kinase. The MAP kinase family of serine/threonine kinase are activated by phosphorylation on either serine/threonine residues and/or tyrosine residues. Since, the 42 kDa MAP-2 kinase is tyrosine phosphorylated by the lymphocyte specific tyrosine kinase $p56^{lck}$, as described above, and by PKC (Nel *et al.*, 1991) the IL-2-stimulated tyrosine phosphorylation step may occur at an early stage of signal transduction leading to the activation of a protein kinase that subsequently affects the DNA binding activity of AP-1, or its ability to interact with the GR by protein–protein interactions.

AP-1-GLUCOCORTICOID RECEPTOR INTERACTIONS

The glucocorticoid and other steroid hormone receptors have also been shown to interact with the components of the AP-1 transcription factor complex (reviewed by Diamond et al., 1990; Miner et al., 1991; Ponta et al., 1992). In the case of the GR a complex pattern of interactions with *fos* and *jun*, the principal AP-1 components, is emerging. This complexity is related to the availability of the various proteins involved (*fos*, *jun*, steroid-occupied GR), their modifications and the nature of the *cis*-acting DNA sequence involved (simple GRE, composite TRE-GRE or simple TRE element). Thus, genes containing either just a TRE or just a GRE as well as those containing a composite GRE-TRE or physically separated GREs and TREs can also be affected.

The mechanism by which an increase in AP-1 DNA binding activity, in response to IL-2 or phorbol esters, represses cell death is not known. Since protein synthesis and, presumably, gene expression is required for IL-2 deprived CTLL cells to undergo apoptosis (Duke and Cohen, 1986) the simplest explanation would be for the direct repression of the gene(s) by AP-1. However, to date, the binding of AP-1 to simple TREs has only been shown to induce gene expression not to repress it. It is possible, of course, that AP-1 induces the expression of a gene product that represses the gene(s) responsible for cell death.

In the presence of IL-2 AP-1 is not only sufficient to repress cell death but also to permit G1 progression and confer resistance to glucocorticoids. When IL-2 is withdrawn, the cells arrest in the cell cycle, AP-1 DNA binding activity is maintained for several hours and then declines. This loss is accelerated by the presence of dexamethasone, becoming undetectable by 8 hours and the rate of cell death is similarly increased. In both cases, therefore, a loss of AP-1 DNA binding activity precedes the onset of DNA fragmentation indicating that the presence of AP-1 represses cell death. Phorbol ester treated cells have high levels of AP-1 DNA binding activity, but remain sensitive to the glucocorticoid analogue, dexamethasone. Similarly, tyrphostin treated cells retain AP-1 DNA binding activity, but become sensitive to dexamethasone. These observations are consistent with a tyrosine phosphorylation-mediated modification to AP-1 being essential for dexamethasone resistance.

For glucocorticoids to induce cell death, gene expression also appears to be required and new genes which are expressed in steroid-treated thymocytes and lymphocytes have been identified (Baughman et al., 1991; Owens et al., 1991). Because of the complexity of the interactions between the GR and AP-1 there are a number of possible mechanisms by which these genes could be controlled. For example, cell death may be induced following an interaction between the steroid-occupied GR and a GRE located upstream of these induced genes. In the presence of IL-2 a fully functional AP-1 transcription factor complex can prevent the GR-induced expression of these genes. To do this the AP-1 must prevent GR binding to its GRE by protein-protein interaction. Since the DNA binding ability of AP-1 is insufficient (i.e. TPA-induced AP-1 cannot repress glucocorticoid-stimulated cell death) AP-1 is likely to interact with the GR through protein–protein interactions that are promoted by the additional modifications to AP-1 which are introduced by IL-2. Since both protein–protein and DNA–protein interactions mediate the complex interactions between the GR and AP-1 it is not possible to definitively predict the nature of the DNA sequence regulating the genes involved in the cell death pathway. Preliminary results indicate that dexamethasone is able to increase the rate of cell death in cells that are depleted of AP-1 DNA binding activity suggesting that the steroid-occupied GR is not simply sequestering AP-1 to relieve repression. The increased rate of cell death is probably due, therefore, to the GR binding to either a composite TRE-GRE or a simple GRE physically separated from the AP-1-regulated site to positively increase the rate of gene expression in the absence of AP-1. In the presence of IL-2 the GR cannot induce gene transcription unless a loss or modification of AP-1 is introduced to permit the GR to bind and increase the rate of transcription of these genes.

In summary, it appears that the PKC and tyrosine kinase-activated signal transduction pathways are both required for IL-2 to generate a functional AP-1 transcription factor complex which is able to maintain the cells in the cell cycle. In the absence of IL-2 the tyrosine phosphorylation-mediated signal declines and the cells will eventually exit the cell cycle and also become sensitive to glucocorticoids. Subsequently all AP-1 DNA binding activity is lost and the cells undergo apoptosis. Although phorbol esters can induce AP-1 DNA binding activity in CTLL cells they cannot effect the modifications to AP-1 that are necessary to confer glucocorticoid resistance and cell cycle progression. Thus, the AP-1 transcription factor complex plays a role not only in controlling the switching between senescence (and possibly quiescence) and the cell cycle, but also in switching between apoptosis and the cell cycle in those cells that cannot become quiescent. This role for AP-1 and the effects of the exogenous stimuli that affect it are summarised in Figure 5.5. The expression of c-*jun* appears to be constitutive in CTLL cells (Walker *et al.*, 1993). The synthesis of c-*fos* on the other hand is induced by IL-2 and TPA, both of which repress cell death by generating the AP-1 DNA binding complex designated $AP-1_A$, presumably by activating an isoform of PKC. $AP-1_A$ can repress cell death by preventing exit from the cell cycle, but cannot promote cell cycle progression and cannot prevent the steroid-occupied GR from degrading it and inducing apoptosis.

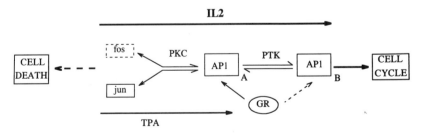

Figure 5.5   Role of AP-1 in switching between the cell death pathway and the cell cycle.

Only IL-2 can induce the further modifications to AP-1_A to generate AP-1_B that is resistant to the GR and can drive G1 progression. The activation of a tyrosine kinase appears essential for this additional modification. At the present time the nature of the additional modification(s) is not known. It may, for example, be further post-translational modifications to *fos* or *jun* proteins or it may be due to the incorporation of additional proteins into either AP-1 or GR complexes.

Finally, these observations point to the fundamental role that the AP-1 transcription factor plays in controlling the transitions of cells between major states such as the cell cycle, cellular quiescence/senescence and now cell death. Moreover, AP-1 interacts with the receptors of a variety of steroid hormones which typically induce differentiation in their target tissues (Herrlich *et al.*, 1990; Ponta *et al.*, 1992) suggesting that it also controls the ability of cells to move into and out of a differentiated state. Further studies on the role of AP-1 in mediating these changes in cellular state are essential for an understanding of the changes in sensitivity, or susceptibility, of cells to apoptosis and how these states may be manipulated by drugs for therapeutic purposes.

REFERENCES

Angel, P., Baumann, I., Stein, B., Delius, H., Rahmsdorf, H. J. and Herrlich, P. (1987) 12-o-tetradecanoyl-phorbol-13-acetate induction of the human collagenase gene is mediated by an inducible enhancer element located in the 5'-flanking region. *Molecular and Cellular Biology*, **7**, 2256–2266.

Araki, S., Shimada, Y., Kaji, K. and Hayashi, H. (1990a) Apoptosis of vascular endothelial cells by fibroblast growth factor deprivation. *Biochemical and Biophysical Research Communications*, **168**, 1194–1200.

Araki, S., Shimada, Y., Kaji, K. and Hayashi, H. (1990b) Role of protein kinase C in the inhibition by fibroblast growth factor of apoptosis in serum-depleted endothelial cells. *Biochemical and Biophysical Research Communications*, **172**, 1081–1085.

Asao, H., Takeshita, T., Nakamura, M., Nagata, K. and Sigamura, K. (1990) Interleukin 2 (IL-2)-induced tyrosine phosphorylation of IL-2 receptor p75. *Journal of Experimental Medicine*, **171**, 637–644.

Clarke, P. G. H. (1990) Developmental cell death: morphological diversity and multiple mechanisms. *Anatomy and Embryology*, **181**, 195–213.

Diamond, M. I., Miner, J. N., Yoshinaga, S. K. and Yamamoto, K. R. (1990) Transcription factor interactions: Positive or negative regulation from a single DNA element. *Science*, **249**, 1266–1272.

Duke, R. C. and Cohen, J. J. (1986) IL-2 addiction: withdrawal of growth factor activates a suicide program in dependent T cells. *Lymphokine Research*, **4**, 289–299.

Ellis, R. E., Yuan, J. and Horvitz, H. R. (1991) Mechanisms and functions of cell death. *Annual Review of Cell Biology*, **7**, 663–698.

Farrar, H. L. and Taguchi, M. T. (1985) Interleukin-2 stimulation of protein kinase C membrane association: evidence for IL-2 receptor phosphorylation. *Lymphokine Research*, **4**, 87–93.

Farrar, W. L. and Anderson, W. B. (1985) Interleukin-2 stimulates association of protein kinase C with membrane. *Nature*, **315**, 233–235.

Farrar, W. L., Ferris, W. L. and Harel-Bellam, A. (1989) The molecular basis of immune cytokine action. *CRC critical reviews in therapeutic drug carrier systems*, **5**, 229–261.

Friere-Moar J., Cherwinski, H., Hwang, F., Ranson, J. and Webb, D. (1991) Expression of protein kinase C isoenzymes in thymocyte subpopulations and their differential regulation. *Journal of Immunology*, **147**, 405–409.

Gillis, S. and Smith, K. A. (1977) Long term culture of tumour-specific cytotoxic T cells. *Nature*, **268**, 154–156.

Hatakeyama, M., Kono, T., Kobayashi, H., Kawahara, A., Levin, S. D., Perlmutter, R. M. and Taniguchi, T. (1991) Interaction of the IL-2 receptor with the *src*-family kinase $p56^{lck}$: Identification of novel intermolecular association. *Science*, **252**, 1523–1528.

Herrlich, P., Jonat, C., Ponta, H. and Rahmsdorf, H. J. (1990) A major decision: proliferation of differentiation. Interactions between pathway-specific transcription factors exemplify a molecular mechanism for the decision. In Bartram, C. R., Munk, K. and Schwab, M., eds. *Oncogenes in cancer diagnosis*. Basel: Karger, pp. 1–9.

Horak, D. J., Gress, R. E., Lucas, P. J., Horak, E. M., Woldmann, T. A. and Bolen, J. B. (1991) T-lymphocyte interleukin 2-dependent tyrosine protein kinase signal transduction involves the activation of $p56^{lck}$. *Proceedings of the National Academy of Science, USA*, **88**, 1996–2000.

Ioannides, C. G., Freedman, R. S., Liskamp, R. M., Ward, N. E., O'Brian, C. A. (1990) Inhibition of IL-2 receptor induction and IL-2 production in the human leukemic cell line Jurkat by a novel peptide inhibitor of protein kinase C. *Cellular Immunology*, **131**, 242–252.

Kanter, P., Leister, K. J., Tomei, L. D., Wenner, P. A. and Wenner, C. E. (1984) Epidermal growth factor and tumour promoters prevent DNA fragmentation by different mechanisms. *Biochemical and Biophysical Research Communications*, **118**, 392–399.

Kizaki, H., Shimada, H. and Ishimura, Y. (1989) 12-O-tetradecanoylphorbol 13-acetate induces DNA cleavage at linker regions in mouse thymocytes. *Journal of Biochemistry*, **105**, 673–675.

Kwast-Welfeld, J., de Belle, I., Walker, P. R., Isaacs, R. J., Whitfield, J.F. and Sikorska, M. (1991) Changes in cyclic adenosine monophosphate-responsive element binding proteions in rat hepatomas. *Cancer Research*, **51**, 528–535.

Larsen, C. S., Christiansen, N. O. and Esmann, V. (1988) Modulation of high-affinity interleukin-2 receptors on activated human T lymphocytes proliferation-related genes. *Scandinavian Journal of Immunology*, **28**, 167–173.

Legrue, S. J. (1988) Interleukin-2 stimulus-response coupling is calcium independent. *Lymphokine Research*, **6**, 1–11.

Lockshin, R. A. and Zakeri, Z. (1991) Programmed cell death and apoptosis. In *Apoptosis: the molecular basis of cell death*. pp. 47–60. Cold Spring Harbour Laboratory Press.

McConkey, D. J., Hartzell, P., Jondal, M. and Orrehius, S. (1989) Inhibition of DNA fragmentation in thymocytes and isolated thymocyte nuclei by agents that stimulate protein kinase C. *Journal of Biological Chemistry*, **264**, 13 399–13 402.

Merida, I., Pratt, J. C. and Gaulton, G. N. (1990) Regulation of interleukin-2-dependent growth responses by glycosylphosphatidylinositol molecules. *Proceedings of the National Academy of Science, USA*, **87**, 9421–9425.

Mills, G. B., Girard, P., Grinstein, S. and Gelfand, E. W. (1988) Interleukin-2 induces proliferation of T lymphocyte mutants lacking protein kinase C. *Cell*, **55**, 91–100.

Miner, J. N., Diamond, M. I. and Yamamoto, K. R. (1991) Joints in the regulatory lattice: composite regulation by steroid receptor-AP-1 complexes. *Cell Growth and Differentiation*, **2**, 525–530.

Munoz, E., Zubiaga, A. M. and Huber, B. T. (1991) Tyrosine phosphorylation is required for protein kinase c-mediated proliferation in T cells. *FEBS Letters*, **279**, 319–322.

Nel, A. E., Hanekom, C. and Hultin, L. (1991) Protein kinase C plays a role in the induction of tyrosine phosphorylation of lymphoid microtubule-associated protein-2 kinase. *Journal of Immunology*, **147**, 1933–1939.

Ojeda, F., Guarda, M. I., Maldonaldo, C. and Folch, H. (1990) Protein kinase C involvement in thymocyte apoptosis induced by hydrocortisone. *Cellular Immunology*, 125, 535–539.

Perotti, M., Toddei, F., Mirabelli, F., Vairetti, M., Bellomo, G., McConkey, J. and Orrenius, S. (1990) Calcium-dependent DNA fragmentation in human synovial cells exposed to cold shock. *FEBS Letters.*, 259, 331–334.

Ponta, H., Cato, A. C. B. and Herrlich, P. (1992) Interference of pathway specific transcription factors. *Biochimica Biophysica Acta*, 1129, 225–261.

Pulverer, B. J., Kyriakis, J. M., Avruch, J., Nikolakaki, E., and Woodgett, J. R. (1991) Phosphorylation of c-*jun* mediated by MAPkinases. *Nature*, 353, 670–674.

Redondo, J. M., Lopez-Rivas, A., Vila, V., Cragoe, E. . Jr. and Fresno, M. (1988) The role of protein kinase C in T lymphocyte proliferation. *Journal of Biological Chemistry*, 263, 17 467–17 470.

Riabowol, K., Schiff, J. and Gilman, M. Z. (1992) Transcription factor AP-1 activity is required for initiation of DNA synthesis and is lost during cellular aging. *Proceedings of the National Academy of Science, USA*, 89, 157–161.

Rodriguez-Tarduchy, G. and Lopez-Rivas, A. (1989) Phorbol esters inhibit apoptosis in IL-2 dependent T lymphocytes. *Biochemical and Biophysical Research Communications*, 164, 1069–1075.

Rossio, J. L., Ruscetti, F. W. and Farrar, W. L. (1986) Ligand-specific calcium mobilization in IL-2 and IL-2 dependent cell lines. *Lymphokine Research*, 5, 163–172.

Roy, C., Brown, D. L., Little, J. E., Valentine, B. K., Walker, P. R., Sikorska, M., Leblanc, J. and Chaly, N. (1992) The topoisomerase II inhibitor teniposide (VM26) induces apoptosis in unstimulated mature murine lymphocytes. *Experimental Cell Research*, 200, 416–424.

Ryseck, R.-P. and Bravo, R. (1991) c-*jun*, *jun* B, and *jun* D differ in their binding affinities to AP-1 and CRE consensus sequences: effect of *fos* proteins. *Oncogene*, 6, 533–542.

Saltzman, E. M., Thom, R. R. and Casnellie, J. E. (1988) Activation of a tyrosine protein kinase is an early event in the stimulation of T lymphocytes by interleukin-2. *Journal of Biological Chemistry*, 263, 6956–6959.

Sato, K., Ido, M., Kamiya, H., Sakurai, M. and Hidaka, H. (1988) Phorbol esters potentiate glucocorticoid-induced cytotoxicity in CEM-C7 human T-leukemia cell line. *Leukemia Research*, 12, 3–9.

Shi, Y., Szalay, M. G., Paskar, L., Boyer, M., Singh, B. and Green D. R. (1990) Activation-induced cell death in T cell hybridomas is due to apoptosis. *Journal of Immunology*, 144, 3326–3333.

Smith, K. A., (1989) The interleukin 2 receptor. *Annual Review of Cell Biology*, 5, 397–425.

Soloff, B. ., Nagle, W. A., Moss, A. J. Jr., Henlie, K. J. and Crawford, J. T. (1987) Apoptosis induced by cold shock *in vitro* is dependent on cell growth phase. *Biochemical and Biophysical Research Communications*, 145, 876–883.

Tenniswood, R. P. M., Welsh, J.-E., Sikorksa, M. and Walker, P. R. (In press) Molecular Biology of Apoptosis. *Proceedings of the Society for Experimental Biology and Medicine*.

Tomei, L. D., Kanter, P. and Wenner, C. E. (1988) Inhibition of radiation-induced apoptosis *in vitro* by tumor promoters. *Biochemical and Biophysical Research Communications*, 155, 324–331.

Turner, B., Rapp, U., App, H., Greene, M., Dobashi, K. and Reed, J. (1991) Interleukin-2 induces tyrosine phosphorylation and activation of p72-74 *raf*-1 kinase in a T-cell line. *Proceedings of the National Academy of Science, USA*, 88, 1227–1231.

Valge, V. E., Wong, J. G. P, Datlof, B. M., Sinskey, A. J. and Rao, A. (1988) Protein kinase C is required for response to T cell receptor ligands but not interleukin-2 in T cells. *Cell*, 55, 101–112.

Walker, P.R., Kwast-Welfeld, J., Gourdeau, H., Leblanc, J., Neugebaur, W. and Sikorska, M. (1993) Relationship between apoptosis and the cell cycle in lymphocytes: roles of protein kinase C, tyrosine phosphorylation and AP-1. *Experimental Cell Research*, **207**, 142–151.

Walker, P. R., Smith, C., Youdale, T., Leblanc, J., Whitfield, J. F. and Sikorska, M. (1991) Topoisomerase II-reactive chemotherapeutic drugs induce apoptosis in thymocytes. *Cancer Research*, **51**, 528–535.

Wyllie, A. H. (1980) Glucocorticoid-induced thymocyte apoptosis is associated with endogenous endonuclease activation. *Nature*, **284**, 555–556.

Chapter 6

INTRACELLULAR ZINC AND THE REGULATION OF APOPTOSIS

Peter D. Zalewski and Ian J. Forbes
Department of Medicine, University of Adelaide,
The Queen Elizabeth Hospital, Woodville, South Australia 5011, Australia

INTRODUCTION

Understanding the regulation of growth requires the elucidation of mechanisms involved in both mitosis and apoptosis, as well as the way in which these separate pathways are coordinately regulated to maintain tissue homeostasis. The numerous examples of stimuli which exert one effect on mitosis and the opposite effect on apoptosis, indicate that mechanisms exist to suppress one pathway when the other is activated. Intranuclear pools of zinc may play a pivotal role in this regulation, on the one hand acting as essential cofactors in multiple events required for mitosis while, on the other hand, suppressing internucleosomal DNA fragmentation and subsequent events in apoptosis. The role of zinc in mitosis has been reviewed elsewhere (Chesters, 1989) and will only be referred to briefly, here. The emphasis in this chapter will be on the involvement of zinc in the regulation of apoptosis.

INDUCTION OF APOPTOSIS IN ZINC DEFICIENCY

In vivo Studies

The essential requirement of zinc for growth, early development, tissue repair and normal function of the immune, nervous, reproductive and endocrine systems has been amply documented (Bach, 1981; Solomons, 1988; Chesters, 1989; Keen and Hurley, 1989). The classic features of zinc deficiency, which include retardation of growth, severe immunodeficiency, characteristic skin lesions, delayed sexual maturation and gastrointestinal problems, reflect the extreme sensitivity of those tissues and organs in which there is rapid turnover of cells (e.g. bone, thymus, epidermis, testis and intestinal crypts) (Solomons, 1988). Similarly, the zinc-deficient foetus develops numerous, severe malformations including abnormalities of the skeleton, fused or missing digits and cleft palate, corresponding to tissues undergoing growth and development at the time when zinc deficiency is imposed (Keen and Hurley, 1989).

In zinc deficiency, there is marked atrophy of thymus, prostate, testis and other affected tissues, due not only to impaired DNA synthesis and cell replication, but also to increased cell death (Elmes, 1977; Elmes and Jones, 1980; Record et al., 1985). The morphology of the dying cells, their presence often alongside apparently healthy cells, and the frequent finding of fragments of dead cells engulfed by neighbouring cells, are all consistent with death by apoptosis rather than necrosis. The finding of depressed activities of some zinc-regulated enzymes in the nuclei of affected tissues in zinc deficiency (Chesters, 1989) indicates that intracellular zinc is depleted in these tissues.

In vitro Studies

A direct relationship between increased apoptosis and a decline in intracellular zinc is supported by studies *in vitro*. Internucleosomal DNA fragmentation and other changes associated with apoptosis are induced in lymphoid and myeloid cell lines when cultured in zinc-free medium (Lennon and Cotter, 1991; Martin et al., 1991); re-addition of zinc to the medium prevents the apoptosis. Apoptosis can also be induced by treatment of cells with zinc chelators. Budzik and colleagues (1982) showed that the divalent cation chelator EDTA mimicked Müllerian-inhibiting substance in causing regression of the Müllerian duct in foetal organ cultures and that zinc was the only metal capable of blocking regression caused by both of these agents. In our laboratory, we have induced rapid apoptosis in various types of lymphoid cells including human chronic lymphocytic leukaemia (CLL) cells (Zalewski et al., 1991), normal human peripheral blood lymphocytes, rat thymocytes and rat splenocytes (Figure 6.1a), by treating them with membrane-permeable zinc chelators such as 1,10 phenanthroline and TPEN (N,N,N'N'-tetrakis-(2-pyridylmethyl)-ethylenediamine). In phenanthroline-treated CLL cells, DNA fragmentation was substantial by 2.5 h, when there was little cell death or morphological change; by 6 h, DNA fragmentation was maximal and about 50% of the cells appeared apoptotic (Zalewski et al., 1991). DNA fragmentation was blocked by the nuclease inhibitor aurin tricarboxylic acid. Typical of cells dying by apoptosis, there was little uptake of trypan blue, except at much higher concentrations of zinc chelator or after prolonged incubation, the cell volume decreased and there was early fragmentation of the nucleus. The rapidity of the onset of DNA fragmentation following addition of chelator would suggest that RNA and protein synthesis are not required although this has not yet been tested.

The apoptosis induced by phenanthroline was not prevented by the simultaneous addition of an excess of metals which do not bind to phenanthroline (including calcium, magnesium, lead, lithium and barium) but it was prevented by zinc and other

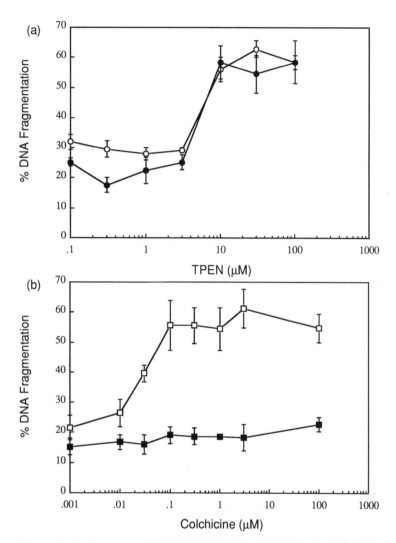

Figure 6.1 Effects of phorbol ester on DNA fragmentation induced by zinc chelator TPEN and colchicine. CLL cells were incubated with varying concentrations of TPEN (a) or colchicine (b), in the absence (○) or presence (●) of 200 nM phorbol dibutyrate. After 18 h, cells were washed, lysed and extent of DNA fragmentation determined by diphenylamine labelling of 15 000 g supernatants. Culture conditions and fragmentation assay were similar to those described in Zalewski et al. (1991). Points are means of triplicates, with error bars indicated.

metals which do bind (including copper, iron, cobalt, nickel and cadmium) (Zalewski et al., 1991). Since copper and iron are biologically-important metals, it cannot be ruled out that induction of apoptosis by phenanthroline is due to chelation of one of these metals rather than, or in addition to, zinc. More specific chelators are needed to establish whether metals other than zinc participate in the suppression of apoptosis.

DNA fragmentation (Figure 6.1a) and cell death induced by intracellular zinc chelators could be distinguished from that caused by other apoptotic stimuli (e.g. colchicine (Figure 6.1b), methylprednisolone and etoposide (Forbes et al., 1992)) in being insensitive to inhibition by phorbol esters. Removal of zinc from a critical target may over-ride the mechanism of inhibition by phorbol esters, which is presumably due to protein kinase C-dependent phosphorylation of a key substrate in the apoptotic pathway. Alternatively, since protein kinase C has two zinc-finger motifs and contains bound zinc atoms (Hubbard et al., 1991), its function may be compromised in a zinc-deficient cell. Studies in our laboratory and others (Csermely and Somogyi, 1989; Zalewski et al., 1990; Forbes et al., 1991) have shown altered subcellular distribution, ligand-binding and functional properties of protein kinase C in cells treated with zinc chelators or zinc ionophores, although the physiological significance of this regulation remains unclear.

Which types of cell are susceptible to zinc deficiency-induced apoptosis needs to be determined. Presumably, only those cells which express the endonuclease and other required gene products will be affected. Consistent with this is the preferential increase in apoptotic cells induced by zinc-deficiency, in those tissues which are rapidly-turning over and where many of the cells are expected to be primed for apoptosis. Some types of cell appear refractory to apoptotic stimuli. Examples that we have studied include mature human peripheral blood lymphocytes and infrequent populations of CLL cells, both of which were resistant to colchicine, methylprednisolone and etoposide. While the CLL cells were also resistant to phenanthroline, the normal peripheral blood lymphocytes were not (Zalewski et al., 1991), indicating different types of blocks in these two populations of cells. Whether the resistant CLL cells lack endonuclease needs to be determined. It would also be interesting to determine whether those cell types which appear to apoptose without undergoing DNA fragmentation are sensitive to zinc chelators.

SUPPRESSION OF APOPTOSIS BY SUPPLEMENTAL ZINC

Changes in the susceptibility of cells to undergo apoptosis may explain the altered resistance of animals to certain toxic agents when zinc intake is increased or decreased, respectively (Dinsdale and Williams, 1977; Bettger and O'Dell, 1981; Waring et al., 1990; Thomas and Caffrey, 1991), particularly since some of these agents (e.g. sporidesmin and colchicine) induce apoptosis in vitro and are antagonized by zinc. There are now numerous examples where presence of high concentrations of extracellular zinc ions (200–5000 μM) suppress internucleosomal DNA fragmentation and subsequent cell death in response to apoptotic stimuli (Table 6.1). In a few examples, inhibition of DNA fragmentation without suppression of cell death has been observed. The effects of zinc on other events in apoptosis such as expression of new gene products and changes in the cytoplasm and cytoskeleton have not yet been reported,

Table 6.1 Inhibition of apoptosis by extracellular zinc.

Stimulus[a]	Cell Type[b]	Reference
Cytotoxic T cells	PB-15 mastocytoma	Duke et al., 1983; Zychlinsky et al., 1991
Dexamethasone	Thymocytes	Cohen, Duke, 1984
Gamma irradiation	Thymocytes	Sellins, Cohen, 1987
Withdrawal of IL-2	T lymphoblasts	Kumar et al., 1987
Tumour necrosis factor	Endothelial cells	Flieger et al., 1990
Sporidesmin	T lymphoblasts	Waring et al., 1990
Cold shock	Fibroblasts	Nagle et al., 1990
Etoposide	HL-60	Shimuzu et al., 1990
Hyperthermia	Burkitt lymphoma	Takano et al., 1991
UV irradiation	HL-60	Martin et al., 1991
Adenosine	Thymocytes	Kizaki et al., 1988
Cytotoxic T cells[c]	YAC-1 lymphoma	Zychlinsky et al., 1991
ATP[c]	Thymocytes	Zheng et al., 1991

a. Agent which induces apoptosis.
b. Cell type in which apoptosis was induced.
c. Inhibition of DNA fragmentation but not cell death; all others, inhibition of both DNA fragmentation and cell death.

although Shimuzu and colleagues (1990) found inhibition by zinc of poly(ADP-ribose) synthesis in etoposide-treated HL-60 cells.

The requirement for high concentrations of extracellular zinc to inhibit apoptosis is probably due to the relatively poor uptake of zinc across the cell membrane. Thus, a much lower concentration of zinc ions (50 μM) will suffice to inhibit calcium-induced DNA fragmentation in isolated nuclei (Cohen and Duke, 1984) and low extracellular concentrations (5–25 μM) of zinc will almost completely suppress internucleosomal DNA fragmentation and cell death in intact cells when given in the presence of zinc ionophores such as pyrithione or di-iodoquinoline (Zalewski et al., 1991) (Figure 6.2). These compounds form lipophilic complexes with zinc and facilitate its uptake across the cell membrane.

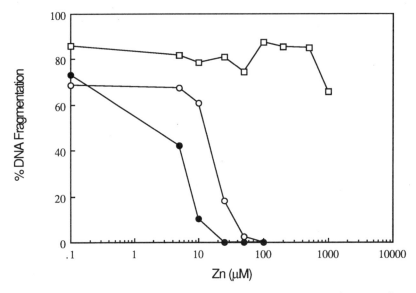

Figure 6.2 Inhibition of colchicine-induced DNA fragmentation by zinc plus ionophore. CLL cells were treated with 50 μM colchicine in the presence of varying concentrations of zinc sulphate, either alone (□), with 1 μM sodium pyrithione (○) or with 10 μM 1,8-dihydroxy iodoquinoline (●). After 18 h, cells were washed and DNA fragmentation determined as in legend to Figure 6.1.

MEASUREMENT OF EXCHANGEABLE INTRACELLULAR POOLS OF ZINC

Intracellular Pools of Zinc

Zinc is primarily an intracellular metal and the bulk of it is tightly bound in metalloenzymes such as carbonic anhydrase and superoxide dismutase. This zinc is essentially non-exchangeable with other pools of zinc, either within or outside the cell (Vallee and Galdes, 1984; Cousins, 1985). Since even in severe zinc deficiency, the total cellular zinc (as measured by atomic absorption spectroscopy) is decreased only slightly or not at all, it has been postulated that the zinc which is most susceptible to depletion and most likely to be involved in growth regulation resides in the minor, readily exchangeable pools (Bettger and O'Dell, 1981; Cousins, 1985; Chesters, 1989; Frederickson, 1989). These include free cytosolic zinc ions, zinc stored within vesicles, zinc loosely-bound in the membrane, to metallothionein or other cytosolic proteins and various pools of zinc within the nucleus (e.g. associated with the nucleolus, chromatin and mitotic spindle). Nuclear zinc, which constitutes about a third of cellular zinc, may be particularly important in regulation of apoptosis. The rapidity of uptake of ^{65}Zn into the nuclei of liver, kidney and spleen cells, shortly after ingestion argues for a direct pathway of exchangeable zinc from extracellular to intranuclear regions (Cousins and Lee-Ambrose, 1992).

Measurement of Intracellular Free or Loosely Bound Zinc with Zinc Fluorophore

Several years ago, we began a systematic search for compounds which might reliably signal fluxes of free or loosely bound intracellular zinc $[Zn^{2+}]_i$, analogous to intracellular fluorescent calcium indicators like Quin-2 and Fura-2. We now have several potentially useful compounds with the desired selectivity and sensitivity for zinc and which are retained by living cells; the one we have studied most intensely, zinquin, is an esterified congener of a compound TS-Q which has previously been used as a fluorescent, histochemical stain for free or loosely-bound zinc in fixed sections of brain and other tissues (Frederickson, 1989), but which has not been useful in studies of zinc in living cells, because of poor cellular retention.

Zinquin is taken up by cells and fluoresces in contact with intracellular zinc, when exposed to low wavelength light. Fluorescence of cells, as determined by microscopy or spectrofluorimetry, is decreased or increased by pretreatment of cells with zinc chelator or zinc ionophore, respectively. Values for $[Zn^{2+}]_i$ can be determined from fluorescence readings before and after lysis of cells in the presence of excess zinc (to obtain maximal fluorescence) and after subsequent addition of excess zinc chelator (to determine minimum fluorescence). Resting values are in the range 150–400 nM depending upon the cell type, although these can only be considered average values since they do not take into account the heterogeneity of fluorescence labelling between individual cells and between different subcellular compartments. The question of subcellular localisation of zinquin is currently being addressed by double-labelling techniques. Synthesis, properties and applications of zinquin to measure $[Zn^{2+}]_i$ will be published in detail elsewhere (Forbes, I.J. *et al.*, unpublished data).

$[Zn^{2+}]_i$ FLUXES AND GROWTH-ASSOCIATED PROCESSES

We are using zinquin (and subsequent versions of the probe as they are developed) to make a systematic investigation of the role of $[Zn^{2+}]_i$ in cell growth. In preliminary experiments, in which we varied $[Zn^{2+}]_i$ in CLL cells, thymocytes or splenocytes by pretreating with a range of concentrations of zinc chelator or zinc plus ionophore to deplete or increase cellular zinc, respectively, the extent of DNA fragmentation induced in these cells either spontaneously or by apoptotic stimuli (colchicine, dexamethasone) correlated inversely with $[Zn^{2+}]_i$. Relatively small changes in $[Zn^{2+}]_i$ were accompanied by marked changes in DNA fragmentation.

There is a need now to define physiological ranges of $[Zn^{2+}]_i$ in different tissues, different cells of the same tissue and cells at different stages of the cell cycle. It will also be important to identify physiological stimuli which increase or decrease $[Zn^{2+}]_i$. Our preliminary experiments with zinquin, indicate up to threefold increases in $[Zn^{2+}]_i$ in rat spleen cells or thymocytes treated for several hours with mitogenic concentrations of lectins in the presence of extracellular zinc. An influx of zinc may be a general feature of mitogenesis, since zinc is a cofactor for several mitogens including phytohaemagglutinin, concanavalin A, serum growth factors and epidermal growth factor (Williams and Loeb, 1975; Grummt et al., 1986; Kobusch and Bock, 1990). Other growth stimuli, such as phorbol esters, may not increase uptake of zinc, but instead cause a redistribution of cellular zinc between compartments (Csermeley and Somogyi, 1989).

Whether fluxes of zinc are a part of the survival signalling mechanism by which many growth stimuli suppress apoptosis is not known. The size of the increase in $[Zn^{2+}]_i$ (detected by zinquin) in concanavalin A-treated rat thymocytes would be sufficient to at least partially suppress DNA fragmentation, provided that zinc is increased in those intracellular (intranuclear?) pools which are involved in the regulation of the endonuclease. The role of zinc in suppression of apoptosis by concanavalin A and other growth stimuli needs to be investigated.

It is also pertinent to question whether the relatively large fluxes of zinc that occur in certain tissues or organs of the body influence the susceptibility of cells in those tissues to apoptose. In many potentially harmful situations such as stress, trauma, burns and infection there is a major redistribution of zinc from plasma to liver, thymus and bone marrow, in part mediated by interleukins and glucocorticoids (Cousins, 1985). There is also an accumulation of zinc in inflamed tissues (Alexander et al., 1982) and a substantial flux of zinc into the reproductive tract during the blastocyst phase of embryonic development (Lutwak-Mann and McIntosh, 1969). Deprivation of zinc is especially deleterious to the early embryo (Keen and Hurley, 1989). The very high concentration of zinc in prostate and seminal fluid (Srivastava et al., 1986) may exert a similar protective role in the male reproductive tract.

POTENTIAL TARGETS OF ZINC

There are numerous potential targets of zinc in the apoptotic cascade, and by analogy with the multiple sites of action of zinc in the mitotic cycle, we might expect zinc to act at more than one site in the apoptotic cascade. The problem will be in distinguishing the physiological, pharmacological and, even, toxicological effects of zinc. The latter is important because incubation of some types of cell for several hours with high concentrations of zinc can trigger events leading to necrosis and, thereby, suppress the important early events leading to apoptosis (Takano et al., 1991). Other types of cells are less sensitive and some T lymphocytes actually mitose when exposed to 200 μM zinc (Reardon and Lucas, 1987).

Calcium Antagonism

In view of the many reported examples of antagonism between the actions of calcium and zinc (Brewer, 1980), it is not surprising that these metals also act in opposite directions in the regulation of apoptosis. If calcium acts to induce apoptosis by several different mechanisms (e.g. induction of gene expression, activation of endonuclease, activation of calmodulin), then zinc may inhibit apoptosis by more than one mechanism. Zinc may also inhibit the cellular or nuclear uptake of calcium, by blocking or competing with calcium for membrane channels or by inhibition of calmodulin (Brewer, 1980).

Ca^{2+}/Mg^{2+} Endonuclease

Several laboratories have now confirmed the initial finding by Duke and colleagues (1983) that zinc is an inhibitor of the Ca^{2+}/Mg^{2+} endonuclease, generally thought to be responsible for the internucleosomal DNA fragmentation in apoptosis. We have found potent inhibition of Ca^{2+}/Mg^{2+} nuclease activity and DNA fragmentation in intact splenocytes, liver cells and CLL cells treated with micromolar concentrations of zinc plus zinc ionophore (Giannakis et al., 1991). Similar concentrations of zinc also strongly inhibited nuclease activity when added directly to cell-free extracts. It remains to be determined whether physiological concentrations of zinc influence the enzyme and whether the endonuclease is affected in zinc deficiency.

Since endonuclease activity was inhibited regardless of whether chromatin or isolated DNA was used as substrate, the action of zinc must be directed, at least primarily, at the enzyme rather than substrate. Nevertheless, zinc also binds to both protein and DNA in chromatin (Walker et al., 1989) and changes in sensitivity of chromatin to digestion by micrococcal nuclease have been reported in zinc-deficient cells (Stankiewicz et al., 1983), although the rate of digestion was decreased, compared with control cells.

Many questions spring to mind. Is the endonuclease normally in a suppressed state because of bound zinc? Does calcium displace zinc? Do phosphorylation, ADP-ribosylation or other covalent modification of the enzyme affect its sensitivity to inhibition by zinc? Does zinc also affect the subcellular distribution, synthesis or degradation of the enzyme? Answers to these questions must wait until we understand better the function and properties of the Ca^{2+}/Mg^{2+} endonuclease.

Zinc Finger Proteins

Other potential targets of zinc include the proteins which possess one or more zinc finger motifs and which interact with the apoptotic cascade. These include glucocorticoid receptors, protein kinase C and poly(ADP-ribose) polymerase (Gradwohl et al., 1990). Zn finger motifs of the type found in glucocorticoid receptors bind one zinc very tightly (K_d in the picomolar range) and a second zinc much more loosely (K_d in the micromolar range) to form zinc-cysteine clusters (Pan et al., 1990); binding of the second zinc will be affected by fluxes of intracellular zinc and could be important in the context of regulation of apoptosis.

Microtubules

Microtubules, which play an important, although poorly understood, role in the regulation of apoptosis are likely targets of zinc. The similar disruption of tubulin assembly that occurs during apoptosis (Martin and Cotter, 1990) and in tissues of zinc-deficient animals (Hesketh, 1981), as well as the prevention by zinc of microtubule depolymerization induced by apoptotic stimuli, such as colchicine and cold shock (Nickolson and Veldstra, 1972; Hesketh, 1984), supports this hypothesis.

Other Targets

Any of the 50 or more zinc metalloenzymes are also potential targets of zinc, although because they bind zinc very strongly, most are expected to be still fully active even in severely zinc-deficient cells and to be unaffected by exogenous zinc. Since zinc induces the expression of some early activation genes (Epner and Herschman, 1991), as well as genes essential for completion of mitosis, it is not unlikely that it may also affect expression of genes that control apoptosis. The labile zinc pathway from extracellular to intranuclear pools clearly alters transcription of at least one gene, that encoding metallothionein (Cousins and Lee-Ambrose, 1992).

We should also not rule out the possibility that zinc has no specific target in suppression of apoptosis, but rather by committing the cell to the mitotic cycle it excludes the possibility of that cell undergoing apoptosis. In this way, zinc deficiency would act like many other inhibitors of DNA synthesis (e.g. anti-cancer drugs) which also cause cells to apoptose.

CONCLUSIONS

There are many hints to the role of zinc in the regulation of apoptosis. Further understanding of the cellular biology of apoptosis and the way in which zinc fluxes are controlled should provide new insights into this role. From a practical viewpoint, we should consider supplementing media with physiologically-relevant concentrations of zinc, bearing in mind that concentrations of zinc in plasma are around 20 μM and may be even higher in some other extracellular fluids in the body. Many culture media contain no added zinc, except that derived from the addition of 5–10% serum or by contamination and under these conditions, cells will, with time, develop zinc-deficiency with a heightened risk of undergoing apoptosis. Some transformed cell lines have apparently overcome this problem by acquiring a greatly decreased dependency on extracellular zinc for growth (Rubin, 1972). With the advent of new tools and probes which can be applied to the study of apoptosis and to the measurement of intracellular zinc, it may now be opportune to reinvestigate the questions surrounding the role of zinc in regulation of apoptosis *in vivo* and the functional implications for zinc-deficient animals and man.

ACKNOWLEDGEMENTS

This work was supported by the National Health and Medical Research Council of Australia, the Anti-Cancer Foundation of South Australia, The Clive and Vera Ramaciotti Foundation and grants from the University of Adelaide.

REFERENCES

Alexander, J., Forre, O., Aaseth, J., Dobloug, J. and Ovrebo, S. (1982) Induction of a metallothionein-like protein in human lymphocytes. *Scandinavian Journal of Immunology*, 15, 217–220.

Bach, J.-F. (1981) The multi-faceted zinc dependency of the immune system. *Immunology Today*, 2, 225–227.

Bettger, W. J. and O'Dell, B. L. (1981) A critical physiological role of zinc in the structure and function of biomembranes. *Life Sciences*, 28, 1425–1438.

Brewer, G. J. (1980) Calmodulin, zinc and calcium in cellular and membrane regulation—an interpretive review. *American Journal of Haematology*, 8, 213–248.

Budzik, G. P., Hutson, J. M., Ikawa, H. and Donahoe, P. K. (1982) The role of zinc in Müllerian duct regression. *Endocrinology*, 110, 1521–1525.

Chesters, J. K. (1989) Biochemistry of zinc in cell division and tissue growth. In *Zinc in Human Biology*, C. F. Mills (ed.), pp. 109–118. London: Springer-Verlag.

Cohen, J. J. and Duke, R. C. (1984) Glucocorticoid activation of a calcium-dependent endonuclease in thymocyte nuclei leads to cell death. *Journal of Immunology*, 132, 38–42.

Cousins, R. J. (1985) Absorption, transport, and hepatic metabolism of copper and zinc: special reference to metallothionein and ceruloplasmin. *Physiology Reviews*, 65, 238–309.

Cousins, R. J. and Lee-Ambrose, L. M. (1992) Nuclear zinc uptake and interactions with metallothionein gene expression are influenced by dietary zinc in rats. *Journal of Nutrition*, 122, 56–64.

Csermely, P. and Somogyi, J. (1989) Zinc as a possible mediator of signal transduction in T lymphocytes. *Acta Physiologica Hungarica*, 74, 195–199.

Dinsdale, D., and Williams, R. B. (1977) The enhancement by dietary zinc deficiency of the susceptibility of the rat duodenum to colchicine. *British Journal of Nutrition*, 37, 135–142.

Duke, R. C., Chervenak, R. and Cohen, J. J. (1983) Endogenous endonuclease-induced DNA fragmentation: An early event in cell-mediated cytolysis. *Proceedings of the National Academy of Science, USA*, 80, 6361–6365.

Elmes, M. E. (1977) Apoptosis in the small intestine of zinc-deficient and fasted rats. *Journal of Pathology*, 123, 219–224.

Elmes, M. E. and Jones, J. G. (1980) Ultrastructural changes in the small intestine of zinc-deficient rats. *Journal of Pathology*, 130, 37–43.

Epner, D. E. and Herschman, H. R. (1991) Heavy metals induce expression of the TPA-inducible sequence (TIS) genes. *Journal of Cellular Physiology*, 148, 68–74.

Flieger, D., Riethmuller, G. and Ziegler-Heitbrock, H. W. L. (1989) Zn^{2+} inhibits both tumor necrosis factor-mediated DNA fragmentation and cytolysis. *International Journal of Cancer*, 44, 315–319.

Forbes, I. J., Zalewski, P. D. and Giannakis, C. (1991) Role for zinc in a cellular response mediated by protein kinase C in human B lymphocytes. *Experimental Cell Research*, 195, 224–229.

Forbes, I. J., Zalewski, P. D., Giannakis, C. and Cowled, P. A. (1992) Induction of apoptosis in chronic lymphocytic leukaemia cells and its prevention by phorbol ester. *Experimental Cell Research*, **198**, 367–372.

Frederickson, C. J. (1989) Neurobiology of zinc and zinc-containing neurons. *Internatioal Review of Neurobiology*, **31**, 145–238.

Giannakis, C., Forbes, I. J. and Zalewski, P. D. (1991) Ca^{2+}/Mg^{2+}-dependent nuclease: tissue distribution, relationship to internucleosomal DNA fragmentation and inhibition by Zn^{2+}. *Biochemical and Biophysical Research Communications*, **181**, 915–920.

Gradwohl, G., Menissier-de-Murcia, J. M., Molinete, M., Simonin, F., Koken, M., Hoeijmakers, J. H. and de-Murcia, G. (1990) The second zinc-finger domain of poly (ADP-ribose) polymerase determines specificity for single-stranded breaks in DNA. *Proceedings of the National Academy of Science, USA*, **87**, 2990–29904.

Grummt, F., Weinmann-Dorsch, C., Schneider-Schaulies, J. and Lux, A. (1986) Zinc as a second messenger of mitogenic induction. Effects on diadenosine tetraphosphate (Ap_4A) and DNA synthesis. *Experimental Cell Research*, **163**, 191–200.

Hesketh, J. E. (1981) Impaired microtubule assembly in brain from zinc-deficient pigs and rats. *International Journal of Biochemistry*, **13**, 921–926.

Hesketh, J. E. (1985) Microtubule assembly in rat brain extracts. Further characterization of the effects of zinc on assembly and cold stability. *International Journal of Biochemistry*, **16**, 1331–1339.

Hubbard, S. R., Bishop, W. R., Kirschmeier, P., George, S. J., Cramer, S. P. and Hendrickson, W. A. (1991) Identification and characterization of zinc-binding sites in protein kinase C. *Science*, **254**, 1776–1779.

Keen, C. L. and Hurley, L. S. (1989) Zinc and reproduction: effects of deficiency on foetal and postnatal development. In *Zinc in Human Biology*, C. F. Mills (ed.), pp. 183–220. London: Springer-Verlag.

Kizaki, H., Shimada, H., Ohsaka, F. and Sakurada, J. (1988) Adenosine, deoxyadenosine, and deoxyguanosine induce DNA cleavage in mouse thymocytes. *Journal of Immunology*, **141**, 1652–1657.

Kobusch, A. B. and Bock, K. W. (1990) Zinc increases EGF-stimulated DNA synthesis in primary mouse hepatocytes. *Biochemical Pharmacology*, **39**, 555–558.

Kumar, S., Baxter, G. D., Smith, P. J., Pemble, L., Collins, R. J., Prentice, R. L. and Lavin, M.F. (1987) DNA fragmentation in childhood T-cell acute lymphoblastic leukaemia. *Molecular Biology in Medicine*, **4**, 111–121.

Lutwak-Mann, C., McIntosh, J. E. A. (1969) Zinc and carbonic anhydrase in the rabbit uterus. *Nature (Lond.)*, **221**, 1111–1114.

Martin, S. J. and Cotter, T. G. (1990) Specific loss of microtubules in HL-60 cells leads to programmed cell death (apoptosis). *Biochemical Society Transactions*, **18**, 299–301.

Martin, S. J., Mazdai, G., Strain, J. J., Cotter, T. G. and Hannigan, B. M. (1991) Programmed cell death (apoptosis) in lymphoid and myeloid cell lines during zinc deficiency. *Clinical and Experimental Immunology*, **83**, 338–343.

Nagle, W. A., Soloff, B. L., Moss, A. J., Jr. and Henle, K. J. (1990) Cultured chinese hamster cells undergo apoptosis after exposure to cold but non-freezing temperatures. *Cryobiology*, **27**, 439–451.

Nickolson, V. J. and Veldstra, H. (1972) The influence of various cations on the binding of colchicine by rat brain homogenates. Stabilization of intact neurotubules by zinc and cadmium ions. *FEBS Letters*, **23**, 309–313.

Pan, T., Freedman, L. P. and Coleman, J. E. (1990) Cadmium-113 NMR studies of the DNA binding domain of the mammalian glucocorticoid receptor. *Biochemistry*, **29**, 9218–9225.

Reardon, C. L. and Lucas, D. O. (1987) Heavy metal mitogenesis: thymocyte activation by Zn^{2+} requires 2-mercaptoethanol and lipopolysaccharide as cofactors. *Immunobiology*, 174, 233–243.

Record, I. R., Tulsi, R. S., Dreosti, I. E. and Fraser, F. J. (1985) Cellular necrosis in zinc-deficient rat embryos. *Teratology*, 32, 397–405.

Rubin, H. (1972) Inhibition of DNA synthesis in animal cells by EDTA and its reversal by Zn. *Proceedings of the National Academy of Science, USA*, 69, 712–716.

Sellins, K. S. and Cohen, J. J. (1987) Gene induction by gamma-irradiation leads to DNA fragmentation in lymphocytes. *Journal of Immunology*, 139, 3199–3206.

Shimuzu, T., Kubota, M., Tanezawa, A., Sano, H., Kasai, Y., Hashimoto, H., Akiyama, Y. and Mikawa, H. (1990) Inhibition of both etoposide-induced DNA fragmentation and activation of poly(ADP-ribose) synthesis by zinc ion. *Biochemical and Biophysical Research Communications*, 169, 1172–1177.

Solomons, N. W. (1988) Zinc and copper. In *Modern Nutrition in Health and Disease*, 7th Edition, M. E. Shils and V. R. Young (eds.), pp. 238–262. Philadelphia: Lea and Febiger.

Srivastava, A., Chowdhury, A. R. and Setty, B. S. (1986) Testicular regulation and subcellular distribution of zinc in the epididymis and vas deferens of rhesus monkey (*Macaca mulatta*). *Acta Endocrinologica (Copenhagen)*, 113, 440–449.

Stankiewicz, A. J., Falchuk, K. H. and Vallee, B. L. (1983) Composition and structure of zinc-deficient *Euglena gracilis* chromatin. *Biochemistry*, 22, 5150–5156.

Takano, Y. S., Harmon, B. V. and Kerr, J. F. R. (1991) Apoptosis induced by mild hyperthermia in human and murine tumour cell lines: a study using electron microscopy and DNA gel electrophoresis. *Journal of Pathology*, 163, 329–336.

Thomas, D. J., and Caffrey, T. C. (1991) Lipopolysaccharide induces double-stranded DNA fragmentation in mouse thymus: protective effect of zinc pretreatment. *Toxicology*, 68, 327–337.

Vallee, B. L. and Galdes, A. (1984) The metallobiochemistry of zinc enzymes. In *Advances in Enzymology*, A. Meister (ed.), Vol 56, pp. 283–430. New York: John Wiley and Sons.

Walker, P. R., LeBlanc, J. and Sikorska, M. (1989) Effects of aluminium and other cations on the structure of brain and liver chromatin. *Biochemistry*, 28, 3911–3915.

Waring, P., Egan, M., Braithwaite, A., Mullbacher, A. and Sjaarda, A. (1990) Apoptosis induced in macrophages and T blasts by the mycotoxin sporidesmin and protection by Zn^{2+} salts. *International Journal of Immunopharmacology*, 12, 445–457.

Williams, R. O. and Loeb, L. A. (1973) Zinc requirement for DNA replication in stimulated human lymphocytes. *Journal of Cell Biology*, 58, 594–601.

Zalewski, P. D., Forbes, I. J. and Giannakis, C. (1991) Physiological role for zinc in prevention of apoptosis (gene-directed death). *Biochemistry International*, 24, 1093–1101.

Zalewski, P. D., Forbes, I. J., Giannakis, C., Cowled, P. A. and Betts, W. H. (1990) Synergy between zinc and phorbol ester in translocation of protein kinase C to cytoskeleton. *FEBS Letters*, 273, 131–134.

Zheng, L. M., Zychlinsky, A., Liu, C. C., Ojcius, D. M. and Young, J. D. (1991) Extracellular ATP as a trigger for apoptosis in programmed cell death. *Journal of Cell Biology*, 112, 279–288.

Zychlinsky, A., Zheng, L. M., Liu, C. C. and Young, J. D. (1991) Cytolytic lymphocytes induce both apoptosis and necrosis in target cells. *Journal of Immunology*, 146, 393–400.

Chapter 7

INDUCTION OF APOPTOSIS BY THE IMMUNOMODULATING AGENT GLIOTOXIN

Paul Waring
Division of Cell Biology, John Curtin School of Medical Research, Australian National University, PO Box 334, Canberra City 2601, Australia

INTRODUCTION

The process of apoptosis or programmed cell death is usually contrasted with necrotic cell death and the latter is often associated in the literature with toxin induced cell death (Kerr *et al.*, 1987; Wyllie *et al.*, 1980). In fact it is now apparent that a large repertoire of toxins of varying and unrelated structure can induce the features commonly observed in classically induced apoptotic cell death—chromatin condensation and internucleosomal DNA fragmentation (Waring *et al.*, 1991). Many of these toxins include anti-neoplastic agents or other molecules in chemotherapeutic use as well as toxins of environmental importance. An understanding of the mechanism(s) by which these agents induce apoptosis in the respective target cell would therefore be of some importance. Table 7.1 indicates examples of toxins known to induce the morphological and/or the biochemical features of apoptotic cell death in a variety of cells. The list is not meant to be comprehensive. It can be seen on initial examination that the structural features of these toxins appear to present no common feature which may allow a rational explanation for their mode of action with respect to induction of apoptosis. A few generalizations however can be made. A number of toxins inducing apoptosis e.g. gliotoxin (Waring *et al.*, 1988), ricin (Griffins *et al.*, 1987; Muldoon *et al.*, 1992), bleomycin (Kuo and Hsu, 1978), menadione (McConkey *et al.*, 1988b) are known to be capable of causing oxidative damage to DNA. Since low level damage to DNA caused by gamma irradiation (Umansky, 1991) is thought to result in apoptosis

Table 7.1 Toxins inducing apoptosis.

Toxin	Properties
Gliotoxin	Protein synthesis inhibitor Capable of single strand DNA damage in cell free conditions
Ricin	Potent protein synthesis inhibitor
Bleomycin	Direct cellular DNA damage
Menadione	Can generate free radicals intracellularly by REDOX cycling
Cycloheximide	Protein synthesis inhibitor
Diphtheria toxin	Protein synthesis inhibitor
Dioxin	Possible free radical generator Increases intracellular calcium
Calcium ionophores	Increases intracellular calcium

in sensitive cells, this may provide an underlying explanation for those DNA damaging toxins inducing apoptosis in exposed cells although the relationship between low level DNA damage and the resulting process which leads to the phenomenon of apoptosis is still unclear. The same may be true of toxins which alkylate DNA (Zwelling et al., 1991).

A second class of toxins or treatments inducing apoptosis in cells are the protein synthesis inhibitors. These include cycloheximide (Waring, 1990; Collins et al., 1991), hyperthermia (Harmon et al., 1990) and gliotoxin (Waring et al., 1989). Cycloheximide is the classic tool which has been used to probe for the requirement of protein synthesis in apoptosis induced by other stimuli such as dexamethasone (Wyllie et al., 1984). A number of other notable inhibitors of protein synthesis have now been shown to induce the features of apoptosis in a variety of cells. The binary toxin ricin was shown to cause the morphology of apoptosis in epithelial cells *in vivo* (Griffins and Leek, 1987). Later it was shown that ricin can also induce the typical DNA fragmentation associated with apoptosis in macrophages and activated T cells (Waring, 1990). Diphtheria toxin has also been shown to induce apoptosis in cells (Chang et al., 1989). These proteinaceous toxins are heterodimers consisting of an A and B chain linked by a disulphide bond. They appear to act by initially binding to the cell by the B chain followed by endocytosis and release of the catalytically active A chain. The latter inhibits protein synthesis by catalytically inactivating the translationary mechanism (Lord et al., 1987). It has been suggested that diphtheria toxin may have intrinsic nuclease activity although this has been disputed (Chang et al., 1989; Bodley, 1990). We have found no evidence for any calcium dependent nuclease activity in ricin (Khan and Waring, unpublished observation). It thus appears that the process of inhibition of protein synthesis *per se* may be sufficient to trigger apoptosis. This is contrary to the generally held dogma that apoptosis requires new protein synthesis in order to proceed although evidence for requirement of protein synthesis in a number of cases is indisputable (Cohen and Duke, 1984; Wyllie et al., 1984; Sellins and Cohen, 1987). These two apparently contradictory aspects to induction of apoptosis have yet to be resolved. One possibility however is that inhibition of survival proteins e.g. Bcl-2 (Sentman et al., 1991) by protein synthesis inhibitors results in apoptotic death by default. It is not clear why this occurs readily in some cells e.g. apoptosis in macrophages induced by cycloheximide but not in others e.g. thymocytes treated with the same concentrations of cycloheximide.

Other toxins may induce apoptosis by mobilizing intracellular calcium since increases in cell calcium have been shown to be associated with apoptosis (McConkey et al., 1989a; McConkey et al., 1989b). The toxin dioxin for example has been shown to induce apoptosis in thymocytes and this occurs with elevated levels of intracellular calcium (McConkey et al., 1988a). Calcium ionophores have also been shown to induce apoptosis (Wyllie et al., 1984)

GLIOTOXIN INDUCED APOPTOSIS

A toxin we have been particularly interested in is the fungal toxin gliotoxin (Taylor, 1971; Mullbacher et al., 1986; Mullbacher et al., 1987; Waring et al., 1990; Waring et al., 1991) which belongs to the epipolythiodioxopiperazine (ETP) class of secondary metabolites. Gliotoxin is a potent immunomodulating agent which appears to be selectively toxic to cells of the haemopeoitic system (Mullbacher et al., 1986). The selective toxicity of gliotoxin is shown in its ability to delete mature T cells from bone marrow cells allowing repopulation of lethally irradiated animals without ensuing graft versus host disease (Mullbacher et al., 1987). Gliotoxin inhibits proliferation of antigen activated T and B cells, macrophage adherence and induction of cytotoxic T cells in a mixed lymphocyte reaction (Mullbacher et al., 1986). It thus appears to inhibit a number of immune functions. The required ED_{50} for inhibition of proliferation in activated T cells is 100 nM. At this concentration there is significant DNA fragmentation typical of apoptosis in treated cells. In comparison, P-815 cells are relatively resistant to the effects of gliotoxin (ED_{50} for inhibition of proliferation is approximately 5 μM and even at these concentrations there is very little apoptotic DNA fragmentation apparent). Similarly, proliferation of the T-cell line WEHI-7, a dexamethasone sensitive cell line, is inhibited by gliotoxin at an ED_{50} of 300 nM. At these concentrations there is massive DNA fragmentation typical of apoptosis (Waring and Mamchack, unpublished data). Inhibition of cell proliferation by gliotoxin therefore seems to be related to its ability to induce apoptosis in the cells. The selective effect of gliotoxin may be related to the known sensitivity of lymphocytes to apoptotic cell death. Sporidesmin, another member of the ETP class of toxins and a number of simple analogs of gliotoxin have also been shown to induce apoptosis in a number of cell types (Waring et al., 1990). Interestingly the anti-phagocytic action of gliotoxin appears to be unrelated to its ability to cause apoptosis in macrophages (Waring et al., 1989).

Mechanism of Action of Gliotoxin Induced Apoptosis

Oxidative stress

The bridged piperazine disulphide ring which defines the ETP class of fungal metabolites appears to be necessary for the biological activity of gliotoxin (Middleton, 1974). Removal of the disulphide ring by reduction and methylation abolishes the immunomodulating activity of these compounds (Mullbacher et al., 1986). Gliotoxin and related analogs in the reduced form have been shown to be capable of causing single strand DNA damage to plasmid DNA in the presence of iron (II) salts (Eichner et al., 1988). The process is inhibited by free radical inhibitors such as ascorbic acid and by catalase and superoxide dismutase. This suggests a role for free radical generation. Agents which cause oxidative stress in hepatocytes have been shown to cause apoptosis in these cells (McConkey et al., 1988) and it is possible that oxidative stress generally may induce apoptosis. Hydrogen peroxide was directly detected during auto-oxidation of the reduced form of a simple analog of gliotoxin (Waring et al., 1989). These data led to the proposal that redox cycling of the oxidized form of the ETP compound by cellular reductants such as glutathione or NADPH followed by oxidation by molecular oxygen may result in oxidative damage to DNA via superoxide and ultimately hydrogen peroxide and hydroxyl radicals. This would be analogous to low level DNA damage induced by gamma irradiation also leading to apoptosis. Sporidesmin, another member of the ETP class of toxins, has been shown to induce oxidative stress in erythrocytes (Munday, 1985) and we have shown that sporidesmin will also cause apoptotic DNA fragmentation in cells of the immune system (see above). Attempts to inhibit apoptosis induced by ETP compounds using a wide range of typical free radical inhibitors have however failed to show any effect of these agents on apoptosis (Waring P., unpublished data). In addition, no change in reduced glutathione levels were observed in cells treated with concentrations of gliotoxin which caused apoptosis (Waring, P., unpublished data). This suggests that gross oxidative stress is not occurring in cells treated with gliotoxin. It still does not rule out the possibility that low level DNA damage caused by gliotoxin but not inhibited by normal free radical inhibitors is occurring in the nucleus. Experiments to test this possibility are under way.

Inhibition of protein synthesis

Gliotoxin is known to be a protein synthesis inhibitor (Eichner et al., 1986; Waring, 1990). Since a number of protein synthesis inhibitors are now known to cause apoptosis by an as yet undefined mechanism (see above), it is possible that gliotoxin is acting in the same way. The problem with this idea is that cycloheximide at concentrations sufficient to inhibit all protein synthesis in thymocytes does not cause apoptosis in these cells whereas gliotoxin induces apoptosis in thymocytes at concentrations as low as 100 nM. Ricin, a potent protein synthesis inhibitor does not induce apoptosis in thymocytes at concentrations known to inhibit protein synthesis although it does cause apoptosis in macrophages (Khan, T. and Waring, P., unpublished data). It would appear that protein synthesis inhibition by gliotoxin may be insufficient to explain apoptosis caused by this toxin.

Calcium mobilization

Because calcium mobilization has been shown to be associated with apoptosis caused by dexamethasone, CD3 or TCR activation as well as apoptosis caused by dioxin (see above), we examined the role played by calcium in gliotoxin induced apoptosis. Work in this laboratory has shown no requirement for extracellular calcium in apoptosis induced by gliotoxin in macrophages (Waring, P. and Sjaarda, A., unpublished data). Treatment of cells with labelled $^{45}Ca^{+2}$ in the presence and absence of gliotoxin in calcium depleted media for 1 h did show a significant increase in uptake of extracellular calcium in treated cells (Figure 7.1). However, EGTA at 500 μM abolishes this uptake but does not inhibit DNA fragmentation—Table 7.2. Similar results were found for other time points. Only at concentrations of EGTA of 8 mM was fragmentation apparently inhibited—however uptake of radiolabelled calcium is inhibited to the same extent by 0.5. 1, 2 or 8 mM EGTA (Waring and Sjaarda, unpublished data). Concentrations of EGTA of 5–8 mM have been reported in the literature as also inhibiting apoptosis caused by dexamethasone, anti-CD3 and dioxin (McConkey et al., 1988a). We used the method of FACS analysis of cells developed by Lyons et al. (1992) to examine apoptotic cells after treatment with EGTA at various concentrations (Figure

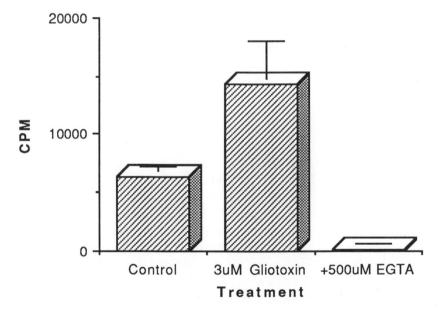

Figure 7.1 Uptake of radiolabelled calcium into macrophages treated with gliotoxin. Macrophages were treated with 3 μM gliotoxin in the presence and absence of 500 μM EGTA in calcium depleted media (Ca^{+2} <100 μM) Cells were pulsed with $^{45}Ca^{+2}$ for 1 h following toxin treatment and radiolabel estimated in the cells after efflux of rapidly exchanging calcium.

Table 7.2 Gliotoxin Induced DNA fragmentation in murine peritoneal macrophages.

Treatment[a]	% DNA Fragmentation[b]
Untreated	< 2
3 μM gliotoxin	67
3 μM gliotoxin in Ca^{+2} depleted media (Ca^{+2} < 100 μM)	77
3 μM gliotoxin in Ca^{+2} depleted media + 500 μM EGTA	74

a. Cells were treated with gliotoxin for 1 h then pulsed with labelled calcium for 1h.
b. DNA fragmentation was estimated by running extracted DNA on 1.5% agarose gels photography under UV light and densitometry of the resulting negative.

7.2). In this technique, ethidium bromide staining of apoptotic cells reveals a discrete population of cells with a small but increased fluorescence (referred to as region 2) which correlates with increased DNA fragmentation (Lyons et al., 1992) and see Figure 7.2, panels A and B. The more overtly staining cells in regions 3 and 4 appear to be associated with the later processes of apoptosis such as apoptotic body formation or secondary necrosis. Cells treated with dexamethasone after 4–6 h for example show up mainly in region 2 while cells treated with dexamethasone for 24 h appear in regions 3 and 4. Treatment of thymocytes with dexamethasone and EGTA at 1 mM and 2 mM shows little effect on apoptosis and this is supported by lack of inhibition of DNA fragmentation (Waring and Sjaarda, unpublished data). This is shown in panels D and E of Figure 7.2 which show the same kind of profiles already described for thymocytes treated with dexamethasone alone. Cells treated with 8 mM EGTA and dexamethasone for only 4 h however display the same profile as cells treated with dexamethasone alone for 24 h, panel F of Figure 7.2. Over 72% of cells in panel F appear either in R3 or R4. If there was protection conferred by EGTA at these concentrations then the profile should resemble panel A. Interestingly, 8 mM EGTA alone has no effect on thymocytes as shown in panel C. This suggests that contrary to inhibiting DNA fragmentation, high concentrations of EGTA in the presence of dexamethasone cause irreversible changes to the cell over the short 4 h period required for the DNA fragmentation to become obvious and the apparent inhibition of DNA fragmentation may simply be a consequence of a direct necrotic type of cell death which normally does not exhibit internucleosomal DNA fragmentation. Caution therefore must be exercised in apparent inhibition of apoptosis by high concentrations of EGTA.

Following treatment of cells for periods of 10–30 min, radiolabelled gliotoxin becomes rapidly cell associated and has been shown to covalently bind to cytosolic proteins (Waring, P. unpublished data). This presumably occurs through mixed

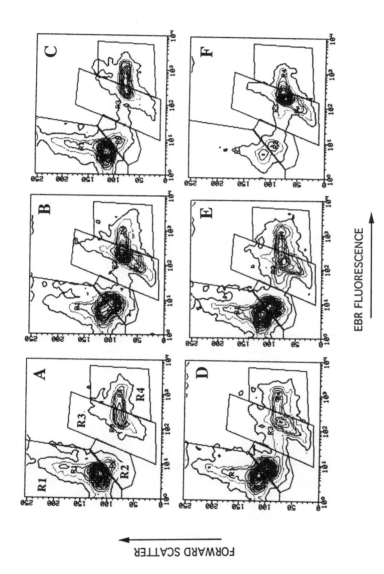

Figure 7.2 FACS Analysis of thymocytes treated with dexamethasone and EGTA. Cells were treated with 1 μM dexamethasone and 1, 2 or 8 mM EGTA for 4 h and stained according to Lyons et al. (1992). Panel A: untreated; panel B: 1 μM Dex.; panel C 8 mM EGTA; panels D-F: 1 μM Dex. with 1,2 and 8 mM EGTA respectively.

disulphide formation between the disulphide of the ETP compound and free thiol groups on proteins. Only a small fraction of the gliotoxin is covalently linked to protein and the interaction is relatively non-specific. Since other thiol specific agents have been shown to covalently inactivate the Ca^{+2} ATPase pump in muscle cell sarcoplasmic reticulum and release calcium from microsomal fractions (Prabhu and Salam, 1990), we examined intracellular calcium levels in cells treated with gliotoxin. Using intracellular calcium sensitive fluorescent dyes we have shown no mobilization of intracellular stores after treatment of thymocytes with gliotoxin between 0.1–3 μm, the range required for apoptosis (Beaver, J. and Waring, P., unpublished data). Interestingly, there does appear to be a rise in cell calcium at concentrations greater than 100 μM. At these concentrations there is no DNA fragmentation and the cells show the features of necrosis.

CONCLUSION

It is clear from these results that we are yet to determine the mechanism of gliotoxin induced apoptosis. This could also be said of the numerous other toxins which induce this form of cell death. It is clear that neither calcium mobilization nor inhibition of protein synthesis—two likely possibilities—provides an obvious answer in the case of gliotoxin. It is almost certain that gliotoxin exerts its effects "downstream" of those caused by stimulation of CD3 or T-cell receptor or even dexamethasone-induced apoptosis since gliotoxin-induced apoptosis in thymocytes is not inhibited by cycloheximide (Waring, 1990). Apoptosis caused by gliotoxin may be a consequence of its anti-proliferative action rather than a cause since cycloheximide causes apoptosis in WEHI-7 cells only at concentrations great enough to inhibit DNA synthesis but not at levels inhibiting protein synthesis (Waring and Mamchack, unpublished data). Investigation of this aspect of gliotoxin toxicity is continuing.

ACKNOWLEDGEMENTS

I would like to acknowledge the following collaborators for their contributions to the study of gliotoxin toxicity: Dr A. Braithwaite, Ms J. Beaver, Mr M. Egan, Dr R. D. Eichner, Mr Ron Tha-La, Ms. Tahira Khan, Dr A. Mullbacher, Mr A. Sjaarda, Dr U. Tiwari-Palni. I would particularly like to thank Dr Bruce Lyons for introducing me to his method of FACS analysis of apoptotic cells.

REFERENCES

Bodley, J. W. (1990) Does diphtheria toxin have nuclease activity? *Science*, **250**, 832.

Chang, M. P., Baldwin, R. L., Bruce, C. and Wisnieski, B. J. (1989) Second cytotoxic pathway of diphtheria toxin suggested by nuclease activity. *Science*, **246**, 1165–1168.

Chang, M. P., Bramhall, J., Gaves, S., Bonavidas, B. and Wisnieski, B. J. (1989) Internucleosomal DNA cleavage precedes diphtheria toxin induced cytolysis. *Journal of Biological Chemistry*, **264**, 15 261–15 267.

Cohen, J. J. and Duke, R. C. (1984) Glucocorticoid activation of a calcium dependent endonuclease in thymocyte nuclei leads to cell death. *Journal of Immunology*, **132**, 38–42.

Collins, R. J., Harmon, B. V., Souvlis, T., Pope, J. H. and Kerr, J. F. R. (1991) Effect of cycloheximide on B-chronic lymphocytic leukaemic and normal lymphocytes *in vitro*—induction of apoptosis. *British Journal of Cancer*, **64**, 518–522.

Eichner, R. D., Al Salami, M., Wood, R. R. and Mullbacher, A. (1986) Effect of gliotoxin on macrophage function. *International Journal of Immunopharmacology*, **8**, 789–797.

Eichner, R. D., Waring, P., Geue, A., Braithwaite, A. W. and Mullbacher, A. (1988) Gliotoxin causes oxidative damage to plasmid and cellular DNA. *Journal of Biological Chemistry*, **263**, 3772–3777.

Griffins, G. D., Leek, M. D. and Gees, D. J. (1987) The toxic plant proteins ricin and abrin induce apoptotic changes in mammalian lymphoid tissue and intestine. *Journal of Pathology*, **151**, 221–224.

Harmon, B. V., Corder, A. M., Collins, R. J., Gobé, G. C., Allen, J., Allan, D. J. and Kerr, J. F. R. (1990) Cell death induced in a murine mastocytoma by 42–47°C heating *in vitro*: evidence that the form of death changes from apoptosis to necrosis above a certain heat load. *International Journal of Radiation Biology*, **58**, 845–858.

Kerr, J. F. R., Searle, J. S., Harmon, B. V. and Bishop, C. J. (1987) Apoptosis in Perspectives. In *Mammalian Cell Death*, C. S. Potten, (ed.), pp. 93–128. New York: Oxford Science Publications.

Kuo, M. Y. and Hsu, T. C. (1978) Bleomycin causes release of nucleosomes from chromatin and chromosomes. *Nature*, **271**, 83–84.

Lord, J. M., Gould, J., Griffiths, D., O'Hare, M., Prior, B., Richardson, P. T. and Roberts, L. M. (1987) Ricin: cytotoxicity, biosynthesis and use in immunoconjugates. In *Progress in Medicinal Chemistry*, G. P. Ellis and G. B. West (eds.), pp. 1–28. Elsevier Science Publishers.

Lyons, A. B., Samuel, K., Sanderson, A. and Maddy, A. H. (In press) Simultaneous analysis of immunophenotype and apoptosis of murine thymocytes by single laser flow cytometry. *Cytometry*.

McConkey, D. J., Hartzell, P., Amador-Perez, J. F., Orrenius, S. and Jondal, M. (1989a) Calcium dependent killing of immature thymocytes by stimulation via the CD3/T-cell receptor complex. *Journal of Immunology*, **143**, 1801–1806.

McConkey, D. J., Hartzell, P., Duddy, S. K., Hakansson, H. and Orrenius, S. (1988a) 2,3,7,8-tetrachlorodibenzo-p-dioxin kills immature thymocytes by Ca mediated endonuclease activation *Science*, **242**, 256–259.

McConkey, D. J., Hartzell, P., Nicotera, P., Wyllie, A. and Orrenius, S. (1988b) Stimulation of endogenous endonuclease activity in hepatocytes exposed to oxidative stress. *Toxicology Letters*, **42**, 123–130

McConkey, D. J., Nicotera, P., Hartzell, P., Bellomo, G., Wyllie, A. H. and Orrenius, S. (1989b) Glucocorticoids activate a suicide process in thymocytes through an elevation of cytosolic Ca^{+2} concentration. *Archives of Biochemistry and Biophysics*, **269**, 365–370.

Middleton, M. C. (1974) The involvement of the disulphide group of sporidesmin in the action of the toxin on swelling and respiration of liver mitochondria. *Biochemical Pharmacology*, **23**, 811–820.

Muldoon, D. F., Hassoun, E. A. and Stohs, S. J. (1992) Ricin induced hepatic lipid peroxidation, glutathione depletion, and DNA single strand breaks in mice. *Toxicology*, **30**, 977–984.

Mullbacher, A., Hume, D., Braithwaite, A. W., Waring, P. and Eichner, R. D. (1987) Selective resistance of bone marrow derived hemopoietic progenitor cells to gliotoxin. *Proceedings of the National Academy of Science, USA.*, **84**, 3822–3825.

Mullbacher, A., Waring, P., Braithwaite, A. W. and Eichner, R. D. (1987) Fungal toxins as immunomodulating agents in *Drugs of the Future*, **12**, 565– 573.

Mullbacher, A., Waring, P., Tiwari-Palni, U. and Eichner, R. D. (1986) Structural relationship of epipolythiodioxopiperazines and their immunomodulating activity. *Molecular Immunology*, **23**, 231–235.

Munday, R. (1985) Studies in the mechanism of toxicity of the mycotoxin sporidesmin IV. *Journal of Applied Toxicology*, **5**, 69–73.

Prabhu, S. D. and Salam, G. (1990) Reactive disulphide compounds induce Ca^{+2} release from cardiac sarcoplasmic reticulum. *Archives of Biochemistry and Biophysics*, **282**, 275–283.

Sellins, K. S. and Cohen, J. J. (1987) Gene induction by gamma irradiation leads to DNA fragmentation in lymphocytes. *Journal of Immunology*, **139**, 3199–3206.

Sentman, C. L., Shutter, J. R., Hockenberry, D., Kanagawa, O. and Korsmeyer, S. J. (1991) bcl-2 inhibits multiple forms of apoptosis but not negative selection in thymocytes. *Cell*, **67**, 879–888.

Taylor, A (1971) The toxicity of sporidesmin and other epipolythiodioxopiperazines. In *Microbial Toxins VII* (S. Kadis, A. L. Ciegler and S. J. Ajl (eds.), pp. 337–371. Academic Press.

Waring, P. (1990) DNA fragmentation induced in macrophages by gliotoxin does not require protein synthesis and is preceded by raised inositol phosphate levels. *Journal of Biological Chemistry*, **265**, 14 476–14 480.

Waring, P., Egan, M., Braithwaite, A., Mullbacher, A. and Sjaarda, A. (1990) Apoptosis induced in macrophages and T blasts by the mycotoxin sporidesmin and protection by Zn^{+2} salts. *International Journal of Immunopharmacology*, **12**, 445–457.

Waring, P., Eichner, R. D. and Mullbacher, A. (1988) Chemistry and biology of the immunomodulating agent gliotoxin. *Medicinal Research Reviews*, **8**, 499–524

Waring, P., Eichner, R. D., Mullbacher, A. and Sjaard, A. J. (1989) Gliotoxin induces apoptosis in macrophages unrelated to its anti-phagocytic properties. *Journal of Biological Chemistry*, **263**, 18 493–18 499.

Waring, P., Kos, F. J. and Mullbacher, A. (1991) Apoptosis or programmed cell death. *Medicinal Research Reviews*, **11**, 219–236.

Wyllie, A., Kerr, J. F. R. and Currie, A. R. (1980) Cell death: the significance of apoptosis. *International Reviews in Cytology*, (G. H. Bourne and J. F. Danielli, eds) Academic Press, **68**, 251–300.

Wyllie, A. H., Morris, R. G., Smith, A. L. and Dunlop, D. (1984) Chromatin cleavage in apoptosis: association with condensed chromatin morphology and dependence on macromolecular synthesis. *Journal of Pathology*, **142**, 67–77.

Zwelling, L. A., Altschuler, E., Cherif, A. and Farquhar, D. (1991) N-(5,5-diacetoxypentyl)doxorubicin: A novel anthracycline producing DNA interstrand cross-linking and rapid endonucleolytic. *Cancer Research*, **51**, 6704–6707.

DNA FRAGMENTATION

Chapter 8

GLUCOCORTICOID-INDUCED PROGRAMMED CELL DEATH (APOPTOSIS) IN LEUKEMIA AND PRE-B CELLS

Emad S. Alnemri and Gerald Litwack

*Department of Pharmacology and the Jefferson Cancer Institute,
Jefferson Medical College of Thomas Jefferson University,
Philadelphia, PA 19107, USA*

INTRODUCTION

There are two main types of cell death, accidental and programmed. In accidental death or necrosis, the death of the cell is due to severe damage, usually to the cell membrane, caused often by mechanical damage, hypoxia, complement-mediated cell lysis and other highly toxic agents. The hallmark of this type of cell death is uncontrolled swelling of the cell (Wyllie, 1987). Thus, in apoptosis or programmed cell death the decision of a cell to die in response to a specific stimulus may represent a favorable adaptation for a tissue which can survive in a changed environment. This situation is observable in tissue modeling processes, normal cell turnover, target cell death induced by cytotoxic T lymphocytes, glucocorticoid-induced thymocyte cell death, some viral infections and others. The hallmark of this type of cell death is chromatin condensation and internucleosomal DNA cleavage (Wyllie, 1987). Although it remains to be seen whether DNA breakdown is the cause or effect of the cell death process, there are increasing observations related to the role of programmed cell death in development, since the developmental process involves the formation of tissues and their replacement by other differentiated tissues, requiring the destruction of the precursor tissue. This is of particular importance in the maturation process of B cells and thymocytes (Cohen, 1991). Consequently, programmed cell death must be a frequent phenomenon in ontogeny. Perhaps the failure of cells to enter the cell death program in response to certain stimuli may produce a candidate for tumour formation (Williams, 1991), a possibility that may focus the interest of cancer researchers. Also, artificially inducing this program may facilitate chemotherapy of tumours.

The process of glucocorticoid-induced programmed cell death is composed of two phases, the cytostatic phase and the cytolytic phase. Both phases together require 3–6 hours for the rat thymocyte and as much as 70 hours or more for human leukemia T and pre-B cells (Figure 8.1). The glucocorticoid hormone diffuses into the cell and binds to the unoccupied cytoplasmic glucocorticoid receptor forming a steroid receptor complex which becomes activated to expose the DNA binding zinc fingers. The activated form translocates to the nucleus through the nucleopore and binds to

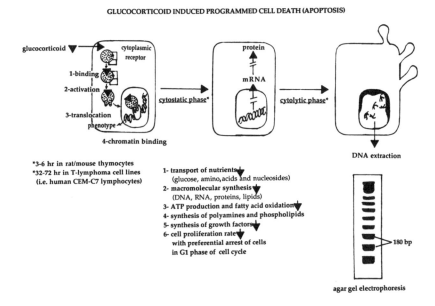

Figure 8.1 Cartoon of the overall mechanism of glucocorticoid-induced programmed cell death in T-derived cells. The glucocorticoid hormone enters the cell (left) possibly by free diffusion and binds to the cytoplasmic glucocorticoid receptor, undergoes a process of activation which dissociates nonhomologous proteins, exposes the DNA binding domain and the activated receptor translocates to the nucleus through the nucleopore. In the nucleus it binds to hormone responsive elements and activates transcription of one or more phenotypes (which, in some cases may be the receptor itself). This leads to a cessation of growth in which macromolecular synthesis is decreased as well as a number of other activities shown in the list in the center. After cessation of growth, DNA cleavage may occur as well as chromatin condensation and cell membrane blebbing (right) and the cell dies. DNA breakdown can be measured by agarose gel electrophoresis and exposure by ethidium bromide (right bottom). Internucleosomal cleavage is indicated by the approximately 180 base pair separation of the breakdown products. In the rat thymocyte, this process takes about 3–6 h, whereas in the CEM-C7 cell, it takes as long as 72 h under cell culture conditions.

hormone responsive elements causing the activation of genes, or in some cases, their suppression. The transcribed mRNAs are translated in the cytoplasm and the increased amounts of specific proteins generate the hormonal response at the cellular level. Among these responses are changes that lead to cytostasis in which macromolecular synthesis ceases. After the cessation of macromolecular synthesis, there is a cleavage of internucleosomal DNA and this is followed by the death of the cell. When, for example, an adrenalectomized rat is injected with dexamethasone, as early as 3 hours following the injection, one can observe the breakdown of DNA in the thymus cell by electrophoresis in 1.8% agarose using ethidium bromide to detect the DNA bands (Alnemri and Litwack, 1989). The bands of DNA are separated by approximately 180 base pairs typical of nucleosomal DNA. Our objectives were to try to understand the glucocorticoid-induced internucleosomal DNA cleavage in human CEM (derived from childhood acute lymphoblastic leukemia) lymphocytes and in rat thymocytes. Later work is extended to pre-B human cells. We also were interested in the mechanism of early glucocorticoid mediated growth arrest and repression of gene expression in human lymphoid cells in particular the participation of proto-oncogenes such as c-*myc* and *bcl*-2.

MECHANISMS OF GLUCOCORTICOID-INDUCED INTERNUCLEOSOMAL DNA CLEAVAGE

Two major theories have been available to explain the glucocorticoid induction of DNA breakdown. These were that the hormone: (a) transcriptionally induces a lysis gene product, possibly an endonuclease (Compton and Cidlowski, 1987), or (b) glucocorticoid action could, in some fashion, activate a constitutive endonuclease. Our early work argued against the induction of a lysis gene product in the form of an endonuclease (Alnemri and Litwack, 1989), and consequently, we focused our attention on the mechanism by which glucocorticoid could activate an endogenous endonuclease (Alnemri and Litwack, 1990). There appeared to be two ways in which this could happen: (a) the hormonal action could provide a suitable form of a co-factor, such as calcium for a calcium-magnesium dependent endonuclease, or (b) it could provide a suitable form of the substrate, producing a conformational change in chromatin which would make it a substrate for a constitutive endonuclease.

THE EFFECTS OF CALCIUM ON HUMAN CEM CELLS VERSUS RAT THYMOCYTES

In these experiments we isolated nuclei from CEM cells and from rat thymocytes that were sensitive to glucocorticoid. These nuclei were subjected to increasing concentrations of calcium and then DNA was prepared and examined for cleavage by agarose gel electrophoresis. In the case of thymocyte nuclei, addition of calcium alone was able to produce DNA breakdown typical of internucleosomal DNA cleavage at an optimal concentration of 1 mM (Figure 8.2). Whereas in the CEM cells, concentrations as high as 10 mM were unable to produce any breakdown of DNA. In the thymocytes, the effective level of calcium could be reduced when 1 mM ATP and 1 mM NAD^+ were added to the nuclei, suggesting that there may be a calcium pumping

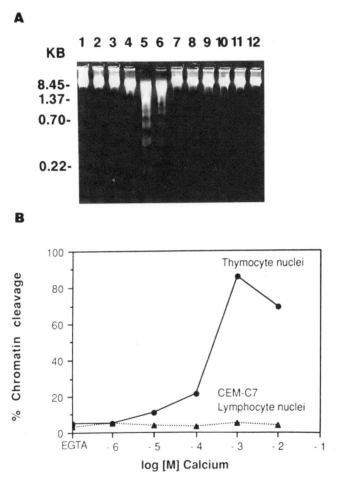

Figure 8.2 Dose-dependence of *in vitro* calcium-induced internucleosomal DNA cleavage in isolated nuclei. Nuclei from rat thymocytes or human CEM-C7 lymphocytes were incubated at 37°C with EDTA or various concentrations of calcium for three hours. A: DNA was isolated and analyzed on 1.8% agarose gel. Lanes 1–6, DNA isolated from rat thymocyte nuclei; lanes 7–12, DNA isolated from human CEM-C7 lymphocyte nuclei. Lanes 1 and 7, 5 mM EGTA, lanes 2 and 8, 1 μM Ca^{2+}; lanes 3 and 9, 10 μM Ca^{2+}. KB, kilobase pair. In B, the percent of chromatin cleavage was determined from the densitometric scan of the photographic negative of the gel shown in A. (Alnemri and Litwack, 1990, reprinted with permission from the *Journal of Biological Chemistry*, © Copyright 1990, The American Society for Biochemistry and Molecular Biology.)

mechanism (Jones et al., 1989; Alnemri and Litwack, 1990), and that physiological amounts of calcium could indeed produce this effect. In contrast, nuclei from CEM-C7 lymphocytes were not at all affected by the combination of calcium with ATP and NAD^+. These experiments led us to conclude that in rat thymocytes there is a constitutive calcium- and magnesium-dependent endonuclease which promotes internucleosomal DNA cleavage. In the case of CEM cells, the endonuclease that cleaves DNA is not sensitive to calcium. Furthermore, the endonuclease in thymocyte nuclei, but not in CEM nuclei, becomes sensitive to near cellular concentrations of calcium in the presence of ATP and NAD^+, suggesting that an ATP dependent nuclear calcium uptake system is involved in the activation of the rat thymocyte endonuclease.

ACTIVATION OF A CONSTITUTIVE ENDONUCLEASE BY PROVIDING A SUITABLE FORM OF THE SUBSTRATE

The assumption is that a DNA conformation-dependent endonuclease is constitutively expressed in the nucleus of CEM-C7 cells. This enzyme would be inactive when chromatin is tightly packaged into a high order structure, but would become active when the linker DNA became accessible for enzyme attack. In order to test this hypothesis we tested a model system with novobiocin which had been shown to interact directly with the histones of chromatin (Sealy et al., 1986). When novobiocin in increasing amounts was added to the medium of CEM-C7 cells for 3 hours, there followed an increased level of DNA breakdown up to a maximum of 400 μg novobiocin/ml of medium. At higher concentrations, up to 1200 μg/ml, there was a decline in internucleosomal cleavage, showing a dose-dependent biphasic effect (Figure 8.3). Possibly, the higher levels inhibited the endonuclease. When cycloheximide was added in similar experiments, it was shown that it had no effect on the ability of novobiocin to generate internucleosomal DNA cleavage. Thus, novobiocin induces internucleosomal DNA cleavage in a dose-dependent manner in CEM-C7 cells, and this effect did not depend on *de novo* protein synthesis (Alnemri and Litwack, 1990).

To examine whether calcium was involved in the novobiocin effect, EGTA was added to a cell culture system, and was demonstrated not to have any influence on the ability of novobiocin to produce programmed cell death. This was shown also for CEM-C1 cells that are resistant to the action of glucocorticoids. The action of the synthetic hormone, triamcinolone acetonide, and novobiocin were compared in viability experiments. It was shown that triamcinolone acetonide produced cytostasis up to about 45 hours and thereafter produced programmed cell death which became highly significant after 70 hours. In the case of novobiocin, cytostasis occurred in about the first 18 hours, and thereafter rapid cell death ensued over a period up to about 60 hours. In these experiments it was clear that the effect of novobiocin was much more rapid than that of triamcinolone acetonide. The same experiments were carried out with hormone resistant cells and novobiocin again gave a similar effect, although the time frame was larger than with the hormone-sensitive cells. During this

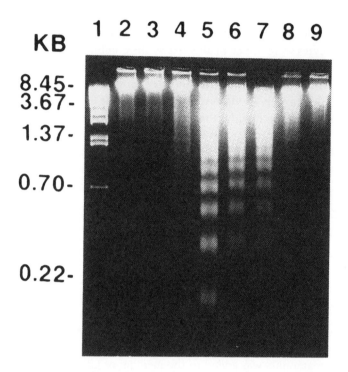

Figure 8.3 Dose-dependence of novobiocin-induced internucleosomal DNA cleavage of CEM-C7 lymphocyte DNA. The cells were incubated in a complete medium (RPMI 1640) for 3 h with or without novobiocin. The DNA was then isolated and analyzed on a 1.8% agarose gel. The following novobiocin concentrations were used: 0 μg/ml (lane 2), 100 μg/ml (lane 3), 200 μg/ml (lane 4), 400 μg/ml (lane 5), 600 μg/ml (lane 6), 800 μg/ml (lane 7), 1000 μg/ml (lane 8), or 1200 μg/ml (lane 9). A lambda DNA-BstEII digest was used as molecular weight markers (lane 1). KB, kilobase pair. (Alnemri and Litwack, 1990, reprinted with permission from the *Journal of Biological Chemistry*, © Copyright 1990, The American Society for Biochemistry and Molecular Biology.)

interval of study, chromatin cleavage was also examined. In the case of the hormone sensitive cells, triamcinolone acetonide at 1 μM produced a steady increase in chromatin cleavage that was significant at 18–32 hours and extended to 72 hours in a linear fashion. By contrast, novobiocin at 200 μg/ml produced immediate cleavage of chromatin that lasted for 3–4 hours, when the effect began to level off. In the hormone resistant cells, a similar experiment was done and the action of novobiocin was also similar except for a longer time frame of action. When the amount of novobiocin was doubled to 400 μg/ml a rapid curve of chromatin cleavage occurred which was superimposable upon the curve obtained using sensitive cells with 200 μg/ml of novobiocin. Preparation of isolated nuclei from CEM-C7 cells was followed by treatment of these nuclei with 200 μg/ml of novobiocin and this was combined with 2 mM calcium or with 1 mM each of ATP and NAD^+ in the presence or absence of calcium. None of these treatments produced DNA breakdown in isolated nuclei, indicating that the action of novobiocin to generate programmed cell death required the intact cell. The effects of novobiocin were also compared to those of triamcinolone acetonide using the electron microscope. Nuclear condensation of chromatin was clearly visualized after 40 h of treatment with triamcinolone acetonide and a similar extent of nuclear condensation of chromatin occurred with novobiocin although in less than 1 h.

The effects of calcium were studied further using ionophore A23187 and other toxic substances, such as sodium azide. These experiments showed that sodium azide had no effect on the ability of novobiocin to produce DNA cleavage and programmed cell death, however the ionophore was able to interfere with the action of novobiocin in causing DNA breakdown in cell cultures. This action was attributed to the ability of the ionophore to damage the cell membrane and is in keeping with the conclusion that the action of novobiocin requires an intact functioning cell. Another possible explanation is that the ionophore may cause activation of protein kinase C so that some possible phosphorylation mechanism might generate this interference with novobiocin's ability to produce DNA damage. Extraction of nuclei, shows that novobiocin action did not result in a change in the integrity of histone proteins compared to the action of triamcinolone acetonide, so that novobiocin must be causing DNA cleavage by actions other than degrading proteins.

MECHANISMS OF NOVOBIOCIN INDUCTION OF INTERNUCLEOSOMAL DNA CLEAVAGE

Three possible mechanisms came to mind for the ability of novobiocin to produce DNA breakdown: (a) inhibition of topoisomerase II activity; (b) novobiocin allows the entrance of an exogenous endonuclease from the cell culture medium (serum) and (c) novobiocin is responsible for the activation of a constitutive endonuclease. Since novobiocin is known to be an inhibitor of topoisomerase II, we examined the type of cleavage induced by treatment with novobiocin. The DNA cleavage activity of topoisomerase II produces a 4-base 5'-overhang and this can be analyzed by polyacrylamide gel electrophoresis. We isolated mono-nucleosomal DNA from breakdown products induced by novobiocin and triamcinolone acetonide, end-labelled the DNA with ^{32}P and analyzed its molecular mass by gel electrophoresis. Thus, if a 5'-overhang occurred through DNA cleavage, there would be an increase in the molecular mass of the DNA band after end-labelling with ^{32}P and treatment with T4 DNA polymerase, since the polymerase would then fill in the 3'-ends. In the case of the 3'-overhang, end-labelling and treatment with T4 DNA polymerase would cleave the 3'-overhang since T4 DNA polymerase has a 3'-exonuclease activity. This would result in a decrease in the molecular mass of the DNA cleavage product resulting in a down shift of the band. Where blunt-ended cleavage occurred there would be no change in the molecular mass resulting in no shift in the electrophoretic mobility of the DNA band after end-labelling with ^{32}P and treatment with T4 DNA polymerase. In all cases, there was no change in the electrophoretic mobility of the DNA bands indicating a blunt-ended cleavage was induced by both triamcinolone acetonide and novobiocin. Furthermore, the cleavage products were examined for the content of 3'- and 5'-groups. Chemical and enzymatic treatment showed that the breakdown products from both hormone and novobiocin produced 5'-phosphate groups and 3'-hydroxyl groups. In addition, other topoisomerase inhibitors were tested and the topoisomerase II inhibitor, AMSA, as well as the topoisomerase I inhibitor, camptothecin, were unable to generate DNA breakdown in CEM-C7 cells. Thus, the type of cleavage induced by these agents was different from that produced by topoisomerase II, leading to the conclusion that DNA cleavage occurs in apoptosis by some other mechanism than through topoisomerase II (or I) activity.

ACTION OF NOVOBIOCIN ON A MODEL MONONUCLEOSOME

Mono-nucleosomes were prepared by using 181 and 239 base pair fragments of an H1 human histone cDNA (van Wijnen et al., 1988) and these fragments were first end-labelled with ^{32}P and then combined with a donor chromatin preparation, either from CEM-C7 cell nuclei or from rat liver nuclei. These were incubated in 1.6 M sodium chloride, and gradually diluted to 80 mM sodium chloride. This treatment caused the proteins of the donor chromatin to dissociate and to decorate the ^{32}P-labelled DNA fragments resulting in the formation of labelled mono-nucleosomes. The effect of novobiocin on the labelled mono-nucleosomes in increasing amounts was then studied by gel retardation analysis. In these experiments increasing amounts of novobiocin caused destabilization of the DNA-histone complex resulting in a decrease in the mono-nucleosomal band and an increase in the free DNA band, regardless of the source of the nuclear proteins. DNase I footprinting analysis showed that 4 regions on the 181 base pair DNA fragment were occupied by proteins. These protected regions gradually returned to a pattern resembling naked DNA when the amounts of novobiocin were increased showing that novobiocin was able to remove these proteins from DNA.

MECHANISM OF INTERNUCLEOSOMAL DNA CLEAVAGE INDUCED BY NOVOBIOCIN AND GLUCOCORTICOIDS

From the foregoing experiments, it appeared that novobiocin was having a direct effect on chromatin, causing the removal of histones and opening the tightly wound DNA that would make the linker accessible to endogenous endonuclease, resulting in DNA cleavage. This process was brief because of the direct action of the agent. In contrast, glucocorticoids require a very long period, about 18–20 hours to achieve significant internucleosomal DNA cleavage, suggesting an indirect action of glucocorticoids. In consequence we decided to examine early genes that might be activated in the process of glucocorticoid-induced DNA cleavage. We measured the levels of glucocorticoid receptor and glutamine synthetase mRNAs. In hormone-sensitive CEM cells we found that treatment with 1 μM triamcinolone acetonide resulted in a ten-fold increase in the glucocorticoid receptor mRNA in less than four hours, which was also reported previously (8), and a gradual induction of glutamine synthetase mRNA. In hormone-resistant cell lines, CEM-C1 and ICR-127, we found a very slight increase in glucocorticoid receptor mRNA, and either no expression of glutamine synthetase or a very slight induction. This induction of glucocorticoid receptor mRNA and glutamine synthesis mRNA was sensitive to camptothecin, a known inhibitor of topoisomerase I. This indicated that the mechanism of up-regulation of mRNA for glucocorticoid receptor and glutamine synthetase was due to a transcriptional effect of the hormone. We also measured the level of c-myc in these cells. c-myc mRNA was rapidly down-regulated after treatment with glucocorticoid, so that by six hours there was a pronounced reduction in the level of c-myc mRNA. These properties were compared in a number of human T and pre-B cell lines. In every case where cells were susceptible to glucocorticoid-induced programmed cell death, which was the case for CEM-C7 cells, MOLT-4 and 697 cells, this susceptibility was accompanied by up-regulation of glucocorticoid receptor mRNA and down-regulation of c-myc mRNA. On the other hand, those cells which did not respond to hormone by programmed cell death, and these included JURKAT, SUPT-1, PEER and CEM-C1 cells, there was no up-regulation of glucocorticoid receptor mRNA and there was no down-regulation of c-myc mRNA. We concluded that the action of the hormone to produce programmed cell death involves the direct effect of the hormone on the up-regulation of the glucocorticoid receptor mRNA. This leads to about a 10-fold increase in the number of molecules of the glucocorticoid receptor which may squelch other transactivating factors which might be necessary for the expression of growth factors critical to the proliferation of the cell, such as c-myc. Further changes at the level of DNA may cause chromatin changes like those observed for novobiocin, which would make accessible the substrate for the resident endonuclease.

PRE-B LEUKEMIA CELLS

We examined the action of glucocorticoid on two pre-B cell lines (Alnemri et al., 1992a), 697 and 380 selected on the basis of their content of the bcl-2 proto-oncogene. In the 697 cell line there is a low level of the cellular proto-oncogene bcl-2 compared to the 380 line, in which bcl-2 is overexpressed through a chromosomal translocation (Tsujimoto et al., 1985). We examined programmed cell death in these two cells lines in response to the glucocorticoid analog triamcinolone acetonide. While inhibition of cell growth is similar between the two cell lines in the presence of the hormone, there are different activities of the hormone. In the case of the 697 cell, the presence of triamcinolone acetonide causes cytostasis and then programmed cell death, which occurs rapidly after about 50 hours. In the case of the 380 cell line, which is overexpressing bcl-2, cytostasis occurs but programmed cell death does not occur within an experimental period of about 100 hours. This was confirmed by examining the state of

DNA in these cells in which 697 cell DNA showed the typical DNA cleavage pattern, whereas the 380 cell line did not show any cleavage when the cells were exposed to triamcinolone acetonide. There was also a diminution of c-*myc* mRNA expression in both cell lines after treatment with glucocorticoid, indicating that Bcl-2 is able to inhibit the DNA cleavage at some point after cytostasis occurs (Figure 8.4) (Alnemri et al., 1992a).

DNA CLEAVAGE; IS IT A CAUSE OR AN EFFECT?

From our work and that of others, DNA cleavage temporally precedes the death of the cell, at least defined by the inability of the cell to resist the entry of certain dyes. Although internucleosomal cleavage has been recognized as a hallmark of apoptosis, there is no firm evidence that it represents the pivotal molecular event, if there is a single such entity. Presumably, apoptosis will be found, in some cases (Ucker et al., 1992), to occur in the absence of DNA cleavage. If so, and in view of the extensive scenarios for inducing apoptosis, there must be many categories of programmed cell death awaiting classification. Our results with rat thymocytes, which in general confirms the conclusions of others (Wyllie, 1980), is that isolated nuclei can be triggered to break down DNA in an apoptotic fashion simply by supplying Ca^{2+} and Mg^{2+} which is needed to activate the endonuclease. This is clearly not the case with CEM-C7. Accordingly, some scenarios of apoptosis induction depend upon the requirements of the endonuclease responsible for DNA breakdown. In addition, there may be an ultimate molecular step, yet to be identified, that is required before endonuclease activation and DNA cleavage.

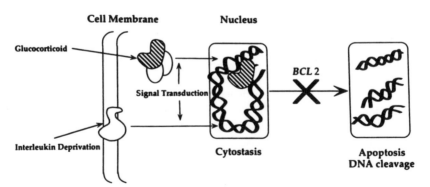

Figure 8.4 Schematic model of the signal transduction pathway involved in glucocorticoid- and IL-3 deprivation-induced apoptosis in pre-B lymphocytes. Glucocorticoid treatment and IL-3 deprivation induce apoptosis in pre-B lymphocytes through a set of signal transduction pathways. Although different mechanisms are utilized initially, evidence suggests that both glucocorticoid treatment and IL-3 deprivation share a final common step between cytostasis and DNA cleavage, which is specifically blocked by Bcl-2. (Alnemri et al., 1992a, reprinted with permission from *Cancer Research*, © Copyright American Association for Cancer Research, Inc.)

OVEREXPRESSION OF *BCL*-2 IN THE BACULOVIRUS SYSTEM

In the baculovirus system Sf9 insect cells are killed after 48 hours due to the viral infection by a process of programmed cell death. However, when *bcl*-2 was overexpressed in these cells, there was a large protective affect against programmed cell death (Alnemri et al., 1992b). Examination of the DNA by agarose gel electrophoresis showed that Bcl-2 prevented the appearance of DNA fragmentation. However, when a deletion mutant of *bcl*-2 which lacks the C-terminal hydrophobic domain was overexpressed there was no protective effect of this overexpressed truncated protein, and the typical DNA cleavage pattern appeared. The subcellular localization was determined biochemically for both full length and truncated *bcl*-2 overexpressed in the Sf9 cell and it was found that little or no protein appeared in the cytosol, but both the full-length and truncated forms appeared in heavy membranes and in progressively purified nuclei. This was confirmed by indirect immunofluorescence in the intact cell with a monoclonal antibody against Bcl-2, in which it appeared that full-length Bcl-2 as well as the truncated form were localized in a band around the perinucleus with some extensions into the cytoplasm. The subcellular localization of the truncated form was more diffuse than that of full length Bcl-2. This perinuclear localization may indicate the means by which the nucleus is protected from apoptotic signals. The ability of Bcl-2 to protect insect cells against programmed cell death suggests a conserved function in evolution.

SUMMARY

The mechanism which controls the activation of internucleosomal DNA cleavage in rat thymocytes differs from that operating in human CEM-C7 cells. Novobiocin's rapid activation of internucleosomal DNA cleavage and chromatin changes in CEM lymphocytes are molecular features of apoptosis or programmed cell death. The mechanism of glucocorticoid- or novobiocin-induced DNA cleavage in CEM lymphocytes involves activation of a constitutive non-calcium-dependent endonuclease. Nuclear chromatin may be generally in a highly compact and charged neutralized state. Disruption of this structure directly by novobiocin, or indirectly by glucocorticoids, may lead to unmasking of internucleosomal DNA regions that are substrates for a constitutive non-calcium-dependent endonuclease. Up-regulation of the glucocorticoid receptor in human T lymphocytes and pre-B cells, in response to glucocorticoids, may cause transcriptional repression of vital genes leading to cytostasis, DNA cleavage and cell death. Overexpression of *bcl*-2 in pre-B cells inhibits DNA cleavage at some point after cytostasis which occurs in response to glucocorticoids. Overexpression of *bcl*-2 in Sf9 cells prolongs the life of these cells, otherwise determined to undergo virus-induced programmed cell death. The protective effect could relate to the subcellular localization of Bcl-2.

ACKNOWLEDGEMENTS

This work was supported by Research Grant DK-13531 to G.L. from the National Institutes of Health.

REFERENCES

Alnemri, E. S., Fernandes, T. F., Haldar, S., Croce, C. M. and Litwack, G. (1992a) Involvement of Bcl-2 in glucocorticoid-induced apoptosis of human pre-B-leukemias. *Cancer Research*, **52**, 491–495.

Alnemri, E. S. and Litwack, G. (1989) Glucocorticoid-induced lymphocytolysis is not mediated by an induced endonuclease. *Journal of Biological Chemistry*, **264**, 4104–4111.

Alnemri, E. S. and Litwack, G. (1990) Activation of internucleosomal DNA cleavage in human CEM lymphocytes by glucocorticoid and novobiocin. *Journal of Biological Chemistry*, **265**, 17 323–17 333.

Alnemri, E. S., Robertson, N. M., Fernandes, T. F., Croce, C. M. and Litwack, G. (1992b) Overexpressed full-length human *bcl*-2 extends the survival of baculovirus-infected Sf9 insect cells. *Proceedings of the National Academy of Science, USA*, **89**, 7295–7299.

Cohen, J. J. (1991) Programmed cell death in the immune system. *Advances in Immunology*, **50**, 55–85.

Compton, M. M. and Cidlowski, J. A. (1987) Identification of a glucocorticoid-induced nuclease in thymocytes. A potential "lysis gene" product. *Journal of Biological Chemistry*, **262**, 8288–8292.

Eisen, L. P., Elsasser, M. S. and Harmon, J. M. (1988) Positive regulation of the glucocorticoid receptor in human T cells sensitive to the cytolytic effects of glucocorticoids. *Journal of Biological Chemistry*, **263**, 12 044–12 048.

Jones, D. P., McConkey, D. J., Nicotera, P. and Orrenius, S. (1989) Calcium-activated DNA fragmentation in rat liver nuclei. *Journal of Biological Chemistry*, **264**, 6398–6403.

Sealy, L., Cotten, M. and Chalkley, R. (1986) Novobiocin inhibits passive chromatin assembly *in vitro*. *The EMBO Journal*, **5**, 3305–3310.

Tsujimoto, Y., Jaffe, E., Cossman, J. and Croce, C. M. (1985) Involvement of the *bcl*-2 gene in human follicular lymphoma. *Science*, **228**, 1440–1443.

Ucker, D. S., Obermiller, P. S., Eckhart, W., Apgar, J. R., Berger, N. A. and Meyers, J. (In press) Genome digestion is a dispensable consequence of physiological cell death mediated by cytotoxic T lymphocytes. *Molecular and Cellular Biology*.

van Wijnen, A. J., Wright, K. L., Massung, R. F., Gerretsen, M., Stein, J. L. and Stein, G. S. (1988) Two target sites for protein binding in the promoter region of a cell cycle regulated human H1 histone gene. *Nucleic Acids Research*, **16**, 571–592.

Williams, G. T. (1991) Programmed cell death: apoptosis and oncogenesis. *Cell*, **65**, 1097–1098.

Wyllie, A. H. (1980) Glucocorticoid-induced thymocyte apoptosis is associated with endogenous endonuclease activation. *Nature*, **284**, 555–556.

Wyllie, A. H. (1987) Cell death. *International Review of Cytology (Supplement)*, **17**, 755–785.

Chapter 9

EVIDENCE FOR THE ROLE OF AN ENDO-EXONUCLEASE IN THE CHROMATIN DNA FRAGMENTATION WHICH ACCOMPANIES APOPTOSIS

Murray J. Fraser, Christine M. Ireland, Stephen J. Tynan and Arthur Papaioannou

Children's Leukaemia and Cancer Research Centre, University of New South Wales, Prince of Wales Children's Hospital, Sydney, NSW 2031, Australia

This report offers a novel suggestion as to the identity of the endonuclease responsible for chromatin DNA fragmentation in apoptosis, namely that it is an activity of one or more forms of endo-exonuclease, an enzyme which is known to function in recombination and in recombinational double-strand break repair in lower eukaryotes. Endo-exonuclease has now been identified as a major nuclease in nuclei of mammalian cells.

NUCLEAR ENDONUCLEASES: A BRIEF REVIEW

Since the early indication of two major types of chromatin-associated endonucleases in mouse and rat liver (Hewish and Burgoyne, 1973), Ca^{2+}, Mg^{2+}-endonuclease and Mg^{2+}-endonuclease, there have been many attempts to purify these enzymes from nuclei, each using different methods. The results of twelve of the more successful purifications, in which the properties of the endonucleases were also relatively well-characterised, are summarised in Table 9.1. It is clear that none of these endonucleases are identical in properties despite the fact that many of them (five of twelve) have been derived from the same source, rat liver nuclei. All of these enzymes have in common Mg^{2+}-dependent activity that releases 5'-phosphoryl-terminated oligonucleotides from DNA and most act optimally in the neutral pH range.

Table 9.1 Chromatin-associated endonucleases.

Source	Size (kDa)[a]	Me^{2+} for Max.Act.[b]	pI	Optimum pH	RNase[c]	Exo[d]	ss/ds[e]	Refs[f]
Rat liver	27N	Mg(Mn)Ca		7			<0.1	1
Rat liver	55N	Mg(Mn)Ca		6–7	+		7	2
Human KB cells	54	Mg	10.3	9.2	+		10	3
Human HeLa cells	22	Mg(Mn)	5.1	7				4
Human KB cells	38N	Mg(Mn)	6.4	9.5	+?		10	5
Rat liver	35N	Mg					1	6
Calf thymus	25-30N	Mg, Ca		6.5–7.5	–?		0.17	7
Bull semen	36,28N	Mg(Mn)Ca		7.4–8	+?		1	8
Hen liver	43	Mg(Mn)	10.2	9.0	+		20	9
Pig liver	29N	Mg, Ca		6.5–8			<0.1	10
Rat liver	46	Mg, Ca	5.7	7.6	+		2.5	11
Rat liver	36(70)	Mg	7.1	7.6		+	2	12,13
Neurospora crassa	31(76)	Mg(Mn)Ca		7–8	+	+	2	14–16
Saccharomyces cerevisiae	72	Mg(Mn)		7–8	+	+	8	17
Monkey CV-1 cells	65	Mg(Mn)Ca		7–8	+	+	2	18

a. The symbol N indicates that only the native molecular weight was determined. In one case this was smaller than the size determined by gel electrophoresis. The values in brackets indicate dimer size (70) or size of largest active polypeptide determined by activity gel analysis (76). All other endonucleases are monomeric.
b. Divalent metal ion(s) required for maximal activity. In most cases Mg^{2+} can be replaced by Mn^{2+}. The valence is not indicated in the text in order to save space.
c. A (+) sign indicates that the purified enzyme has RNase as well as DNase activity.
d. A (+) sign indicates that the purified enzyme has exonuclease as well as endonuclease activity.
e. ss/ds refers to the ratio single-strand/double-strand DNase activity.
f. Ishida *et al.*, 1974; 2. Cordis *et al.*, 1975; 3. Wang *et al.*, 1978; 4. Fischman *et al.*, 1979; 5. Wang & Rose, 1981; 6. Machray & Bonner, 1981; 7. Nakamura *et al.*, 1981; 8. Hashida *et al.*, 1982; 9. Tanigawa & Shimoyama, 1983; 10. Stratling *et al.*, 1984; 11. Hibino *et al.*, 1989; 12. Hibino *et al.*, 1988; 13. Hibino *et al.*, 1991; 14. Chow & Fraser, 1983; 15. Ramotar *et al.*, 1987; 16. Fraser *et al.*, 1989; 17. Chow & Resnick, 1987; 18. Couture & Chow, 1992.

The dependence on Ca^{2+} of the so-called Ca^{2+}, Mg^{2+}-endonucleases is problematical since the most highly purified enzyme in this class, that from bull semen (Hashida *et al.*, 1982), is clearly not dependent on but rather only stimulated by Ca^{2+}. In most other cases, the supposed dependence on Ca^{2+} has not been adequately demonstrated. The sometimes dramatic stimulations observed with Ca^{2+} at different stages of purification seem to be highly variable and could well be influenced indirectly by the interactions of a Mg^{2+}-dependent endonuclease with Ca^{2+}-binding proteins. In short, the description Ca^{2+}, Mg^{2+}-endonuclease may apply only to a minor activity in nuclei such as that isolated from thymocytes (Nakamura et al., 1981). For the major chromatin-bound activites, it is misleading and its use should be abandoned (see section on CEM cell activities below). Taking these doubts to the extreme, there has even been a recent proposal that the nuclease responsible for chromatin DNA fragmentation in apoptosis does not require a divalent metal ion at all, namely that it is DNase II (Barry and Eastman, 1992). However, this enzyme is lysosomal in origin and much more likely to be released during the autolysis associated with necrotic cell death. In any case, DNase II makes breaks in DNA with 5'-OH termini as opposed to the 5'-P termini observed in chromatin DNA fragments (Alnemri and Litwack, 1990).

RNase activity has been found most often in association with the Mg^{2+}-endonucleases (Table 9.1), but also with two of the so-called Ca^{2+}, Mg^{2+}-endonucleases (Cordis et al., 1975; Hibino et al., 1989). RNAse activity is relatively single-strand-specific and is not easily detected when assayed by following the release of acid-soluble material from transfer and ribosomal RNAs because of their inaccessible secondary and tertiary structures in the presence of divalent metal ions. These substrates have been commonly employed by the investigators who have purified the enzymes listed in Table 9.1. Polyriboadenylate (poly rA) in the presence of 0.5–4.0 mM Mg^{2+} seems to be the best substrate for the chromatin-bound RNase activity (see below). Thus, it is not clear whether the endonucleases which are reported to lack RNase activity are sugar non-specific or are truly DNA specific.

The chromatin-associated endonucleases also exhibit a wide range of strand specificities for DNA (Table 9.1), from those which appear to be essentially double-strand specific in the Ca^{2+}, Mg^{2+}-endonuclease class, e.g. two from liver (Ishida et al., 1973; Stratling et al., 1984) and one from thymus (Nakamura et al., 1981), to those which exhibit variable single-strand specificity strongly affected by changes in salt concentrations, e.g. that from hen liver (Tanigawa and Shimoyama, 1983).

The isolation of double-strand specific species of Ca^{2+}, Mg^{2+}-endonucleases, that also apparently lack RNase activity, has led to the use by others of another potentially misleading term, "DNase I-like endonuclease". One of these (Nakamura et al., 1981), was neither inhibited by g-actin nor by a polyclonal antibody that specifically and completely inhibited bovine pancreatic DNase I. We have confirmed that the DNase activities extracted from CEM cell nuclei are not inhibited by such an antibody raised independently in rabbits. A previous study has shown also that DNase I was found in detectable amounts in mammals only in secretory tissue such as the pancreas and salivary glands (Lacks, 1981).

Finally, the fact that both classes of chromatin-associated endonuclease have been found associated with different-sized polypeptides, some of which were determined to have different isoelectric points (Table 9.1) shows that each purified activity is likely to be associated with a different protein. This raises yet another problem: How is it that there are so many different "major" chromatin-associated endonucleases? As many as eight different isoelectric forms have been detected in nuclei of human leukaemic and mouse melanoma cells (Lambert et al., 1982). Of course, some of these isoelectric forms may result from different covalent modifications such as phosphorylation or ADP-ribosylation. The endonuclease from bull semen has been demonstrated to be inhibited by monoADP-ribosylation in vitro (Tanaka et al., 1984), however, there have not been any reports of covalent modifications of active forms.

Nuclear Endo-Exonucleases

An alternative explanation for the presence of multiple endonucleases is suggested by the recent finding of endo-exonuclease in mammalian cells (Couture and Chow, 1992). This chromatin-associated enzyme has been studied to date mainly in lower eukaryotes, Neurospora crassa (Chow and Fraser, 1983; Ramotar et al., 1987; Fraser et al., 1989), Aspergillus nidulans (Koa et al., 1990) and the yeast Saccharomyces cerevisiae (Chow and Resnick, 1987). The endo-exonucleases also have Mg^{2+} (Mn^{2+})-dependent activities which generate 5'-P-terminated products, a relatively single-strand specific endonuclease activity with both DNA and RNA and an exonuclease activity with DNA that is not strand-specific and requires 5'-P-termini (Chow and Fraser, 1983; Fraser et al., 1989). The latter activity is much more salt-sensitive than the endonuclease activity as is the case for the mammalian Mg^{2+}-endonucleases (Table 9.1). Exonuclease activity was also recently detected (Hibino et al., 1991) in association with one of the endonucleases from rat liver nuclei.

In Neurospora, endo-exonuclease accounts for at least 90% of the total nuclear DNase activity and the enzyme occurs in extracts of chromatin as multiple active polypeptides. These were detected by activity gel analysis (Ramotar et al., 1987) and were all inhibited and immunoprecipitated by a specific antibody raised to the purified 31 kDa form of the enzyme (Fraser et al., 1986). The largest active polypeptide in nuclei was 76 kDa in size, associated with the nuclear matrix, while 43 and 37 kDa forms were predominant in chromatin extracts (Ramotar et al., 1987). The latter are very similar in size to several of the endonucleases isolated from mammalian nuclei (Table 9.1).

In addition, an inactive but trypsin-activatable, 93 kDa form of endo-exonuclease has also been detected in Neurospora (Kwong and Fraser, 1978) and may be a precursor of the active enzyme. It is associated with chromatin in a ratio of about 2:1, inactive to active forms (Ramotar et al., 1987). It is exquisitely sensitive to proteolysis as is the exonuclease activity of the active enzyme (Chow and Fraser, 1983). The inactive forms of Neurospora and Aspergillus endo-exonuclease and active endo-exonucleases of Neurospora, Aspergillus and Saccharomyces have all been shown to be immunochemically related (Fraser et al., 1986; Koa et al., 1990). The finding that the originally isolated single-strand specific endonuclease of Neurospora (Linn and Lehman, 1965 a, b) was also immunochemically cross-reactive with endo-exonuclease (Fraser et al., 1986) indicates additionally that the multiple forms of enzyme observed are very likely derived from the inactive percursor by limited proteolysis either in vivo or in vitro during extraction and isolation.

The above findings with *Neurospora* endo-exonuclease and our recent results with human leukaemic CEM-cell nuclei outlined below, suggest the possibility that the very similar $Mg^{2+}(Mn^{2+})$-dependent endo-exonuclease recently isolated from mammalian nuclei (Couture and Chow, 1992) may have also arisen by limited proteolysis of a precursor. The CV-1 endo-exonuclease is known to be related to the endo-exonucleases of lower eukaryotes since it was purified on an immunoaffinity column generated by cross-linking antibody raised to the *Neurospora* endo-exonuclease to protein A-Sepharose (Couture and Chow, 1992). It is activated about 2-fold by Ca^{2+}, as is the exonuclease activity of the purified *Neurospora* enzyme (Chow and Fraser, 1983). The *Neurospora* endo-exonuclease has also been found to be strongly inhibited by 1 mM Zn^{2+} and by aurin tricarboxylic acid, ATA (Chow and Fraser, 1983), inhibitors of DNA fragmentation in apoptosing thymocytes and HL-60 cells (McConkey *et al.*, 1989; Shimizu *et al.*, 1990).

The normal functions of endo-exonuclease, at least in the lower eukaryotes, have been shown to be in recombination and recombinational double-strand break repair (Resnick *et al.*, 1984; Chow and Resnick, 1988). The nuclear endo-exonuclease activity is very poorly expressed in *rad52* mutants of the yeast which are deficient in these processes (Chow and Resnick, 1988) and is also partially deficient (10–12% activity) in nuclei of the repair deficient *uvs-3* mutant of *Neurospora* (Ramotar *et al.*, 1987). It is possible that the double-stand break repair function will prove to be the most relevant to a role in chromatin DNA fragmentation in mammalian cells. The purified *Neurospora* endo-exonuclease can generate double-strand breaks in supercoiled DNA (Chow and Fraser, 1983) and has a site-specific cleavage activity of moderate base specificity with linear double-strand DNA which generates "ladders" of DNA fragments under appropriate conditions (Fraser *et al.*, 1989). It might be expected that endo-exonuclease would normally be activated in the G2 phase of the cell cycle for post-replication repair and that unregulated activation could lead to DNA degradation. Many DNA damaging agents induce G2 arrest and, at cytotoxic doses, this leads to chromatin DNA fragmentation and apoptosis. On the other hand, during excision repair, which precedes post-replication repair in damaged cells, any active nuclease present in nuclei might be expected to be inhibited, for example by ADP-ribosylation.

NUCLEASE ACTIVITIES IN HUMAN LEUKAEMIC CEM CELL NUCLEI

We have compared the major nuclease activities in nuclei of human leukaemic CEM-f2 cells with those found previously in *Neurospora* nuclei (Ramotar *et al.*, 1987). Strong resemblances have been found between them. As can be seen in Table 9.2, the CEM cell ss- and ds-DNase activities are present in nuclear fractions in a ratio of 2:1, about 60% in nucleoplasm, 20% released from chromatin with 0.5 M NaCl and the remaining 20% bound tightly to the nuclear matrix as in *Neurospora* nuclei (Ramotar *et al.*, 1987). Trypsin-activatable ss- and ds-DNase activities were also present in each of these fractions at about 2 times the levels of active enzyme as in *Neurospora* (data not

Table 9.2 Distribution of single-strand DNase (ss-DNase), double-strand DNase (dsDNase) and RNase activities in nuclear fractions of human leukaemic CEM-f2 cells.[a]

Nuclear Fraction[b]	Activity[c] ($nmoles/min/10^8 cells$)		
	ss-DNase	ds-DNase	RNase
Nucleoplasm	5.7, 5.1, 4.2, 5.2	2.4, 1.3, 2.6	17, 19, 20
Chromatin I	2.2, 1.6	2.3, 0.5	12, 24
Chromatin II	0, 0	0, 0	5, 9
Matrix	2.0, 2.1	1.0, 1.5	0.8, 0.8

a. The cells used for preparation of the nuclei were tetraploid after many passages in culture. In two experiments sonicates of these cells were found to have 2.0 times the level of these activities as sonicates of diploid cells.

b. Nucleoplasm is defined here as the fraction extracted from purified nuclei in 20 mM Tris-Cl, pH 7.5 containing 5 mM Na_2 EDTA (TE-buffer). The other fractions were derived by serial salt extractions in this buffer. Chromatin I is derived from the insoluble pellet after contrifuging for 10 min at 15,000 × g by extraction with 0.5 M NaCl, chromatin II by extraction with 2.0 M NaCl. Matrix is defined operationally as the insoluble pellet remaining after extensive shearing of the viscous chromatin suspension in TE-buffer-2.0 M NaCl by passage through 18-23 gauge needles and centrifugation.

c. Assays were carried out with 400 μg/ml heat-denatured DNA (ss-DNase) and native DNA (ds-DNase) and 300 μg/ml polyrA (RNase) each in the presence of 50 mM Tris-Cl, pH 7.5 containing 4 mM $MgCl_2$, 5 mM NaCl, 1 mM dithiothreitol and 0.1 mM Na_2 EDTA for 0, 15 and 30 min at 37°C. Reactions were stopped by addition of an equal volume of 0.2 M perchloric acid and the nucleotides in acid-soluble fractions determined from the absorbances at 260 nm.

shown). This was demonstrated by pre-treating each fraction with an optimal concentration of trypsin for 30 min at room temperature, in the range 10–200 μg/ml, depending on the protein concentration.

RNase activity was also detected in all fractions in variable ratios with the DNase activities. It was found that a fraction of this RNase activity but not the DNase activities, was specifically inhibited by heated CEM cell cytosol, Δ Cyt (Table 9.3). In the presence of an excess of Δ Cyt, the ratios of ss-DNase: ds-DNase: RNase were close to 2:1:1 in all nuclear fractions, very similar to those found for the *Neurospora* endo-exonuclease. As for the latter enzyme also, all three activities were strongly inhibited by 25 μM ATA (Table 9.3). At 50–100 μM ATA, the inhibitions were complete.

All three activities also showed complete dependence on excess Mg^{2+} in the presence of a low concentration of EDTA and exhibited no activity in the presence of EDTA alone. Optimal activity was found in the range 2–4 mM Mg^{2+} but was only slightly less at 10 mM Mg^{2+}. The effects of Ca^{2+} added in the presence of Mg^{2+} were highly variable for these crude nuclear fractions. For example, with nucleoplasm in the presence of 2 mM Mg^{2+}, the addition of 1 and 2 mM Ca^{2+} abolished all RNase activity, reduced the ss-DNase activity to 48% and 16% respectively but the ds-DNase activity was reduced to only 80% and 63% repectively with 1 and 2 mM Ca^{2+}. Thus, in the

Table 9.3 Effects of aurin tricarboxylic acid, ATA (25 μM) and excess heated cytosol[a] (Δ Cyt) on the nuclease activities in nuclear fractions of CEM-f2 cells.

Nuclear Fraction	+ATA			% Activity[b] −Cyt		
	ss-DNase	ds-DNase	RNase	ss-DNase	ds-DNase	RNase
Nucleoplasm	35	21	2	96	104	76
Chromatin I	2	22	2	100	–	45
Chromatin II	–	–	1	–	–	69
Matrix	0	0	12	96	94	60

a. Heated cytosol was prepared from a lysate of washed CEM-f2 cells made in the presence of Nonidet P-40 and centrifuged to sediment out crude nuclei. The supernatant was heated at 100°C for 10 min and then centrifuged to remove coagulated protein.
b. See footnotes of Table 9.2 for assays.

presence of 2 mM concentrations of both Mg^{2+} and Ca^{2+}, the nucleoplasmic activity appeared to be specific for DNA and exhibited more activity with ds- than with ss-DNase. Whether such responses to divalent metal ions will be shown by the purified CEM endo-exonuclease remains to be determined.

That at least a major proportion of the CEM cell activity is related to the endo-exonuclease of *Neurospora* is made likely by the observation that up to 70% of the ss-DNase activity in nuclear extracts and cell sonicates bound to an affinity column generated by cross-linking antibody raised to *Neurospora* endo-exonuclease with protein A-Sepharose. The enzyme was recovered by elution with 3.5 M $MgCl_2$, followed by rapid dialysis. At least 80% of the ss-DNase activity in cell sonicates was inhibited by 100 μM ATA. In addition, as for *Neurospora* endo-exonuclease in crude extracts (Chow and Fraser, 1983), 90% of the ss-DNase activity in CEM cell sonicates bound reproducibly to DEAE-Sepharose and was eluted with 0.1–0.4 M NaCl in a linear 0–1.0 M NaCl gradient; this activity, in turn, bound to ss-DNA-cellulose. The activity eluted from DEAE-Sepharose was inhibited by 90% or more with antibody raised to the purified monkey CV-1 cell endo-exonuclease (Couture and Chow, 1992) kindly supplied by T. Y.-K. Chow. Preliminary results indicate that the CEM activity purified by immunoaffinity chromatography was also very sensitive to inhibition by this antibody and by ATA.

To investigate whether or not this major CEM nuclear activity might be involved in the chromatin DNA fragmentation that accompanies apoptosis, we assayed the levels of ss-DNase, ds-DNase and RNase (with poly rA) activities in nuclear fractions derived from CEM cells that had been pre-treated with 1 μM dexamethasome (DEX) for 48 h or with 17 μM etoposide (VP-16) for 3 h, treatments known to induce apoptosis in these cells. In four experiments, instead of the expected activation of nuclease activity, a reduction of the three activities (to 45% activity on average) was observed in all nuclear fractions from DEX-treated cells relative to those from DEX solvent (DMSO) controls. The same result was observed in one experiment with VP-16-treated CEM cells. The latter result has now been repeated using a simplified procedure that monitored the total ss-DNase activity in sonicates of untreated and VP-16-treated CEM cells over a 4 h time course. In this case, the activity also dropped in the 1, 2, 3 and 4 h treatments respectively to 86%, 63%, 70% and 54% of the untreated 4 h drug solvent (DMSO) control.

These losses in ss-DNase activity may result from ADP-ribosylation, proteolysis or both, since these processes are also triggered by apoptotic stimuli including VP-16 treatment (Kaufmann, 1989; Shimizu et al., 1990). If any conversion of the inactive to active endo-exonuclease had occurred in response to VP-16, it was likely to have been obscured by this inactivation. Proteolysis has already been implicated in the rapid turnover of the endogenous endonuclease in thymocytes (McConkey et al., 1990). Direct evidence for inactivation of the ss-DNase activity by ADP-ribosylation in vivo was obtained by incubating untreated control and VP-16 treated cells (1.5 h treatment) without and with the polyADP-ribosyltransferase inhibitor, 3-aminobenzamide, 3-AB (3 mM). In this experiment, VP-16 alone had little or no effect in 1.5 h on the ss-DNase activity extracted by sonication (99% activity), while 3-AB treatment alone caused a 36% increase in ss-DNase activity of control cells and a 63% increase in VP-16-treated cells. Similar effects have been observed at earlier (45 min) and later (2.5 h) times but the maximum activation seen so far in VP-16-treated cells with 3-AB was 1.6-fold.

VP-16 treatment results in the production of 100–1000 times the number of irreversible breaks in chromatin DNA as those induced transiently and reversibly at the target sites, topoisomerase II-DNA complexes (Kaufmann, 1989). Presumably, the irreversible breaks result from the activation and release of chromatin-bound nuclease and trigger polyADP-ribosyltransferase activity. Evidently, by blocking ADP-ribosylation, the expected activation of the nuclease in response to VP-16 can be detected. ADP-ribosylation may be a step in the normal turnover of the enzyme since active enzyme also accumulated significantly in control cells treated with 3-AB. Thus, the combined use of 3-AB and proteolysis inhibitors (McConkey et al., 1990) might permit the detection of even higher activation of the endogenous CEM endo-exonuclease in response to VP-16 treatment. A maximum 3-fold increase would be expected if all of the trypsin-activatable form were activated.

CLONING AND SEQUENCING OF THE ENDO-EXONUCLEASE GENE

The cloning and sequencing of the Saccharomyces, Neurospora and human nuclear endo-exonuclease genes has been undertaken in the laboratory of T.Y.-K. Chow by screening the appropriate cDNA libraries in the λgt11 expression vector using antibody raised to Neurospora endo-exonuclease to detect expressed immunoreactive polypeptides. In each case, positive clones have been isolated and this has led or is leading to the isolation of the corresponding genes.

The yeast gene has been shown (Chow et al., 1992) to encode a predicted polypeptide of 485 amino acids which is very unusual in having a region near the N-terminus (residues 73–187) which shows 50% homology to the human *rho* oncogene products, small GTPases related to the *ras* proteins. Four putative GTP-binding sequences are highly conserved and the yeast endo-exonuclease has been found to bind to GTP-agarose. It may be relevant that the GTP-binding sites in some of the *rho* proteins are also targets for inactivation by ADP-ribosylation, in this case mediated by various bacterial toxins (Bourne et al., 1990).

The cloned human DNA expresses a polypeptide which not only cross-reacts with antibody raised to the *Neurospora* endo-exonuclease, but also cross-reacts with antibodies raised to the yeast nuclear and monkey CV-1 cell endo-exonucleases (T.Y.-K. Chow, personal communication). Preliminary sequence data on the cloned human DNA indicates that it also contains a *rho*-like sequence. It is not yet known whether the yeast and human enzymes have GTPase activity or are autophosphorylated. This site may prove to be a regulatory site governing the switching of the nuclease activity "off" and "on", "off" possibly by ADP-ribosylation and "on" possibly by phosphorylation.

We have synthesized oligonucleotides (21-mers) corresponding to sequences respectively near the termini of the coding and non-coding strands of the cloned human DNA provided to us by Dr Chow. These have been used to amplify two cDNA sequences detected in methotrexate-treated, but not in untreated, HL-60 cells using PCR techniques. The amplified DNAs have been purified and are being sequenced to confirm that they correspond to the cloned human DNA isolated by Chow's group. These DNAs will be used to prepare probes for the study of the expression of the human endo-exonuclease gene in human leukaemic cell lines in response to agents which induce apoptosis including those used in chemotherapy.

SUMMARY

Endo-exonuclease has been identified as a major chromatin-matrix-associated nuclease in mammalian nuclei. Several active polypeptides are present in nuclear extracts and may be derived by limited proteolysis from a larger inactive form *in vivo* or *in vitro* during extraction and isolation. This limited proteolysis may account for the previous isolation and purification of most of the different endonucleases from mammalian nuclei.

By analogy with its known roles in lower eukaryotes, the normal functions of this enzyme may be in recombination and in post-replication repair in the G2 phase of the cell cycle. In addition, in cells exposed to cytotoxic doses of DNA-damaging agents, endo-exonuclease participates in the genomic DNA degradation that accompanies apoptosis. This is formally analogous to the "reckless" DNA degradation in bacteria that is carried out by the major recombination nuclease in heavily DNA-damaged *Escherichia coli* when the repair systems are overwhelmed. This DNA degradation also occurs in mammalian cells in response to other stimuli that induce apoptosis such as glucocorticoid treatment of human leukaemic CEM and HL-60 cells and of thymocytes. CEM cell endo-exonuclease is inhibited by 1 mM Zn^{2+} and 25–100 μM aurin tricarboxylic acid, inhibitors of chromatin DNA fragmentation and apoptosis in intact cells.

A cytotoxic dose of the topoisomerase II inhibitor, etoposide (VP-16), triggered a rapid loss of extractable endo-exonuclease in CEM cells. This was partially blocked by an inhibitor of ADP-ribosylation, a treatment which permitted activation of endo-exonuclease to be observed in response to VP-16 treatment. Finally, cloning and sequencing of the human endo-exonuclease gene is in progress and has already indicated that the nuclease contains a *rho*-like GTP-binding domain which might be a target for ADP-ribosylation.

REFERENCES

Alnemri, E. S. and Litwack, G. (1990) Activation of internucleosomal DNA cleavage in human CEM lymphocytes by glucocorticoid and novobiocin. Evidence for a non-Ca^{2+}-requiring mechanism(s). *Journal of Biological Chemistry*, 265, 17323–17333.

Barry, M. A. and Eastman, A. (1992) Endonuclease activation during apoptosis: The role of cytosolic Ca^{2+} and pH. *Biochemical and Biophysical Research Communications*, 186, 782–789.

Bourne, H. R., Sanders, D. A. and McCormick, F. (1990) The GTPase superfamily: a conserved switch for diverse cell functions. *Nature*, 348, 125–132.

Chow, T. Y.-K. and Fraser, M. J. (1983) Purification and properties of single strand DNA-binding endo-exonuclease of *Neurospora crassa*. *Journal of Biological Chemistry*, 258. 12 010–12 018.

Chow, T. Y.-K., Perkins, E. and Resnick, M. A. (In press) Yeast RNC1 encodes a chimeric protein, RhoNUC, with a human *rho* motif and deoxryibonuclease activity. *Nucleic Acids Research*.

Chow, T. Y.-K. and Resnick, M. A. (1987) Purification and characterization of an endo-exonuclease from *Saccharomyces cerevisiae* that is influenced by the *rad52* gene. *Journal of Biological Chemistry*, 262, 17 659-17 667.

Chow, T. Y.-K. and Resnick, M. A. (1988) An endo-exonuclease activity of yeast that requires a functional *rad52* gene. *Molecular and General Genetics*, 211, 41–48.

Cordis, G. A., Goldblatt, P. J. and Deutscher, M. P. (1975) Purification and characterization of a major endonuclease from rat liver nuclei. *Biochemistry*, 14, 2596–2603.

Couture, C. and Chow, T. Y.-K. (In press) Purification and characterization of a mammalian endo-exonuclease. *Nucleic Acids Research*.

Fischman, G. J., Lambert, M. W. and Studzinski, G.P. (1979) Purification and properties of a nuclear DNA endonuclease from HeLa S_3 cells. *Biochimica Biophysica Acta*, 567, 464–471.

Fraser, M. J., Chow, T. Y.-K., Cohen, H. and Koa, H. (1986) An immunochemical study of *Neurospora* nucleases. *Biochemistry and Cell Biology*, 64, 106–116.

Fraser, M. J., Hatahet, Z. and Huang, X. (1989) The actions of *Neurospora* endo-exonuclease on double strand DNAs. *Journal of Biological Chemistry*, 264, 13 093–13 101.

Hashida, T., Tanaka, Y., Matsunami, N., Yoshihara, K., Kamiya, T., Tanigawa, T., and Koide, S. S. (1982) Purification and properties of bull seminal plasma Ca^{2+}, Mg^{2+}-dependent endonuclease. *Journal of Biological Chemistry*, 257, 13 114–13 119.

Hewish, D. R., Burgoyne, L. A. (1973) The calcium-dependent endonuclease activity of isolated nuclear preparations. Relationships between its occurrence and the occurrence of other classes of enzymes found in nuclear preparations. *Biochemical and Biophysical Research Communications*, 52, 475–481.

Hibino, Y., Iwakami, N. and Sugano, N. (1991) A nuclease from rat-liver nuclei with endo- and exonucleolytic activity. *Biochemica Biophysica Acta*, 1088, 305–307.

Hibino, Y., Yamamura, T. and Sugano, N. (1989) Purification and properties of an endonuclease endogenous to rat liver nuclei. *Biochemica Biophysica Acta*, 1008, 287–292.

Hibino, Y., Yoneda, T. and Sugano, N. (1988) Purification and properties of a magnesium-dependent endodeoxyribonuclease endogenous to rat liver nuclei. *Biochemica Biophysica Acta*, 950, 313–320.

Ishida, R., Akiyoshi, H. and Takahashi, T. (1974) Isolation and purification of calcium- and magnesium-dependent endonuclease from rat liver nuclei. *Biochemical and Biophysical Research Communications*, 56, 703–710.

Kaufmann, S. H. (1989) Induction of endonucleolytic DNA cleavage in human acute myelogenous leukaemic cells by etoposide, camptothecin and other cytotoxic anticancer drugs: a cautionary note. *Cancer Research*, 49, 5870–5878.

Koa, H., Fraser, M. J. and Kafer, E. (1990) Endo-exonuclease of *Aspergillus nidulans*. *Biochemistry and Cell Biology*, 68, 387–392.

Kwong, S. and Fraser, M. J. (1978) Neurospora endo-exonuclease and its inactive (precursor?) form. *Canadian Journal of Biochemistry*, 56, 370–377.

Lacks, S. A. (1981) Deoxyribonuclease I in mammalian tissues. Specificity of inhibition by actin. *Journal of Biological Chemistry*, 256, 2644–2648.

Lambert, M. W., Lee, D. E., Okorodudu, A. O. and Lambert, W. C. (1982) Nuclear deoxyribonuclease activities in human lymphoblastoid and mouse melanoma cells. *Biochemica Biophysica Acta*, 699, 192–202.

Linn, S. and Lehman, I. R. (1965a) An endonuclease from *Neurospora crassa* specific for polynucleotides lacking an ordered structure. I. Purification and properties of the enzyme. *Journal of Biological Chemistry*, 240, 1287–1293.

Linn, S. and Lehman, I. R. (1965b) An endonuclease from *Neurospora crassa* specific for polynucleotides lacking an ordered structure. II. Studies of enzyme specificity. *Journal of Biological Chemistry*, 240, 1294–1304.

Machray, G. C. and Bonner, J. (1981) Purification and some properties of a deoxyribonucleic acid endonuclease endogenous to rat liver chromatin. *Biochemistry*, 20, 5466–5470.

McConkey, D. J., Hartzell, P., Nicotera, P. and Orrenius, S. (1989) Calcium-activated DNA fragmentation kills immature thymocytes. *FASEB Journal*, 3, 1843–1849.

McConkey, D. J., Hartzell, P., Nicotera, P. and Orrenius, S. (1990) Rapid turnover of endogenous endonuclease activity in thymocytes: Effects of inhibitors of macromolecular synthesis. *Archives of Biochemistry and Biophysics*, 278, 284–287.

Nakamura, M., Sakaki, Y., Watanabe, N. and Takagi, Y. (1981) Purification and characterization of the Ca^{2+} plus Mg^{2+}-dependent endodeoxyribonuclease from calf thymus chromatin. *Journal of Biochemistry*, (Tokyo) 89, 143–152.

Ramotar, D., Auchincloss, A. H. and Fraser, M. J. (1987) Nuclear endo-exonuclease of *Neurospora crassa*. Evidence for a role in DNA repair. *Journal of Biological Chemistry*, 262, 425–431.

Resnick, M. A., Sugino, A., Nitiss, J. and Chow, T. (1984) DNA polymerases deoxyribonucleases, and recombination during meiosis in *Saccharomyces cerevisiae*. *Molecular and Cellular Biology*, **4**, 2811–2817.

Shimizu, T., Kubota, M., Tanizawa, A., Sano, H., Kasai, Y., Hashimoto, H., Akiyama, Y. and Mikawa, H. (1990) Inhibition of both etoposide-induced DNA fragmentation and activation of poly (ADP-ribose) synthesis by zinc ion. *Biochemical and Biophysical Research Communications*, **169**, 1172–1177.

Stratling, W. H., Grade, C. and Horz, W. (1984) Ca/Mg-dependent endonuclease from porcine liver. *Journal of Biological Chemistry*, **259**, 5893–5898.

Tanaka, Y., Yoshihara, K., Itaya, A., Kamiya, T. and Koide, S. S. (1984) Mechanism of the inhibition of Ca^{2+}, Mg^{2+}-dependent endonuclease of bull seminal plasma induced by ADP-ribosylation. *Journal of Biological Chemistry*, **259**, 6579–6585.

Tanigawa, Y. and Shimoyama, M. (1983) Mg^{2+}-dependent/poly (ADP-ribose) sensitive endonuclease. *Journal of Biological Chemistry*, **258**, 9184–9191.

Wang, E. C., Furth, J. J. and Rose, J. A. (1978) Purification and characterization of a DNA single strand specific endonuclease from human cells. *Biochemistry*, **17**, 544–549.

Wang, E. C. and Rose, J. A. (1981) Deoxyribonuclease and single-stand-specific endonucleases in human cells: Partial purification of a salt-resistant endonuclease with an acidic isoelectric point. *Biochemistry*, **20**, 755–758.

Chapter 10

CRITICAL CHANGES IN APOPTOSIS IN THYMOCYTES PRECEDE ENDONUCLEASE ACTIVATION

Gerald M. Cohen, Xiao-Ming Sun, Roger T. Snowden, Michael G. Ormerod and David Dinsdale

MRC Toxicology Unit, Medical Research Council Laboratories, Woodmansterne Road, Carshalton, Surrey SM5 4EF, UK

Apoptosis is a major form of cell death occurring in a wide variety of different biological systems including normal cell turnover, embryonic and T-cell development, metamorphosis, hormone-dependent atrophy, immune and toxic cell killing (Arends and Wyllie, 1991; Ellis et al., 1991; Cohen et al., 1992b). It is an active form of self-destruction, which is characterized by condensation and fragmentation of nuclear chromatin, accompanied by compaction of cellular organelles, dilatation of the endoplasmic reticulum and a marked reduction in cell volume (Wyllie et al., 1980; Arends and Wyllie, 1991). Volume reduction is such a prominent feature of apoptosis that initially this type of cell death was known as shrinkage necrosis (Kerr, 1971). The morphological changes of apoptosis have been associated with endonuclease cleavage of DNA into nucleosomal length fragments of 180–200 base pairs or multiples thereof (Wyllie, 1980). Here we report that key morphological changes of apoptosis in thymocytes, precede and can be dissociated experimentally from DNA fragmentation produced as a result of endonuclease activity. Thus internucleosomal cleavage of DNA is a late event in apoptosis.

In addition, we describe the isolation of pre-apoptotic thymocytes, which are intermediate in size and density between normal and apoptotic thymocytes and have sharply defined clumps of heterochromatin abutting onto the nuclear membrane. These morphological changes precede internucleosomal cleavage of DNA. Subsequent dramatic changes in the nuclei of these cells result in both the typical morphology of apoptotic cells and internucleosomal cleavage of DNA. Thus critical biochemical and morphological changes in apoptosis occur prior to endonuclease activation. The identification of the enzyme(s) responsible for these early nuclear changes would represent an important contribution to our current understanding of apoptosis.

SEPARATION OF NORMAL AND APOPTOTIC THYMOCYTES

The induction of apoptosis in immature thymocytes by glucocorticoids, such as dexamethasone and methylprednisolone, is one of the best characterized systems for inducing apoptosis both *in vivo* and *in vitro* (Wyllie, 1980; Cohen and Duke, 1984; Cohen et al., 1992b). Recently we have described a novel flow cytometric method to separate and quantify normal and apoptotic thymocytes (Sun et al., 1992). Thymocytes were isolated from immature rats and incubated with dexamethasone (0.1 μM) for periods up to 4 h. Subsequent incubation with the vital bisbenzimidazole dye Hoechst 33342 and the DNA intercalating agent propidium iodide enabled three distinct populations of cells to be identified. Nonviable cells fluoresced red due to propidium iodide, whereas normal and apoptotic cells fluoresced blue due to Hoechst 33342. Apoptotic cells were distinguished from normal thymocytes both by their higher intensity of blue fluorescence and their smaller size as determined by a reduction in forward light scatter (Figure 10.1). Cells from these populations were purified by fluorescence activated cell sorting (FACS). The larger cells, with low blue fluorescence, showed normal thymocyte ultrastructure and the absence of any DNA fragmentation, as measured by agarose gel electrophoresis. In contrast, the smaller cells showed both the morphological characteristics of apoptosis and extensive internucleosomal fragmentation of DNA to multiples of approximately 180 base pairs.

An illustration of the flow cytometric method is shown in Figure 10.1 where the induction of apoptotic cells (region 1) by dexamethasone is clearly seen (Figure 10.1b). The formation of these apoptotic cells was inhibited by cycloheximide and actinomycin D (Figure 10.1c and 10.1d) inhibitors of protein and mRNA synthesis respectively. The inhibition of thymocyte apoptosis by these agents is in agreement with other studies (Cohen and Duke, 1984; Wyllie et al., 1984) and provides further validation for our new flow cytometric method.

ZINC PREVENTS SOME BUT NOT ALL OF THE NUCLEAR CHANGES INDUCED BY DEXAMETHASONE

Zinc has also been reported to prevent apoptosis, an effect assumed to be due to the ability of the metal ion to inhibit a Ca^{2+}/Mg^{2+}-dependent endonuclease (Cohen and Duke, 1984; Waring et al., 1990). It was therefore surprising that zinc did not decrease the formation of apoptotic cells, as judged by the property of high blue fluorescence (region 1 in Figure 10.1e), under conditions where zinc almost completely inhibited

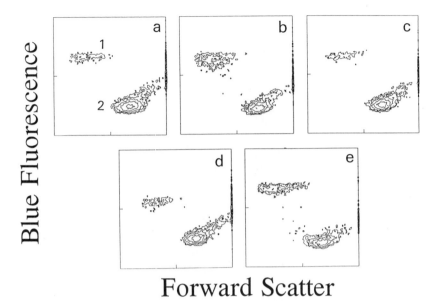

Figure 10.1 Zinc does not prevent the formation of apoptotic thymocytes by dexamethasone. Freshly isolated immature rat thymocytes were incubated (20 × 10^6 cells/ml in RPMI-1640 containing 10% foetal bovine serum) for 4 h either alone (Figure 10.1a) or with dexamethasone (0.1 μM) in the absence (Figure 10.1b) or in the presence of either cycloheximide (10 μM) (Figure 10.1c), actinomycin D (1 μM) (Figure 10.1d), or zinc acetate dihydrate (1000 μM) (Figure 10.1e). Normal (region 2) and apoptotic (region 1) thymocytes were separated by flow cytometry following incubation for 10 min with the bisbenzimidazole dye Hoechst 33342 (1 μg/ml) and propidium iodide (Sun et al., 1992). Non-viable cells, which included propidium iodide, were gated out. Apoptotic cells, with high blue fluorescence and low forward light scatter, were distinguished from normal thymocytes, with low blue fluorescence and high forward light scatter. Forward scatter gives an indication of size i.e. apoptotic cells were smaller than normal cells.

both DNA fragmentation and DNA laddering induced by dexamethasone (Cohen et al., 1992a).

Previous studies have utilized isopycnic centrifugation on discontinuous Percoll gradients to separate and purify normal and apoptotic thymocytes (Wyllie and Morris, 1982). Applying this system to cells treated with dexamethasone (0.1 μM) alone, we obtained a discrete cell fraction, which had both a smaller mean diameter and higher modal density than normal thymocytes. This fraction was examined both by electron microscopy and agarose gel electrophoresis. Following incubation of thymocytes with dexamethasone alone, the high density cells separated by Percoll gradients, exhibited high blue fluorescence with Hoechst 33342, and showed extensive DNA laddering (Figure 10.2). On ultrastructural examination, they showed the distinct morphological features of apoptosis (Figure 10.3a) including condensed heterochromatin usually coalesced against one pole of the nuclear membrane (Wyllie et al., 1980; Arends and

Figure 10.2 Zinc inhibits dexamethasone-induced DNA laddering in small, high density apoptotic thymocytes. Apoptotic cells of high modal density and small diameter were examined by agarose gel electrophoresis (Sorenson et al., 1990) after separation by isopycnic centrifugation on discontinuous Percoll gradients following incubation of thymocytes for 4 h with dexamethasone (0.1 μM) either alone (lane 3) or in the presence of zinc (1000 μM) (lane 2). Lane 1 contains molecular size standards of multiples of 123 bp.

Wyllie, 1991). Strikingly different results were noted in the high density cells obtained following incubation with dexamethasone in the presence of zinc (1000 μM). These thymocytes were also shrunken, with dilation and "bubbling" of the smooth endoplasmic reticulum, but they showed no evidence of DNA laddering (Figure 10.2). The heterochromatin in these cells was condensed and arranged in several sharply defined clumps, which were contiguous with the nuclear membrane (Figure 10.3b). Similar changes have been described, in various cell types, as the earliest signs of apoptosis (Kerr et al., 1987; Walker et al., 1988). Thus in the presence of zinc, the cells have probably been arrested at an early stage of apoptosis prior to the effects of the endonuclease. The cells (Figure 10.3b) showed the earliest ultrastructural features of apoptosis, indicating that, in the presence of zinc, dexamethasone induces many key features of apoptosis without any evidence of DNA laddering. Similar results were observed in thymocytes treated with etoposide, a DNA topoisomerase II inhibitor, in the presence of zinc (unpublished results). These results suggest that in the induction of apoptosis, critical changes occur both in nuclear DNA and in the cytoplasm prior to endonuclease cleavage of DNA into nucleosomal fragments. Whilst our results are

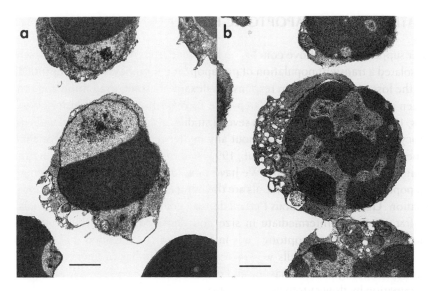

Figure 10.3 Zinc arrests apoptotic changes in dexamethasone-treated thymocytes. Thymocytes, treated with dexamethasone, demonstrated the characteristic chromatin condensation and cytoplasmic contraction of apoptosis (Figure 10.3a). The heterochromatin was condensed and usually coalesced against one pole of the nuclear membrane. Cells treated with dexamethasone in the presence of zinc showed only the earliest signs of apoptosis (Figure 10.3b). Discrete clumps of condensed chromatin were abutted against the nuclear membrane, a further clump was present in the centre of many of these nuclei. The nuclear membrane, although usually intact, was often convoluted. The euchromatin retained its normal density but often included one or more clusters of intensely-stained nucleolar remnants. Bars represent 1 μm.

consistent with a role for zinc in inhibiting the endonuclease (Cohen and Duke, 1984; Waring et al., 1990), they do not exclude other possibilities such as zinc binding to the DNA. Zinc may also modify apoptosis by activating protein kinase C or by inhibiting phosphorylases associated with inositol phosphate metabolism (Waring et al., 1991).

This to our knowledge is the first time, in thymocytes, that a clear dissociation has been observed between the morphological features of apoptosis and DNA laddering. A similar absence of DNA laddering has been observed in some cases of programmed cell death, for example in insect metamorphosis and in normal limb development (Lockshin and Zakeri, 1991). Our results with zinc support the hypothesis that the induction of the earliest morphological changes of apoptosis involves enzymes other than the Ca^{2+}/Mg^{2+}-dependent endonuclease and that this endonuclease is not involved until the later stages, when it is responsible for internucleosomal fragmentation of the DNA (Wyllie, 1980).

ISOLATION OF A PRE-APOPTOTIC POPULATION OF THYMOCYTES

Further support for the above conclusion has come from our recent studies, where we have isolated a transient population of pre-apoptotic thymocytes. In apoptotic thymocytes, the loss of cell volume in response to dexamethasone, irradiation and etoposide has been associated with a single step-wise increase in buoyant density (Wyllie and Morris, 1982; Walker et al., 1991). In several studies of apoptosis, normal and apoptotic thymocytes have been isolated without any evidence of cells of an intermediate size (Wyllie and Morris, 1982, Walker et al., 1991) and thus by implication, no intermediate population (Cohen et al., 1992b). We have now identified and isolated such a transitional population of cells. These cells are the immediate precursors of an apoptotic cell population. Using discontinuous Percoll density gradients, we isolated a population of cells (fraction 3-F3) intermediate in size and density between the normal cells in fraction 2 (F2) and the apoptotic cells in fraction 4 (F4) (Cohen et al., unpublished data). This population of cells was increased by a number of agents, which induce apoptosis, including dexamethasone and etoposide.

Examination by flow cytometry of the different populations of cells from the Percoll gradients, after incubation with Hoechst 33342, showed that cells in F1 and F2 exhibited high forward light scatter and low blue fluorescence, typical of normal cells (Figure 10.4a and 4b). Cells in F3 appeared to be heterogeneous containing both smaller cells with high blue fluorescence (region 3, Figure 10.4c), as well as apparently normal cells (region 1, Figure 10.4c), which exhibited low blue fluorescence but were somewhat smaller (lower mean forward light scatter) than the corresponding cells from F1 and F2. A third group of cells, present within F3 (region 2, Figure 10.4c) had properties intermediate between the other two populations. Cells in F4 were predominantly of a smaller size (lower forward light scatter) and exhibited high blue fluorescence with Hoechst 33342 (Figure 10.4d).

Agarose gel electrophoresis of the cells, from each Percoll fraction obtained following exposure to etoposide, showed little or no DNA laddering in the normal cells in F1 and F2, whereas extensive laddering was observed in the purported apoptotic cells in F4. Significant laddering was also observed in cells from F3, but it was less than that observed from F4.

Cells from the four different Percoll fractions, prepared after treatment with etoposide for 4 h, were processed for electron microscopy. The morphology of the majority (>85%) of the cells in F1 and F2 was indistinguishable from that of most thymocytes isolated from untreated rats. The nuclei in both fractions were rounded and characterised by relatively condensed perinuclear heterochromatin, which lined the nuclear membrane except for small regions around the nuclear pores. This heterochromatin extended as several, ill-defined projections into the euchromatin. The population of highest buoyant density in F4 was primarily apoptotic, most of which had accumulations of condensed chromatin forming dense apical caps within the nucleus. Discontinuities in the nuclear membrane were often evident within the cytoplasm of these cells and nucleolar remnants, particularly the dense fibrillar

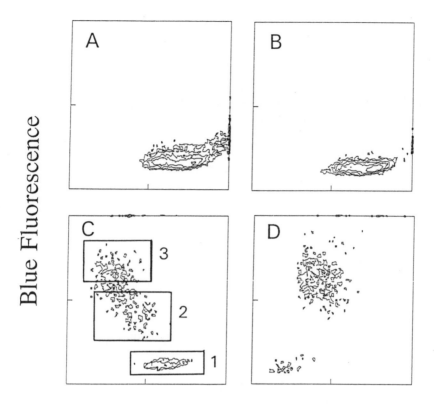

Figure 10.4 Flow cytometric analysis of cells fractionated from Percoll. Thymocytes were incubated with etoposide (10 μM) for 4 h. The cells were separated on Percoll and the four fractions of increasing density from F1 to F4 (A to D respectively) incubated with Hoechst 33342 at 37°C for a further 10 min and then examined by flow cytometry as described in the legend to Figure 10.1. Similar results were obtained following dexamethasone (0.1 μM) treatment of cells for 4 h.

components, were prominent. The cisternae of the endoplasmic reticulum were often dilated and many of the resulting vacuoles fused with the cell membrane, usually towards one pole of the cell.

Unlike the other three fractions, F3 was strikingly heterogeneous, it contained a few apoptotic bodies (<5%) and many profiles (40–50%) exhibited the features of cells in F4. The remaining profiles were cells with a distinctly different morphology. These cells had a rather dense, granular cytoplasm containing normal organelles, except for a few small vacuoles or dilated cisternae of the endoplasmic reticulum. The most striking feature of these cells was the compaction of the perinuclear heterochromatin into dense, sharply-defined clumps abutting onto the nuclear membrane. Unusually large regions of this membrane were devoid of heterochromatin and the membrane itself was often convoluted. The euchromatin and fibrillar centres of these cells were apparently normal.

EVIDENCE THAT F3 CONTAINS AN INTERMEDIATE POPULATION OF PRE-APOPTOTIC THYMOCYTES

In order to test the hypothesis that the cells in F3 were an intermediate population between the normal (F1 and F2) and the apoptotic thymocytes (F4), thymocytes were incubated with etoposide (10 μM) for 4 h and following Percoll separation the cells in F3 were further sorted by flow cytometry into the population of cells of either low blue fluorescence (region 1, Figure 10.4c), or of intermediate and high blue fluorescence (regions 2 and 3, Figure 10.4c). On examination by agarose gel electrophoresis, no laddering was observed with cells from region 1, whereas significant laddering was observed with cells from regions 2 and 3. Ultrastructural examination of the cells from region 1 by electron microscopy showed that this population was homogeneous (Figure 10.5) and characterised by distinctive perinuclear accumulations of condensed heterochromatin. The sharp demarcation between these clumps and the euchromatin was accentuated by the apparent loss of the perichromatin fibrils, which permeated this region in both untreated cells and in the etoposide-treated cells collected in F1 and F2. In marked contrast, most of the cells (>95%) in regions 2 and 3 showed the ultrastructural features associated with apoptotic cells (Figure 10.3a). In addition, when a homogeneous population of cells from F3 was isolated and then incubated for a further 2 h in the presence or absence of etoposide (10 μM), the cells became smaller, exhibited higher blue fluorescence with Hoechst 33342 and DNA ladders developed.

Thus in these studies, we have identified and partially characterized a transitional population of cells, intermediate in size between normal and apoptotic thymocytes and exhibiting a distinctive morphology. No internucleosomal cleavage of DNA was observed in these transitional cells until the dramatic changes of nuclear morphology, characteristic of apoptosis were observed. This observation provides strong evidence for a causative link between these two events. Cells in F1 and F2 had normal morphology, whilst those in F4 were typically apoptotic. All these fractions were fairly homogeneous as judged by both fluorescence characteristics with Hoechst 33342 and electron microscopy. In marked contrast, after 4 h incubation with etoposide, cells in F3 were heterogeneous both by flow cytometry (Figure 10.4c) and electron microscopy (results not shown). Separation of these heterogeneous cells by FACS showed subtle, but distinct, changes in the distribution of heterochromatin in most cells from region 1 (Figure 10.5). These changes resemble those observed in thymocytes in which the process of dexamethasone-induced apoptosis has been arrested by the presence of zinc (Cohen et al., 1992a). This peripheral coalescence of heterochromatin has been recognised by several workers as an early sign of apoptosis (Kerr et al., 1987; Walker et al., 1988; Bursch et al., 1992). It probably indicates major changes in the DNA

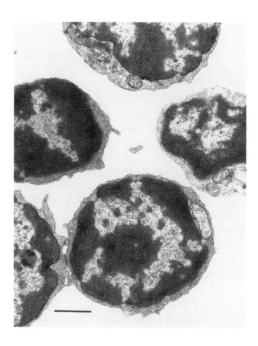

Figure 10.5 Ultrastructure of pre-apoptotic thymocytes. Thymocytes were incubated for 4 h in the presence of etoposide and separated into four fractions (F1–F4) by isopycnic Percoll centrifugation. At least two types of cell were evident in F3 which were subsequently sorted, by flow cytometry, according to their fluorescence intensity with Hoechst 33342. Cells with high blue fluorescence were apoptotic, similar to those in F4. The pre-apoptotic population, with low blue fluorescence, showed aggregation of heterochromatin into discrete clumps, which were sharply delineated from the remaining nucleoplasm.

but, in the absence of any laddering in this fraction, it clearly precedes internucleosomal cleavage. In contrast, cells from regions 2 and 3 (Figure 10.4c) exhibited a dramatic change in morphology demonstrating all the characteristic changes associated with apoptosis, together with DNA laddering. These data provide strong evidence that internucleosomal cleavage of DNA is responsible for the most characteristic morphological changes in the nucleus of apoptotic thymocytes, in agreement with the suggestion of Arends et al. (1990).

Thus in two different series of experiments, firstly in the studies with dexamethasone in the presence of zinc and secondly in the isolation of the pre-apoptotic population of cells, we have identified cells with marked nuclear changes (Figures 10.3b and 10.5) in the absence of internucleosomal cleavage of DNA. These results strongly suggest that key enzyme(s), other than the endonuclease, are implicated in the earliest stages of both dexamethasone- and etoposide-induced apoptosis in thymocytes. The isolation of this pre-apoptotic population of cells will greatly facilitate the identification of critical early changes in apoptosis.

REFERENCES

Arends, M. J., Morris, R. G. and Wyllie, A. H. (1990) Apoptosis: the role of the endonuclease. *American Journal of Pathology*, 136, 593–608.

Arends, M. J. and Wyllie, A. H. (1991) Apoptosis: Mechanisms and roles in pathology. *International Review of Experimental Pathology*, 32, 223–254.

Bursch, W., Oberhammer, F. and Schulte-Hermann, R. (1992) Cell death by apoptosis and its protective role against disease. *Trends in Pharmacological Sciences*, 13, 245–251.

Cohen, G. M., Sun, X.-M., Snowden, R. T., Dinsdale, D. and Skilleter, D. N. (1992a) Key morphological features of apoptosis may occur in the absence of internucleosomal DNA fragmentation. *Biochemistry Journal*, 286, 331–334.

Cohen, J. J. and Duke R. C. (1984) Glucocorticoid activation of a calcium-dependent endonuclease in thymocyte nuclei leads to cell death. *Journal of Immunology*, 132, 38–42.

Cohen, J. J., Duke, R. C., Fadok, V. A. and Sellins, K. S. (1992b) Apoptosis and programmed cell death in immunity. *Annual Review of Immunology*, 10, 267–293.

Ellis, R. E., Yuan, J. and Horvitz, H. R. (1991) Mechanisms and functions of cell death. *Annual Review of Cell Biology*, 7, 663–698.

Kerr, J. F. R. (1971) Shrinkage necrosis: a distinct mode of cellular death. *Journal of Pathology*, 105, 13–20.

Kerr, J. F. R., Searle, J., Harmon, B. V. and Bishop, C. J. (1987) Apoptosis. In *Perspectives on Mammalian Cell Death*, C. S. Potten (ed.), pp. 93–128. Oxford: Oxford University Press.

Lockshin, R. A. and Zakeri, Z. (1991) Programmed cell death and apoptosis. In *Apoptosis: The Molecular Basis of Cell Death*, L. D. Tomei and F. O. Cope (eds), pp. 47–60. New York: Cold Spring Harbor Laboratory Press.

Sorenson, C. M., Barry, M. A. and Eastman, A. (1990) Analysis of events associated with cell cycle arrest at G2 phase and cell death induced by cisplatin. *Journal of the National Cancer Institute*, 82, 749–755.

Sun, X.-M., Snowden, R. T., Skilleter, D. N., Dinsdale, D., Ormerod, M. G. and Cohen, G. M. (1992) A flow cytometric method for the separation and quantitation of normal and apoptotic thymocytes. *Analytical Biochemistry*, 204, 351–356.

Walker, N. I., Harmon, B. V., Gobe, G. C. and Kerr, J. F. R. (1988) Patterns of cell death. *Methods and Achievements in Experimental Pathology*, 13, 18–54.

Walker, P. R., Smith, C., Youdale, T., Leblanc, J., Whitfield, J. F. and Sikorska, M. (1991) Topoisomerase II-reactive chemotherapeutic drugs induce apoptosis in thymocytes. *Cancer Research*, 51, 1078–1085.

Waring, P., Egan, M., Braithwaite, A., Mullbacher, A. and Sjaarda, A. (1990) Apoptosis induced in macrophages and T blasts by the mycotoxin sporidesmin and protection by Zn^{2+} salts. *International Journal of Immunopharmacology*, 12, 445–457.

Waring, P., Kos, F. J. and Mullbacher, A. (1991) Apoptosis or programmed cell death. *Medicinal Research Reviews*, 11, 219–236.

Wyllie, A. H. (1980) Glucocorticoid-induced thymocyte apoptosis is associated with endogenous endonuclease activation. *Nature*, 284, 555–556.

Wyllie, A. H., Kerr, J. F.R. and Currie, A. R. (1980) Cell death: The significance of apoptosis. *International Review of Cytology*, 68, 251–306.

Wyllie, A. H. and Morris, R. G. (1982) Hormone-induced cell death: purification and properties of thymocytes undergoing apoptosis after glucocorticoid treatment. *American Journal of Pathology*, 109, 78–87.

Wyllie, A. H., Morris, R. G., Smith, A. L. and Dunlop, D. (1984) Chromatin cleavage in apoptosis: association with condensed chromatin morphology and dependence on macromolecular synthesis. *Journal of Pathology*, 142, 67–77.

Chapter 11

STRUCTURAL ANALYSIS OF DNA RELATED TO APOPTOSIS
Evidence for a Specific Lesion Associated with DNA Replication

*Bernard W. Stewart, Robert J. Sleiman,
Daniel R. Catchpoole and Sally M. Pittman*
Children's Leukaemia and Cancer Research Centre, University of
New South Wales, Prince of Wales Children's Hospital,
Sydney, NSW 2031, Australia

In mammalian cells, fragmentation of DNA into nucleosomal lengths is widely recognized as the molecular criterion for apoptosis (Bursch *et al.*, 1990). The phenomenon is of significance both as a means of monitoring apoptosis and providing a rationale, in terms of the activation of endogenous nuclease, for relating morphological to molecular change (Jones *et al.*, 1989; Arends *et al.*, 1990; Sorenson *et al.*, 1990). The precise digestion pattern evident in preparations of DNA isolated from cells undergoing apoptosis due to a variety of stimuli is also a criterion for distinguishing so-called 'programmed cell death' from necrosis (Corcoran and Ray, 1992). The latter is characterized by random digestion of DNA mediated by lysosomal enzymes.

Apart from cell death, structural change in DNA is recognized under a variety of circumstances. Transcription necessitates strand separation, the process being mediated by a variety of transcription factors and associated proteins. Several modes of repair necessitate elimination of modified bases from DNA through the re-synthesis of a polynucleotide strand thereby maintaining informational integrity. Replication of DNA necessitates strand separation in such a manner as to permit the entire genome to be duplicated. Granted the scope of structural change known to occur in DNA, any such change must be precisely characterized to allow its identification with a particular biological process. This chapter is concerned with structural change in DNA which may be specifically related to apoptosis. The data suggest a specific structural lesion indicative of an association between disruption of DNA replication and apoptosis.

BD-CELLULOSE AS A MEANS OF ANALYSING DNA

Fractionation of DNA by column chromatography on derivatized DEAE-cellulose, specifically benzoylated DEAE- (BD-)cellulose or benzoylated naphthoylated DEAE-(BND-)cellulose, was initially introduced as a means of isolating replicative DNA, an effect subsequently developed for the purpose of assaying DNA repair (Strauss, 1981). The methodology is based on the principle that totally double stranded DNA may be eluted from such media in the presence of approximately 1 M NaCl, whilst recovery of double stranded DNA containing single stranded regions (i.e. unpaired nucleotides) requires addition of caffeine or formamide to the salt solution.

Many assays of DNA repair (e.g. determination of non-semiconservative synthesis by density labelling) are predicated on distinguishing between structural change associated with repair on the one hand and *de novo* replication of DNA on the other. Specifically, models for both excision repair and replicative synthesis of DNA envisage the transient generation of single stranded regions in DNA. Accordingly, identification of any single-stranded character in DNA with a specific biological process requires a degree of molecular detail as a basis for classification.

A caffeine gradient method for elution of BD-cellulose, to permit sizing of single stranded regions in bound DNA, was developed in this laboratory. Initial studies using single stranded DNA indicated that, after logarithmic transformation, polynucleotide length (at least up to 50 kb) was proportional to the eluting caffeine concentration, although such a principle is not applicable to BND-cellulose (Haber and Stewart, 1981). These studies were extended using appropriate constructs to confirm that the elution characteristics of double stranded DNA containing single stranded regions are independent of the associated double standed regions (Norris and Stewart, 1988). Accordingly, this chromatographic procedure provides a means of sizing such single stranded regions in otherwise double stranded DNA fragments.

STRUCTURAL CHANGE ASSOCIATED WITH REPLICATION

DNA isolated from mammalian cells during S phase exhibits single stranded character as variously indicated by single strand-specific nuclease digestion, chromatography and electron microscopy (Henson, 1978; Collins, 1979; Micheli *et al.*, 1982). There have been few attempts to determine the polynucleotide length or configuration of single stranded regions generated during eukaryotic DNA replication. Single stranded regions of approximately 200 nucleotides in immediate association with the replicating fork of Simian virus (SV) 40 were described by Herman *et al.* (1981) based upon S1 nuclease digestion.

Replicating DNA isolated from mammalian cells is bound to BD- or BND-cellulose (Henson, 1978; Stewart et al., 1979). We have recently reported analysis, by caffeine-gradient elution, of the size of single stranded regions in DNA isolated from proliferating human leukaemic (CCRF-CEM) cells (Stewart et al., 1991). After pulse-labelling for periods between 30 seconds and 10 minutes, two classes of structural defect were evident in DNA sheared to a modal length of 20 kb: a uniform population of approximately 200 nucleotides and a disperse population containing single stranded regions from 900 to approximately 4000 nucleotides. Most newly incorporated radioactivity was associated with the latter structural defect, a finding not readily correlated with current concepts of the 'replicating fork' (Spadari et al., 1989). Rather, the strand-separation model of DNA replication proposed by Gaudette and Benbow (1986) offers a basis for interpreting these and similar findings.

STRUCTURAL CHANGE IN DNA DURING NECROSIS

It has been generally accepted that biological interpretation of structural change in DNA would be confounded by concomitant necrosis. DNA degradation, and hence strand breakage, is considered to be inevitably associated with cell death and tissue destruction. Thus, for example, in studies of carcinogen-induced DNA damage and repair in intact animals, treatment regimes have typically been adjusted to preclude tissue necrosis (Cox et al., 1973). Despite wide acceptance of the proposition that extensive structural DNA damage is co-incident with necrosis, few studies have specifically examined structural change in DNA following toxic tissue injury (Corcoran and Ray, 1992).

To characterise single stranded regions in DNA consequent upon cell injury, Haber and Stewart (1985) utilized centrilobular necrosis induced in rat liver by carbon tetrachloride intoxication (1 ml/g body wt in corn oil by gavage). Within 4 hours of dosing with the halogenated alkane, marked disruption of cellular function is detectable at the biochemical level, whilst histological evidence of injury is apparent within 24 hours. Despite these considerations, no increase in the proportion of DNA exhibiting single stranded character is evident by stepwise elution from BD-cellulose. By 48 hours after treatment, gross haemorrhagic injury to the liver is apparent and chromatographic analysis indicates a three-fold increase in the proportion of DNA containing single stranded regions. Such structural change was found to be heterogeneous both in terms of the length of single stranded regions and the reduced size overall of DNA fragments subject to analysis. There was no indication from the these studies of a specific structural lesion, as distinct from progressive degradation of DNA.

STRUCTURAL CHANGE ASSOCIATED WITH APOPTOSIS

Evidence of structural change in DNA possibly associated with apoptosis arose in the course of monitoring replicative intermediates in the CCRF-CEM cell line. Taylor and

Hodson (1984) reported that this line is characterised by accumulation of S phase cells under conditions of limiting growth. If such conditions, which include limiting cell density, extremes of pH or depletion of media, are not adjusted, apoptosis occurs. In our laboratory, comparison was made of the distribution of newly incorporated ^{3}H-thymidine (2 minute pulse – 3 minute chase) within DNA isolated from populations of CCRF-CEM cells in both the mid-log phase of cell growth and at confluence. In these, and all other experiments described herein, the 'pre-existing' (or 'total' DNA) was labelled by exposure of cells to ^{14}C-thymidine 24 hours prior to 'zero' time. After ^{3}H-labelling, DNA was isolated from the respective pulse-labelled populations, sheared and subjected to stepwise eluation from BD-cellulose. In respect of ^{3}H-radioactivity, the labelling of DNA from mid-log cells was consistent with the dynamic pattern previously described (Haber et al., 1986): viz. radioactivity is rapidly lost from the thymidine triphosphate pool (eluted during loading with 0.3 M NaCl), is transiently associated with DNA replicative intermediates eluted with caffeine solution and is ultimately recovered in double stranded DNA in 1.0 M NaCl. The latter fraction having been labelled by incorporation of ^{14}C-thymidine 24 hours previously, accounts for 80% (or more) of such radioactivity recovered.

The dynamics of radiolabelling was markedly different in confluent cells. As might have been anticipated, less intracellular ^{3}H-thymidine-derived radioactivity was recovered, and most of this was present as intracellular TTP, with only a minor proportion incorporated into DNA as either replicative intermediates or mature double stranded DNA (Table 11.1). Not anticipated was the distribution of ^{14}C-radioactivity. The proportion of acid-soluble activity (0.3 M NaCl) was increased more

Table 11.1 Distribution of radioactivity following BD-cellulose chromatography of DNA isolated from mid-log and confluent cultures of CCRF-CEM cells.[a]

	Mid-log		Confluent	
	^{14}C	^{3}H	^{14}C	^{3}H
Total radioactivity				
(dpm × 10^3)	196	691	45	217
% recovered eluted with				
0.3 M NaCl (TTP)	0.03	0.2	13.8	65.5
1.0 M NaCl (dsDNA)	80.0	55.1	60.1	10.7
2% caffeine (DNA with ss regions)	19.7	44.7	26.1	13.8

a. Cells were pre-labelled with ^{14}C-thymidine 24 h prior exposure to ^{3}H-thymidine for 2 minutes. The cells were re-suspended in free media for a further 3 minutes after which DNA was isolated, sheared and subjected to BD-cellulose chromatograpy. Recovery of radioactivity following stepwise elution with 0.3 M NaCl, 1.0 M NaCl and 2% caffeine is expressed as percentage of total radioactivity, the respective fractions containing thymidine triphosphate (TTP), double stranded DNA (dsDNA) and DNA with single stranded (ss) regions.

than 100-fold by comparison with the mid-log result, and that eluted with caffeine was also increased in terms of the same comparison. Whilst degradation of prelabelled DNA might account for loss of ^{14}C-radioactivity from double-stranded DNA, there is no obvious basis for attributing increased labelling of the caffeine-eluted fraction to a degradative pathway.

A possible relationship between structural change in DNA and growth to near confluence of CCRF-CEM cells was examined by monitoring these two parameters over a 3-day period under varying conditions. When cells were monitored at low initial density, their growth rate indicated a doubling time of less than 24 hours. Under these conditions, there was no variation in the proportion of pre-labelled DNA eluted with caffeine from BD-cellulose. However after 48 hours, a decrease in growth rate occurred and by 72 hours a marked increase in the proportion of DNA exhibiting single stranded character was evident (Figure 11.1, upper).

When the same study was initiated at higher cell density, cell numbers did not progressively increase during the ensuing 3 days. In one instance, increase in cell number occured over the first 24 hours, after which decline occurred. The proportion of caffeine-eluted DNA increased markedly once growth declined (Figure 11.1, middle). Likewise, when virtually no cell growth occurred, the proportion of caffeine-eluted DNA increased successively over the 3-day course of the study (Figure 11.1, lower).

A STRUCTURAL LESION ASSOCIATED WITH LIMITING GROWTH

The structural defect responsible for increased single stranded character of DNA isolated from CCRF-CEM cells approaching apoptosis consequent upon limiting cell density was determined by caffeine-gradient elution of BD-cellulose. In these experiments, structural change specifically associated with replication was identified by pulse-labelling with ^{3}H-thymidine for 5 minutes before DNA isolation. Typical results are illustrated in Figure 11.2. At 'zero' time, DNA fragments exhibiting long single stranded regions are evident consequent upon incorporated ^{3}H-thymidine: the result noted earlier. In the same preparation, there was no evidence of any discrete structural change associated with ^{14}C-labelling. DNA isolated from similar cells maintained in culture for a further 72 hours also exhibited the 'replicative intermediates' labelling pattern in respect of ^{3}H-radioactivity. However, the ^{3}H-elution profile was now almost duplicated by the distribution of ^{14}C-radioactivity. This ^{14}C-elution profile is most reasonably attributable to structural change associated with DNA replication. Such an increasing amount of pre-labelled DNA is suggestive of a blockage during DNA replication which is affecting an increasing proportion of cells.

The structural lesion evident in confluent cultures of CCRF-CEM cells is considered to precede, rather than accompany, apoptosis for the following reasons. Morphological assessment of the respective cultures provided no evidence of an increasing number of markedly apoptotic cells over the 3-day course of the study. Whilst cell

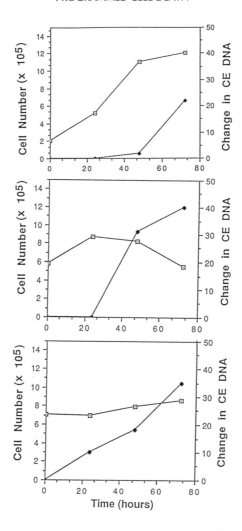

Figure 11.1 Relationship between cell number and proportion of DNA subsequently eluted from BD-cellulose in caffeine solution (CE-DNA) following maintenance of CCRF-CEM cells in culture over 3 days. The results of three independent experiments are shown. Zero time was initiated at different cell densities and preceded by 24 hour exposure to ^{14}C-thymidine which served as the basis for monitoring DNA elution from BD-cellulose. Variation in the amount of CE-DNA is indicated in arbitrary units calculated by subtracting the proportion (expressed as a percentage) recovered at zero time from results obtained on successive days of the experiment.

numbers may not be increasing, the cultures are not characterized by large numbers of dead or dying cells relative to the situation at 'zero' time. More importantly, there is evidence that rapid causation of apoptosis in the same cells by drug treatment does involve such effects.

Etoposide (VP-16) is known to cause endonucleolytic cleavage of DNA in leukaemic cells (Kaufmann, 1989). In our laboratory, degradation of internucleosomal

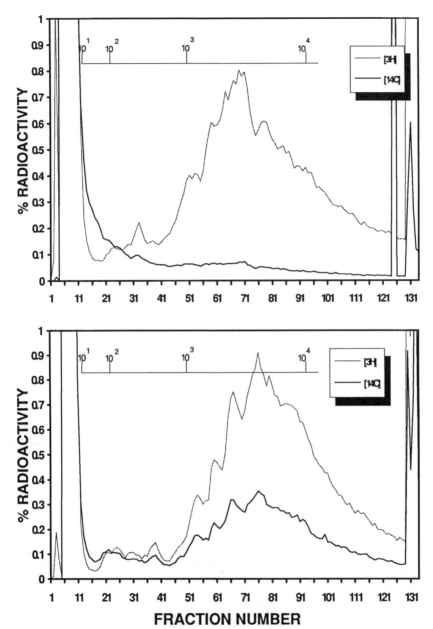

Figure 11.2 Caffeine gradient elution of DNA isolated from CCRF-CEM cells, the growth characteristics of which are shown in Figure 11.1, middle panel. DNA was isolated at zero time (proliferating cells, upper panel) and after 3 days culture (confluent cells, lower panel) and, after pulse labelling with ^{3}H-thymidine as described in the text, subjected to caffeine-gradient elution from BD-cellulose (Norris and Stewart, 1988). Double stranded DNA is eluted prior to fraction 12 and the caffeine gradient covers subsequent fractions. The polynucleotide length of single stranded regions eluted is indicated. The prominent peak which was similar in both preparations was generated by ^{3}H-radioactivity, the heavier line indicating ^{14}C-radioactivity.

DNA, as indicated by agarose gel electrophoresis, was apparent within 3 hours of exposure of CCRF-CEM leukaemic cells to the drug and morphological evidence of apoptosis was also apparent. Under these conditions, at concentrations up to 0.1 mM, there was no change in the proportion DNA eluted from BD-cellulose in caffeine solution. Gradient analysis suggested a separate structural lesion (a peak at fractions 20–31) may be consequent upon drug treatment (D. Catchpoole and B. W. Stewart, unpublished data). Finally, the present findings (Figure 11.2) are not readily attributable to apoptosis *per se* because the specific structural lesion—increased single stranded regions—is not anticipated on the basis of endonuclease activity associated with apoptosis (see Fraser *et al.*, Chapter 9).

The molecular lesion characteristic of confluent CCRF-CEM cells is indicative of a blockage in DNA replication subsequent to nucleotide polymerization. It is possible that such structural aberrations could render chromatin more susceptible to constitutive endonucleases (Alnemri and Litwack, 1989). The precise stuctural basis remains to be clarified, but is more easily reconciled to models of DNA replication at a level of organization involving, for example, matrix attachment rather than the structure of a replicating fork (Mirkovitch *et al.*, 1984; Huberman, 1987). The findings recorded may be specific for CCRF-CEM cells insofar as the particular stage (S phase) of the cell cycle which is blocked. The effect, however, is evocative of the suggestion by Schimke and his colleagues (Kung *et al.*, 1990) that dissociation of normally integrated cell cycle events is responsible for cell death.

REFERENCES

Alnemri, E. S. and Litwack, G. (1989) Glucocorticoid-induced lymphocytolysis is not mediated by an induced endonuclease. *Journal of Biological Chemistry*, **264**, 4104–4111.

Arends, M. J., Morris, R. G. and Wyllie, A. H. (1990) The role of endonuclease. *American Journal of Pathology*, **136**, 593–608.

Bursch, W., Kleine L. and Tenniswood, M. (1990) The biochemistry of cell death by apoptosis. *Biochemistry and Cell Biology*, **68**, 1071–1074.

Collins J. M. (1979) Transient structure of replicative DNA in normal and transformed human fibroblasts. *Journal of Biological Chemistry*, **254**, 10167–10172.

Corcoran G. B. and Ray, S. D. (1992) The role of the nucleus and other compartments in toxic cell death produced by alkylating hepatotoxicants. *Toxicology and Applied Pharmacology*, **113**, 167–183.

Cox, R., Damjanov, I., Abanobi, S. E. and Sarma, D. S. R. (1973) A method for measuring DNA damage and repair in the liver *in vivo*. *Cancer Research*, **33**, 2114–2121.

Gaudette, M. F. and Benbow, R. M. (1986) Replication forks are underrepresented in chromosomal DNA of xenopus laevis embryos. *Proceedings of the National Academy of Science, USA*, **83**, 5953–5957.

Haber, M., Kavallaris, M. and Stewart, B. W. (1986) Two-stage incorporation of thymidine triphosphate into mammalian DNA as indicated by chromatography on benzoylated DEAE-cellulose. *Journal of Chromatography*, **382**, 127–134.

Haber, M. and Stewart, B. W. (1981) Sizing of DNA fragments by preparative, benzoylated DEAE-cellulose chromatography. *FEBS Letters*, **133**, 72–74.

Haber, M. and Stewart, B. W. (1985) Patterns of structural change in DNA during tissue necrosis indicated by benzoylated DEAE-cellulose chromatography. *Chemico-Biological Interactions*, **53**, 247–255.

Henson P. (1978) The presence of single-stranded regions in mammalian DNA. *Journal of Molecular Biology*, **119**, 487–506.

Herman, T. M., Depamphilis, M. L. and Wassarman, P. M. (1981) Structure of chromatin at deoxyribonucleic acid replication forks: Location of the first nucleosomes on newly synthesized simian virus 40 deoxyribonucleic acid. *Biochemistry*, **20**, 621–630.

Huberman, J. A. (1987) Eukaryotic DNA replication: A complex picture partically clarified. *Cell*, **48**, 7–8.

Jones, D. P., McConkey D. J. Nicotera, P. and Orrenius, S. (1989) Calcium-activated DNA fragmentation in rat liver nuclei. *Journal of Biological Chemistry*, **264**, 6398–6403.

Kaufmann, S. H. (1989) Induction of endonucleolytic DNA cleavage in human acute myelogenous leukemia cells by etoposide, camptothecin, and other cytotoxic anticancer drugs: A cautionary note. *Cancer Research*, **49**, 5870–5878.

Kung, A. L., Zetterberg, A., Sherwood, S. W. and Schimke, R. T. (1990) Cytotoxic effects of cell cycle phase specific agents: Result of cell cycle perturbation. *Cancer Research*, **50**, 7307–7317.

Micheli, G., Baldari, C. T., Carri, M. T., Cello, G. D. and Buongiorno-nardelli, M. (1982) An electron microscope study of chromosomal DNA replication in different eukaryotic systems. *Experimental Cell Research*, **137**, 127–140.

Mirkovitch, J., Mirault, M-E. and Laemmli, U. K. (1984) Organisation of the higher-order chromatin loop: Specific DNA attachment sites on nuclear scaffold. *Cell*, **39**, 223–232.

Norris, M. D. and Stewart, B. W. (1988) Sizing of single-stranded regions in double-stranded DNA by preparative benzoylated DEAE-cellulose chromatography. *FEBS Letters*, **228**, 223–227.

Sorenson, C. M., Barry, M. A. and Eastman, A. (1990) Analysis of events associated with cell cycle arrest at G_2 phase and cell death induced by cisplatin. *Journal of the National Cancer Institute*, **82**, 749–755.

Spadari, S., Montecucco, A., Pedrali-Noy, G. and Ciarrocchi, G. (1989) A double-loop model for the replication of eukaryotic DNA. *Mutation Research*, **219**, 147–156.

Stewart, B. W., Huang, P. H. T. and Brian, M. J. (1979) Endogenous single-stranded regions in comparison with damage induced *in vivo* by a carcinogen. *Journal of Biochemistry*, **179**, 341–352.

Stewart, B. W., Kavallaris, M., Catchpoole, D. and Norris, M. D. (1991) Two classes of single-stranded regions evident in deproteinized preparations of replicating DNA isolated from mammalian cells. *Experimental Cell Research*, **192**, 639–642.

Strauss, B. S. (1981) Use of benzoylated naphtholylated DEAE cellulose. In *DNA Repair. Laboratory Manual of Research Procedures.*, edited by Friedberg, E. C. and Hanawalt, P. C. New York: Marcel Dekker Inc., pp. 319–339

Taylor, I. W. and Hodson, P. J. (1984) Cell cycle regulation by environment pH. *Journal of Cellular Physiology*, **121**, 517–525.

GENETIC REGULATION

Chapter 12

REGULATION OF PROGRAMMED CELL DEATH

J. John Cohen

Department of Microbiology and Immunology, University of Colorado School of Medicine, Denver, CO 80262, USA

Programmed cell death (PCD) is a term commonly applied to cell death when it is a normal part of the life of a metazoan. The programmed death of a cell should be predictable. In the nematode *Caenorhabditis elegans*, the entire complement of cells has been fate-mapped and death of individual cells can be exactly predicted (Sternberg, 1990). This death is cell autonomous, that is, it will happen essentially independently of the environment the cell finds itself in, and thus is likely to be regulated by genes that are expressed at a critical time within the cell itself (Yuan and Horvitz, 1990). Perhaps this is the only case of cell death so far known where we are really justified in using the term programmed. In the mammalian examples to be considered here, the process is more accurately called physiologic cell death, in that it will occur with a high degree of probability, but it is nevertheless stochastic. Within the pool of developing B lymphocytes, for example, we know that at least 95% will die before maturity (Deenen et al., 1990), but before they die we can not say which ones are so fated.

In spite of this relative uncertainty, the term PCD has been adopted by students of vertebrate cell biology. As will be clear to readers of this volume, PCD has unfortunately come to have two definitions. The first, and more justifiable, is similar to that used by *C. elegans* scientists, and refers to cell death that can reasonably be thought of as desirable or essential for the form or function of the body. There are a number of examples, including morphogenetic death during development, death from hormone or growth factor removal, death from exposure to hormones, and death from inappropriate receptor stimulation. In each case a reasonable case may be made that the cell's death is good for the rest of the cellular community. The second definition that has accreted to PCD is apoptosis. This is obviously because many examples of PCD (using our first, functional definition) are apoptosis by morphology. But the functional and the morphological definitions are uneasy partners. There are many instances in which cells die by what is certainly apoptosis, although it is difficult to see in what way that death might be programmed. As a single example, thymocytes exposed to the dioxin TCDD, an insecticide developed in the last 50 years, die by apoptosis (McConkey et al., 1988); how can this response be considered "programmed", when it would not have occurred if the chemical had never been invented? It is useful, therefore, to broaden the definition of PCD to include cell death that follows a particular biochemical program, namely that characterized by the morphologically-defined event apoptosis, regardless of the teleological explanation for the death. In some cases, we will eventually find out the "reason" the apoptotic cell died. In others, as we have speculated elsewhere, apoptosis may be a nonspecific suicide response to injury (Sellins and Cohen, 1987; Cohen et al., 1992).

In a multicellular creature like the human, maintaining the steady state requires that cell division must be exactly matched by cell loss or death, or a tissue or organ system will get larger. We often think of malignancy as a disorder of cell growth; but it is just as possible, at least mathematically, that it could result from inadequate cell death. In the adult human, about 500 000 polymorphonuclear neutrophils die each second; this is programmed and normal cell death, not the result of a suicidal activation in response to pathogens. If this death rate decreases without a concomitant decrease in production, the number of neutrophils will rise, just as it would if production were increased. Thus the size of most tissues and organ systems must be regulated by balanced production and death, and the two processes must be informationally linked. This hints that the death of cells in these circumstances will not be completely "cell autonomous", but may be regulated within broader or narrower limits by factors external to them. The survival of neutrophils, for example, although programmed to be brief, can be moderately extended by colony-stimulating factors (Colotta et al., 1992).

In fact, although a cell might die for purely internal reasons, its survival could depend on conditions in its environment. The question is not whether the cell is "bad" or "should die", but rather whether it is bad or should die in a particular context. For example, a developing T lymphocyte expresses its receptor for antigen based on a random rearrangement of the genes for the receptor's structural elements. Such a rearrangement could constitute anti-self reactivity in one individual, but not in others whose self-defining molecules are different. The lymphocyte is dangerous in one case, innocuous in the other. When immature self-reactive (potentially autoimmune) lymphocytes engage self-determinants, the resultant internal signals activate a suicide pathway and the cell is destroyed (Finkel et al., 1989; Smith et al., 1989; Hasbold and Klaus, 1990). Cells that survive this process of negative selection, and complete their maturation, respond to the same sorts of signalling by positive, immunologically suitable activation.

There are other sorts of environmental signals to commit suicide. Generically, these can be humoral or contact-mediated. In the first category is death mediated by exposure to, or withdrawal from, hormones. The most widely studied example of programmed cell death in mammals is currently the rodent thymus. Most cells of this organ are responsive to glucocorticoids, and curiously, their response is death (Cohen, 1989). As mentioned above, when a developing T lymphocyte expresses its surface receptor for antigen, some will have anti-self reactivity and have to be aborted. A few will fulfil the very rigorous requirement that a useful T lymphocyte must recognize both antigen and self together, but neither separately. The majority, probably 99% of all cells that develop, will be neither of use nor harmful to the host. One possibility is that these cells die when the daily plasma concentration peak of glucocorticoid washes through the thymus (Cohen, 1991). Mature T lymphocytes, whether still within the thymus or already exported, are resistant to the lethal effects of glucocorticoids (Cohen, 1989). Thus the glucocorticoid-induced death of thymocytes is not only apoptosis but is likely also to be programmed in our first sense, and is a proper model for studying these processes.

As mentioned above, if an immature lymphocyte (T or B) is stimulated via its receptor for antigen, it will usually commit suicide, whereas if allowed to mature, it will now respond positively to the same stimulus, and divide or differentiate. One explanation for this difference might be that the receptor is connected to different parts of the cellular biochemical machinery in the immature and mature states, and there is in fact evidence that signalling differs at these two stages (Finkel *et al.*, 1989). One might imagine that the immature cell is not fully ready to respond normally to stimulation, and the result is that some but not all of the biochemical changes associated with activation take place. The ensuing chaos leads to the activation of a suicide program. There is support for this general concept. If a lymphocyte is damaged by low-dose ionizing radiation or by brief exposure to elevated temperature, it will respond by committing suicide (Sellins and Cohen, 1987, 1991). These insults produce a degree of damage or disorder that other cells readily repair; the lymphocyte's response is suicide. Perhaps the reason lymphocytes (as far as we know, uniquely among cells) do so is that they are extremely dangerous cells if incorrectly or persistently activated. For the good of the community of cells, it seems to be preferable that a damaged lymphocyte—however mild the damage—kill itself rather than attempt repair. We sometimes call this the "better dead than wrong" paradigm.

Several instances have been published of cell suicide following hormone removal. Again taking examples from the immune system, activated helper or inducer T lymphocytes secrete short-range hormone-like mediators called lymphokines (such as interleukins 2, 3 and 4). Other cells, including B cells and cytotoxic T cells, respond to these lymphokines by proliferation or differentiation. In many cases the responding cell becomes dependent on the lymphokine not only for these functions, but for its very survival. It is addicted to the growth factor, and will commit apoptotic suicide if denied it (Duke and Cohen, 1986). Death by suicide upon hormone withdrawal is also seen in the endometrium (Rotello et al., 1992), prostate (Kerr and Searle, 1973), among leukocytes (Williams et al., 1990), and in the central nervous system (Rich, 1992).

This brief and selective review is meant to support the idea that PCD is an essential part of the design of multicellular organisms, and as such is probably regulated with a degree of complexity similar to that of other essential and widespread processes, such as mitosis (it may actually be less complex than mitosis, as anyone who has compared the construction of an office block to its demolition would agree). The genetic regulation of apoptosis is fascinating to consider because of its incredible conceptual depth. Are there "death genes", that is, genes that code for a suicide protein? This would be the simplest sort of arrangement: in live cells the genes would be transcriptionally silent, and they would be activated only in response to a definitive signal coming from within or without the cell. This may be the case, but it is a risky design, as it requires that the death genes be in a transcriptionally inactive but readily available state, not safely extinguished by, for example, methylation. A variant of this model would hold that death genes are expressed at all times, but death does not result because the level of expression is too low, or because there are "death inhibitor" proteins which represent the true locus of apoptosis regulation. Alternatively, death genes might simply be normal genes abnormally expressed. For example, the nuclear breakdown associated with apoptosis might be caused by the activation of genes whose normal function is to remodel the nucleus during mitosis. Finally, apoptosis may not involve any specific genes at all: it may simply represent the most probable biochemical response of an injured or confused cell, when the most vulnerable pathway has become exhausted. Although most workers have their preferences for one or another of these patterns, the definitive answer will have to wait until death genes are cloned, or until the complete biochemical pathway of apoptosis is elucidated.

Greg Owens, Bill Hahn and I decided that the evidence we had obtained indicating the requirement for new mRNA and protein synthesis in several normal cell systems of apoptosis was convincing enough that we should try to identify the genes involved. The cloning procedures we used have been published (Owens et al., 1991; Owens and Cohen, 1992). Our approach was to induce apoptosis in thymocytes with dexamethasone and isolate mRNA. This turned out to be difficult, as mRNA degradation is an early event in apoptosis (Cidlowski, 1982) and our yields were too small to work with. We decided to add cycloheximide to the dexamethasone, so that gene induction could take place but the expression of the apoptotic program would be blocked. This allowed us to isolate enough mRNA. If there were "late" death genes we would miss them, but our evidence is that there are no essential genes in that category. During this work we found that cycloheximide treatment alone increased the steady-state levels of several messages that we later identified as death-associated. This suggests that there is, in fact, constitutive expression of some death genes, at least in normal immature thymocytes. Because of this, isolation of death-associated messages is not simple, especially by the method we used: subtractive hybridization. We prepared probes by subtracting "housekeeping" messages from those in the apoptotic cDNA library. The remaining cDNA, about 2% of the total, was used as a probe in differential screening of apoptotic cDNA libraries. About 20 independent inserts showed increased expression in dexamethasone-treated thymocytes. Since some of these might be dexamethasone-response or thymus specific, we screened candidate genes on Northern blots of RNA from unrelated cells undergoing apoptosis for other reasons; this assured us that any positive clones were "death-related", if not actually part of the death pathway. We are still screening the library, but some of our early results are encouraging, in that RP-8 and RP-2 are expressed in a variety of cell systems in the body, but apparently primarily in cells that are undergoing programmed cell death.

In contrast to the examples considered above, where the induction of apoptosis is blocked if mRNA or protein synthesis is inhibited, there are other examples of apoptosis where cycloheximide, for example, actually causes apoptosis (Martin et al., 1990; Ledda-Columbano et al., 1992). Since all the changes that we use to characterize apoptosis take place in the presence of inhibitors, this must mean that in these cell types at least, all the proteins necessary for apoptosis are already available. It is possible that blocking macromolecular synthesis results in the preferential loss of a death inhibitor. Cells with this mechanism in place, then, are in a curious metastable state: they have already activated the death program but are holding it in abeyance with a death-inhibiting protein (one is tempted to think of bcl-2, although there is at present no evidence to support it). There is a fascinating parallel in bacteria: the F plasmid of E. coli codes for two proteins, a poison called CcdB and its antidote, CcdA (Bernard and Couturier, 1992; Maki et al., 1992). CcdA has a very short half-life. If the bacterium contrives to get rid of the plasmid (which it would prefer to do, as the extra DNA is a metabolic drag), the CcdA protein rapidly decays and the CcdB toxin is free to kill the cell. As the saying goes, the cell that rides a tiger will find it difficult to dismount.

If inhibition of apoptosis is a common phenomenon, it may be that in at least some examples of apoptosis the trigger may actually act as, or induce the synthesis of, a repressor of the inhibitor's gene expression. Thus the critical message would be decreased in a cDNA library from apoptotic cells, rather than (as we assumed) increased. Our subtraction strategy would have missed such genes, although the cloning vectors we used would make it relatively straightforward to reverse the direction of the subtraction.

When cytotoxic T cells or other killer cells interact with their appropriate target cells, changes take place in the targets that are indistinguishable morphologically from apoptosis (Cohen et al., 1985; Berke, 1991; Cohen et al., 1992). In most examples of this sort of killing, there is no need on the target's part for new protein synthesis. This means either that all the proteins necessary for apoptosis are present in virtually all cells; or that the killer cell physically transfers some or all of the necessary molecules. This issue is hotly debated, and quite reasonable data have been offered in support of both sides. The outcome is eagerly awaited by students of apoptosis, as it may tell us whether, or how, we should continue our search for death genes.

A completely different strategy for identifying death genes is to look for vertebrate genes homologous to those already identified in C. elegans, the ced series (Ellis and

Horvitz, 1986). The direct approach, using *ced* probes to screen mammalian cDNA libraries, has not yet met with any published success. However, it has recently been found (Vaux *et al.*, 1992) that the mammalian apoptosis-suppressing gene *bcl*-2 can inhibit programmed cell death in transgenic *C. elegans*. In this it resembles the nematode gene *ced*-9, and since *ced*-9 is thought to inhibit the actions of genes *ced*-3 or *ced*-4, a search of the databases for vertebrate genes homologous to these may well prove to be rewarding.

REFERENCES

Berke, G. (1991) Lymphocyte-triggered internal target disintegration. *Immunology Today*, 12, 396–399.

Bernard, P. and Couturier, M. (1992) Cell killing by the F plasmid CcdB protein involves poisoning of DNA-topoisomerase II complexes. *Journal of Molecular Biology*, 226, 735–745.

Cidlowski, J. A. (1982) Glucocorticoids stimulate ribonucleic acid degradation in isolated rat thymic lymphocytes *in vitro*. *Endocrinology*, 111, 184–190.

Cohen, J. J. (1989) Lymphocyte death induced by glucocorticoids. In *Anti-inflammatory Steroid Action: Basic and Clinical Aspects*, R. P. Schleimer, H. N. Claman and A. Oronsky (eds), pp. 110–131. San Diego, CA: Academic Press, Inc.

Cohen, J. J. (1991) Programmed cell death in the immune system. *Advances in Immunology*, 50, 55–85.

Cohen, J. J., Duke, R. C., Chervenak, R., Sellins, K. S. and Olson, L. K. (1985) DNA fragmentation in targets of CTL: an example of programmed cell death in the immune system. In *Mechanisms of Cell-Mediated Cytotoxicity II*, edited by P. Henkart and E. Martz, pp. 493–506. New York: Plenum Press.

Cohen, J. J., Duke, R. C., Fadok, V. A. and Sellins, K. S. (1992) Apoptosis and programmed cell death in immunity. *Annual Review of Immunology*, 10, 267–293.

Colotta, F., Re, F., Polentarutti, N., Sozzani, S. and Mantovani, A. (1992) Modulation of granulocyte survival and programmed cell death by cytokines and bacterial products. *Blood*, 80, 2012–2020.

Deenen, G. J., van Balen, I., Opstelten, D. (1990) In rat B-lymphocyte genesis sixty percent is lost from the bone marrow at the transition of nondividing pre-B cell to sIgM+ B lymphocyte, the stage of Ig light chain gene expression. *European Journal of Immunology*, 20, 557–564.

Duke, R. C., Cohen, J. J. (1986) IL-2 addiction: withdrawal of growth factor activates a suicide program in dependent T cells. *Lymphokine Research*, 5, 289–299.

Ellis, H. M., Horvitz, H. R. (1986) Genetic control of programmed cell death in the nematode *C. elegans*. *Cell*, 44, 817–829.

Finkel, T. H., Marrack, P., Kappler, J. W., Kubo, R. T., Cambier, J. C. (1989) αβ T-cell receptor and CD3 transduce different signals in immature T cells. Implications for selection and tolerance. *Journal of Immunology*, 142, 3006–3012.

Hasbold, J. and Klaus, G. G. (1990) Anti-immunoglobulin antibodies induce apoptosis in immature B cell lymphomas. *European Journal of Immunology*, 20, 1685–1690.

Kerr, J. F. R. and Searle, J. (1973) Delection of cells by apoptosis during castration-induced involution of the rat prostate. *Virchows Archive* [B], 13, 87–102.

Ledda-Columbano, G. M., Coni, P., Faa, G., Manenti, G. and Columbano, A. (1992) Rapid induction of apoptosis in rat liver by cycloheximide. *Amererican Journal of Pathology*, 140, 545–549.

Maki, S., Takiguchi, S., Miki, T. and Horiuchi, T. (1992) Modulation of DNA supercoiling activity of *Escherichia coli* DNA gyrase by F plasmid proteins. Antagonistic actions of LetA (CcdA) and LetD (CcdB) proteins. *Journal of Biological Chemistry*, 267, 12244–12251.

Martin, S. J., Lennon, S. V., Bonham, A. M. and Cotter, T. G. (1990) Induction of apoptosis (programmed cell death) in human leukemic HL-60 cells by inhibition of RNA or protein synthesis. *Journal of Immunology*, 145, 1859–1867.

McConkey, D. J., Hartzell, P., Duddy, S. K., Hakansson, H. and Orrenius, S. (1988) 2,3,7,8-Tetrachlorodibenzo-p-dioxin kills immature thymocytes by Ca^{2+}-mediated endonuclease activation. *Science*, 242, 256–259.

Owens, G. P. and Cohen, J. J. (1992) Identification of genes involved in programmed cell death. *Cancer and Metastasis Reviews*, 11, 149–156.

Owens, G. P., Hahn, W. E. and Cohen, J. J. (1991) Identification of mRNAs associated with programmed cell death in immature thymocytes. *Molecular and Cellular Biology*, 11, 4177–4188.

Rich, K. M. (1992) Neuronal death after trophic factor deprivation. *Journal of Neurotrauma*, 9, S61–S69.

Rotello, R. J., Lieberman, R. C., Lepoff, R. B. and Gerschenson, L. E. (1992) Characterization of uterine epithelium apoptotic cell death kinetics and regulation by progesterone and RU 486. *American Journal of Pathology*, 140, 449–456.

Sellins, K. S. and Cohen, J. J. (1987) Gene induction by gamma-irradiation leads to DNA fragmentation in lymphocytes. *Journal of Immunology*, 139, 3199–3206.

Sellins, K. S. and Cohen, J. J. (1991) Hyperthermia induces apoptosis in thymocytes. *Radiation Research*, 126, 88–95.

Smith, C. A., Williams, G. T., Kingston, R., Jenkinson, E. J. and Owen, J. J. T. (1989) Antibodies to CD3/T-cell receptor complex induce death by apoptosis in immature T cells in thymic cultures. *Nature*, 337, 181–184.

Sternberg, P. W. (1990) Genetic control of cell type and pattern formation in *Caenorhabditis elegans*. *Advances in Genetics*, 27, 63–116.

Vaux, D. L., Weissman, I. L. and Kim, S. K. (1992) Prevention of programmed cell death in *Caenorhabditis elegans* by human *bcl*-2. *Science*, 258, 1955–1957.

Williams, G. T., Smith, C. A., Spooncer, E., Dexter, T. M. and Taylor, D. R. (1990) Haemopoietic colony stimulating factors promote cell survival by suppressing apoptosis. *Nature*, 343, 76–79.

Yuan, J. Y. and Horvitz, H. R. (1990) The *Caenorhabditis elegans* genes *ced*-3 and *ced*-4 act autonomously to cause programmed cell death. *Developmental Biology*, 138, 33–41.

Chapter 13

MACROMOLECULAR SYNTHESIS, C-*MYC*, AND APOPTOSIS

Douglas R. Green[1] and Thomas G. Cotter[2]

[1]*Division of Cellular Immunology, La Jolla Institute for Allergy and Immunology, La Jolla, CA 92037, USA*
[2]*The Department of Biology, St. Patrick's College, Maynooth, Co. Kildare, Ireland*

INTRODUCTION: HOW GENES CAN CONTROL DEATH

Twenty years ago, Kerr *et al.* (1972) outlined the morphological evidence that underpins the concept of apoptosis, that cells dying under physiological conditions can do so in a manner that is quite distinct from that of necrosis. Although apoptosis is often observed under pathological conditions, much of the interest in this phenomenon stems from its involvement in normal developmental processes. This vital role of focal cell death in the morphogenesis of normal embryos was clearly defined more than 40 years ago (Glucksmann, 1951), and in most cases such cell death can be clearly shown to have the characteristics of apoptosis. Subsequently, Wyllie, Kerr, and Currie (1980) brought the concept of apoptosis to the general attention of biologists and provided the only unambiguous marker of apoptosis (besides morphology) that is currently available—the fragmentation of DNA into multiples of 200 base pairs (Wyllie, 1980). Nevertheless, it was several years before the subject of apoptosis "crossed-over" into mainstream biology to become a subject of widespread and considerable interest.

At least part of the attraction in studying apoptosis is the application of new techniques to the identification of genes that regulate cell death. Apoptosis appears to be a specific "goal" of a physiological process and not simply a secondary consequence of other events. This suggests that apoptosis is a result of the expression of specific genes whose products direct the process of cell death and account for its several characteristic features. Thus, active cell death should represent the sequential interactions of defined (or definable) gene products, and as in the events that regulate the cell cycle, the elucidation of the events controlling cell death will have far reaching consequences for our understanding of many diverse biological phenomena.

In this overview we consider the evidence for and against the existence of unique "death genes" and a central, regulated "apoptotic pathway" of biochemical interactions controlling the process in all cases (regardless of cell type or mode of induction). In the next section we will discuss the evidence from the perspective of the effects of pharmacologic inhibition of RNA or protein synthesis. On *a priori* grounds, though, at least four types of real and hypothetical gene products in the process of apoptosis can be classified by function, and we will consider these before proceeding to the evidence and arguments surrounding them.

The first class of "death genes" are those that "set up" conditions leading to apoptosis, for example by transducing signals, triggering other molecules, and/or driving the cell cycle to a required point for exit into the death pathway. We might expect molecules in this class to be required for apoptosis under specific rather than all conditions. A trivial example might be glucocorticoid receptors required for dexamethasone-induced apoptosis. Another likely member of this class, though less obvious, is the c-Myc protein, which is required for some but not all forms of apoptosis. We will consider *myc* in detail in the last section of this review.

Once apoptosis is "set up", it presumably proceeds via the action of gene products that are essential in all cases. A candidate for this class is the putative endogenous endonuclease responsible for DNA fragmentation during apoptosis (Duke *et al.*, 1983; Cohen and Duke, 1984; Arends *et al.*, 1990). Although it is not clear that only one endonuclease is involved in all apoptotic phenomena, and attempts to identify this protein have not yet met with complete success. Evidence that endonucleases in general are required for apoptosis has been presented (Cohen and Duke, 1984; McConkey *et al.*, 1989). In contrast, though, some evidence exists that apoptosis can proceed in the absence of oligonucleosomal fragmentation (Lockshin and Zakeri, 1991; Tomei, 1991), and therefore this molecule might not represent a central player in all forms of apoptosis.

Perhaps the best evidence for a single apoptotic pathway comes from studies in the nematode, *C. elegans*, in which genes whose functions appear to be required for all forms of developmental death have been identified (Ellis *et al.*, 1991). While some questions remain as to whether the cell death controlled by these genes is apoptosis (Robertson and Thomson, 1982), there is little doubt that regulated cell death occurs via a genetically defined pathway. If homologues of these genes can be found in other organisms, and/or if techniques can be established for the study of cell lines in nematodes, we stand to learn a great deal about the molecular process of apoptosis. Until that time, some caution should be exercised in extending these observations to conclusions about the process of apoptosis in all organisms.

The search for gene expression associated with apoptosis in vertebrates has not been without success, and several genes expressed during apoptosis have been identified. While many investigators hope that these will fall into the second class of essential molecules for the process of apoptosis, it is likely that most of these are members of a third class. This class represents the products of genes that are expressed in response to cell stress or damage, and which sometimes account for features of apoptosis that are characteristic but are not necessarily essential for the process. Two examples of these are transglutaminase (Fesus et al., 1987) and a protein with several names including TRPM-2, SGP-2, and clusterin (Bursch et al., 1990). The first is responsible for the formation of a detergent-insoluble cellular envelope formed from extensively cross-linked protein in many apoptotic cells (Fesus et al., 1987). The second functions to block the lytic action of the complement pathway on apoptotic bodies (Jenne and Tschopp, 1989). Another candidate for this class of molecules is the structure identified on apoptotic cells by vitronectin receptors on phagocytes (Savill et al., 1990). These and other molecules of this class are undoubtedly important in determining the characteristics of apoptotic cells, and are probably critical in preventing secondary necrosis and inflammation in response to the apoptotic bodies. It is unlikely, however, that the expression of such genes is a requirement for death, per se.

The final class of gene products result from the expression of "anti-apoptosis genes", and their involvement is to prevent, rather than cause apoptosis. One well-known example is Bcl-2, as discussed elsewhere in this volume, and others probably also exist. We will consider the possible roles of such gene products in the induction or inhibition of apoptosis in more detail later in this review.

THE PARADOXICAL RESULTS OF PHARMACOLOGIC STUDIES WITH MACROMOLECULAR SYNTHESIS INHIBITORS

The central evidence that expression of gene products may control apoptosis comes from pharmacologic studies in which protein or RNA synthesis inhibitors block the induction of cell death in some systems. From the outset it is important to note that such evidence does not prove that gene expression in response to apoptotic signals is a requirement for death, nor does it prove that the genes expressed must be novel. If inhibition of gene expression blocks apoptosis in some cases, it may be through loss of a labile protein that is constitutively present in the cell. If a required protein and its RNA have short half-lives, then blockade of macromolecular synthesis will result in its rapidly being lost from the cell. If its function is in any way required for the process of apoptosis (or its "set up", see above), then apoptosis will not occur.

Many examples of the inhibition of apoptosis by cycloheximide or actinomycin D have been reported. These include developmental apoptosis in palatal epithelial cells (Pratt and Greene, 1976) and the metamorphosing tadpole tail (Tata, 1966); apoptosis induced by nitrogen mustard or cytosine arabinoside in intestinal epithelial cells (Lieberman, et al., 1970) or bone marrow cells (Ben-Ishay et al., 1985); apoptosis induced in thymocytes by glucocorticoid (Cohen and Duke, 1984; Wyllie et al., 1984), the calcium ionophore A23187 (Wyllie et al., 1984), irradiation (Yamada and Ohyama, 1988), or dioxin (McConkey et al., 1988); apoptosis induced by withdrawal of growth factors (Duke and Cohen, 1986); and activation-induced apoptosis in T-cell hybridomas (Ucker et al., 1989; Odaka et al., 1990; Shi et al., 1990), to name a few. Such examples are sufficiently diverse that some investigators have contended that a requirement for macromolecular synthesis is a hallmark of apoptosis. Further, the endonuclease activity in isolated nuclei that is often considered to be responsible for DNA fragmentation during apoptosis is rapidly lost from cells treated with protein or RNA synthesis inhibitors (McConkey et al., 1990), and these observations provided additional support for the view that continued synthesis is necessary for active cell death.

The interpretation of results of experiments utilizing these inhibitors must be viewed with caution. As Carson and colleagues have pointed out (Carson et al., 1988) the inhibition of protein synthesis might prolong the life of a cell under conditions where such synthesis represents a considerable drain on energy resources. Alternatively, inhibition might act to deplete proteins that fall into the first class of mediators mentioned above, that is, those that allow or "encourage" apoptosis to occur under some circumstances without necessarily playing a direct role in the process itself. Thus, ongoing protein synthesis might be necessary to trigger apoptosis in some cases, but this may not be a general feature of the apoptotic process.

Several situations have been described where apoptosis does not appear to depend on ongoing RNA or protein synthesis. For example, CTL-induced apoptosis in target tumour cells can occur in the absence of macromolecular synthesis (Duke et al., 1983). We clearly demonstrated that in the human promyelocytic leukaemia cell line HL-60 that apoptosis induced by a number of agents occurred in the absence of macromolecular synthesis (Martin et al., 1990). Another example is apoptosis induced by mild hyperthermia (Sellins and Cohen, 1991). Thus, since apoptosis can apparently occur in the absence of RNA or protein synthesis, such a requirement should not be considered a hallmark of the phenomenon.

Such cautious interpretation is especially important in light of the paradoxical observation that protein and RNA synthesis inhibitors can themselves *induce* apoptosis under some circumstances. For example, apoptosis in leukaemic myeloblasts, intestinal crypt cells and some tumour cells can be induced by actinomycin D or cyclohexímde (Heine et al., 1966; Baxter et al., 1989; Ijiri and Potten, 1987). We demonstrated that both actinomycin D and cycloheximide readily induced apoptosis in HL-60 cells and also in a number of other human tumour cell lines (Martin et al., 1990). This effect was both dose-dependent and highly reproducible. Similarly, spontaneous apoptosis in a lymphocytic leukaemia cell line was enhanced by macromolecular synthesis inhibitors rather than prevented (Collins et al., 1991). Finally, the ability of tumour necrosis factor to induce apoptosis in some cells is enhanced rather than inhibited by blocking of macromolecular synthesis (Rubin et al., 1988). Clearly, if inhibitors of macromolecular synthesis can trigger and/or enhance apoptosis, it seems extremely unlikely that apoptosis necessarily requires RNA or protein synthesis.

The paradoxical abilities of macromolecular synthesis inhibitors to either induce or prevent apoptosis requires some explanation. Of course, it may be that the conflicting results might be due to species differences (e.g. human cells versus rodent cells), but it is difficult to conceive how a process as fundamental as apoptosis would have evolved quite distinct mechanisms in related species. Alternatively, it may well be that different cell types, with quite distinct physiological functions, have differerent pathways leading to apoptosis. For example, a specific cell type might possess some or all of the necessary proteins to allow it to undergo apoptosis, which then might be prevented by limiting factors and/or active inhibitory mechanisms. Thus, in such a situation addition of macromolecular synthesis inhibitors might have no effect or might effectively induce apoptosis by removing a labile inhibitor.

While this idea is attractive and may in some cases be correct, there are additional problems. In particular, apoptosis may be either inhibited or induced by cycloheximide or actinomycin D in a single cell line, depending upon the dose of the drug used and the time period analyzed. For example, apoptosis can be induced in a T-cell hybridoma, Al.1 cells by exposure to anti-CD3 antibodies or other signals that activate the cell (Shi et al., 1990) as has been observed in other hybridomas (Ucker et al., 1989; Odaka et al., 1990). This activation-induced apoptosis can be inhibited if the cells are pretreated with small amounts of either actinomycin D (1 μg/ml) or cycloheximide (5 μM) (Shi et al., 1990), and this has similarly been observed in the other hybridomas referenced above. This dependence on macromolecular synthesis is also seen when stimuli such as glucocorticoids or a combination of phorbol myristate acetate plus ionomycin are used. There is clearly a requirement for RNA and protein synthesis for apoptosis in this system. However, a number of agents, including camptothecin, etoposide, and aphidicolin all induce apoptosis in A1.1 cells in a macromolecular synthesis-independent manner (Cotter et al., 1992). Thus, while some forms of apoptosis depend upon macromolecular synthesis in this cell line, other do not. It is particularly interesting that when these cells are exposed to levels of either actinomycin D or cycloheximide that are gradually increased there is a parallel increase in the levels of apoptosis (Figure 13.1). Although both these agents inhibit apoptosis at low concentrations they clearly enhance or induce it at higher levels (Figure 13.1). These observations thus demonstrate that several pathways leading to apoptosis can co-exist within the same cell type.

The simplest explanation for these and related findings is that in cells such as A1.1, the necessary machinery for apoptosis is effectively "in place" in the cells. In some cases, such as those of activation or glucocorticoid exposure, the induction of the apoptotic process appears to depend on the continuation of protein synthesis. In other cases it clearly does not, and the pathway must be induced by other means. At least one important difference between these two modes of apoptosis induction is that the former agents (antibodies to the T-cell receptor, glucocorticoids, PMA plus ionomycin) are not inherently toxic, while the others function to damage cells by virtue of their pharmacologic properties. Thus, apoptosis induced by such toxic effects might bypass the requirement for macromolecular synthesis in these cells, while that induced by otherwise nontoxic agents must rely on the effects of gene expression to trigger the apoptotic pathway.

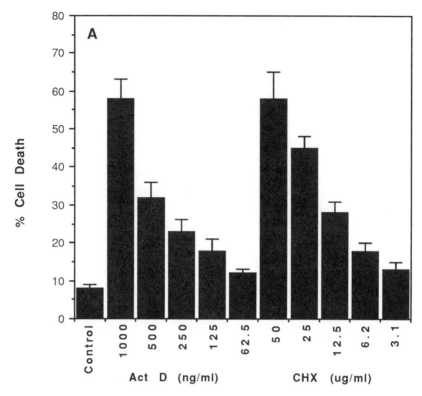

Figure 13.1 (A) A1.1 cells were induced to undergo apoptosis by incubating for 15 h at 37°C in the presence of either actinomycin D or cycloheximide at the concentrations indicated. Cell death was assessed by the MTT assay.

This is not to say that the gene expression in these cases must be novel. As we have already pointed out, inhibition of constitutive gene expression could have important effects if the products have short half-lives in the cells. One example is the c-Myc protein, which we discuss in the next section, and the continued expression of this protein appears to be a requirement for activation-induced apoptosis (but not glucocorticoid-induced apoptosis) (Shi et al., 1992).

The possibility that the induction of apoptosis may not depend upon *de novo* gene expression is suggested by recent studies on the glucocorticoid receptor. Receptors lacking regions required for dimerization and transactivation of transcription can nevertheless function to induce cell death (Nazareth et al., 1991). While this observation requires more thorough analysis (for example, it was not demonstrated that the death induced via these defective receptors was apoptotic) the result is tantalizing. If correct, this will provide the first direct evidence that glucocorticoid-induced gene expression does not include genes required for apoptosis. On the other hand, the ability of glucocorticoid receptors to induce death does depend upon their ability to

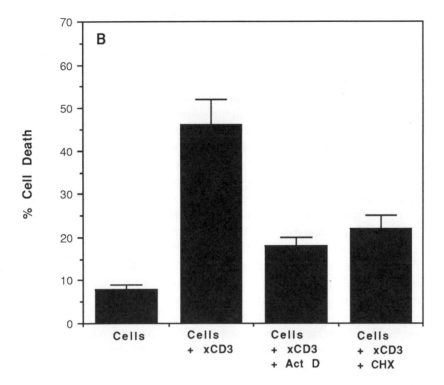

Figure 13.1 (B) A1.1 cells were cultured for 15 h at 37°C in microtitre plate wells which had been coated with an anti-CD3 monoclonal antibody. Both actinomycin D and cycloheximide were potent inhibitors of the cell death induced at concentrations of 100 ng/ml and 5 µg/ml respectively. Untreated cells were used as a control.

bind to DNA (Dieken and Miesfeld, 1992) and this may indicate that inhibition of gene expression (rather than direct induction) plays a role in glucocorticoid-induced apoptosis.

C-MYC AND APOPTOSIS

In the course of our studies on activation-induced cell death in T-cell hybridomas, we have discovered that expression of c-*myc* is a requirement for apoptosis in this system (Shi *et al.*, 1992). Our approach was to use antisense oligodeoxynucleotides to specifically control the function of this gene in a T-cell hybridoma, A1.1, which undergoes activation-induced apoptosis (Shi *et al.*, 1990). Antisense oligonucleotides corresponding to c-*myc* dramatically blocked activation-induced cell death, while neither nonsense oligonucleotides (having the same base composition) nor antisense oligonucleotides corresponding to c-*fos* showed any effect. When we analyzed cell lysates for c-Myc, we found that the ability of the c-*myc* antisense to block cell death correlated with a loss of c-Myc protein from these cells. Further, DNA fragmentation and morphological changes associated with apoptosis were dramatically reduced when cells were activated in the presence of the c-*myc* antisense. Finally we examined the effect of c-*myc* antisense on another response parameter of T-cell activation through the T-cell receptor, i.e. the release of lymphokines such as interleukin-2. Neither antisense nor control oligonucleotides affected the amount of lymphokine released following activation by anti-CD3, and therefore inhibition of c-*myc* did not influence activation *per se*, but only the apoptotic consequence.

More recently, a similar effect of c-*myc* antisense on activation-induced apoptosis in a B lymphoma line has been observed (Fisher *et al.*, unpublished data). This line, WEHI-231, undergoes cell cycle arrest and apoptosis upon cross-linking of its surface immunoglobulin (Benhamou *et al.*, 1990). Not only did the presence of c-*myc* antisense oligonucleotides prevent apoptosis in this system, but it also prevented the cell cycle arrest. This, in turn correlated with the phosphorylation status of the retinoblastoma (RB) protein in these cells: activation via the surface immunoglobulin led to the appearance of only nonphosphorylated RB protein and cell cycle arrest, while in cells activated in the presence of the c-*myc* antisense only the phosphorylated form of RB was detected. Direct interactions between Myc and RB have been observed (Rustgi *et al.*, 1991), and it is therefore possible that an interaction between c-Myc protein and the RB protein is important in activation-induced apoptosis.

As outlined above, apoptosis is blocked in some cases by the inhibition of RNA or protein synthesis and these observations have led to the concept of "death genes" required for the apoptosis process. Although many other laboratories have provided evidence for genes associated with apoptosis, our results represent the first demonstration of a specific gene requirement in any form of apoptosis. However, while expression of c-*myc* appears to be required for activation-induced apoptosis in these systems, we found that glucocorticoid-induced apoptosis (Shi *et al.*, 1992) as well as apoptosis induced by cytotoxic agents (unpublished observations) were unaffected by treatment with c-*myc* antisense oligonucleotides. Thus, Myc protein is essential in some but not all forms of apoptosis.

A role for the c-*myc* protooncogene in some other forms of apoptosis has been suggested from a number of interesting studies. Wurm *et al.* (1986) engineered Chinese hamster ovary (CHO) cells carrying an amplified construct in which c-*myc* is expressed under the control of a heat shock promoter. They found that upon heat treatment *myc* expression was induced, and this resulted in a dramatic loss of viability in these cells. The extent of cell death was found to correlate with the amount of c-Myc protein expressed in the cells (Pallavicini *et al.*, 1990). Subsequently, we have shown that this cell death is via apoptosis (Bissonnette *et al.*, 1992). Similarly, Wyllie and colleagues (1987) found an increased rate of apoptosis in fibroblast lines that had been transfected with c-*myc*. Thus, it is clear that *myc* can induce apoptosis in some cells, but how?

The answer to this question (at least in part) was provided by studies that initially addressed a different issue. For many years it has been known that while some cells simply come to rest upon withdrawal of growth factors, other cells instead undergo apoptosis (and thus, such cells have been said to be "addicted" to their growth factor (Duke and Cohen, 1986)). An obvious problem, then, was to define the essential differences between these two very different responses to growth factor withdrawal. Two studies provided an important insight into this problem by providing a role for the c-*myc* proto-oncogene in the process of growth factor addiction. A cell line that was maintained in IL-3 was found to arrest cell division without apoptosis upon growth factor withdrawal, but when these cells were transfected with c-*myc*, IL-3 withdrawal rapidly led to apoptosis (Askew et al., 1991). Similarly, RAT-1 cells expressing a chimaeric, estrogen-dependent c-Myc/estrogen receptor protein (Eilers et al., 1989) underwent apoptosis upon serum deprivation only if c-*myc* was active (Evan et al., 1992). A relationship between this and the CHO cell system discussed above was provided by the observation that at very low levels of *myc*-induction in the CHO cells, apoptosis was observed at low but not high concentrations of serum (unpublished observations). These lines of evidence support the idea that c-*myc* can produce a state of growth factor dependency such that factor withdrawal induces apoptosis.

The observation that c-*myc* and probably other genes can induce a state of growth factor addiction can be viewed in a slightly different way. That is, some genes (such as c-*myc*) induce apoptosis which can be inhibited by the addition of growth factors. An important consequence of this viewpoint is the possibility that the survival signal provided by growth factors can be replaced by other signals, such as might be provided by other genes products. If so, then this would resolve the paradox of how a gene that is normally associated with cell growth and transformation (e.g. *myc*) can participate in apoptosis—a second survival signal converts apoptotic induction into a signal to proliferate.

This two signal idea suggests that activation of *myc* induces *either* apoptosis or growth, depending upon the presence (growth) or absence (apoptosis) of the survival signal. If *myc* expression is absent or reduced, then the third option of cell cycle arrest (rest) in the absence of growth factors becomes available. This simple idea is somewhat complicated by the observation that growth factors may induce *myc* expression (Shibuya et al., 1992) in addition to providing the putative survival signal. Nevertheless, it provides a framework for understanding the protective effects of other genes in the presence of c-*myc* induction.

Direct evidence that a survival signal can prevent c-*myc*-induced apoptosis was provided in our studies using c-*myc* and *bcl*-2 (Bissonnette et al., 1992). The ability of *bcl*-2 to block apoptosis in some systems is reviewed in detail elsewhere in this volume. In the CHO system mentioned above, we found that expression of the *bcl*-2 gene under the control of a constitutive promoter (Hockenbery et al., 1990) had no effect on the levels of *myc* expression observed following induction, but dramatically inhibited c-*myc*-induced apoptosis. Other oncogene products appear to be capable of blocking c-*myc*-induced apoptosis as well. The increased rate of apoptosis observed by Wyllie and colleagues (1987) in fibroblasts upon transfection with *myc* was not seen upon co-transfection with Ha-*ras*, and thus the Ha-*ras* oncogene might provide a survival signal. Another source for a survival signal might be found in the *abl* gene product. A myeloid cell line expressing v-*myc* and a temperature-sensitive v-*abl* underwent cell death at the nonpermissive temperature (Vogt et al., 1987), and we have recently found that this death is apoptotic (unpublished observations). Further, in a preliminary study, apoptosis was significantly enhanced in K562 cells when they were treated with antisense oligonucleotides corresponding to *bcr-abl* (unpublished observations) Figure 13.2. Thus, we propose that *bcl*-2, Ha-*ras*, *abl*, *bcr/abl*, and other oncogenes provide signals that inhibit *myc*-induced apoptosis. The importance of this idea is underscored by the ability of each of these oncogenes to cooperate with *myc* in cell transformation (Wyllie et al., 1987; Vaux et al., 1988; Sawyers et al., 1992).

Thus, the balance of apoptotic and anti-apoptotic signals in a cell will determine its survival, growth, or death, and under normal circumstances this is likely to be under very rigorous control. Loss of regulation will usually lead to cell death, and this "default" is likely to represent a defense mechanism against such deregulation. However, if both the growth/death-inducing signal (e.g., *myc*) and an anti-apoptotic signal (e.g., *bcl*-2) become constitutively active, the result can be transformation. If all transformation events include such anti-apoptotic signals, then therapy directed at removing such signals should provide a route towards favoring apoptosis in tumour cells, and thus provide a new avenue to successful treatment of cancer.

In this brief overview, we have attempted to bring together the conflicting evidence supporting a role for macromolecular synthesis in apoptosis. While it is unlikely that all apoptotic processes are dependent upon ongoing protein synthesis, clearly some

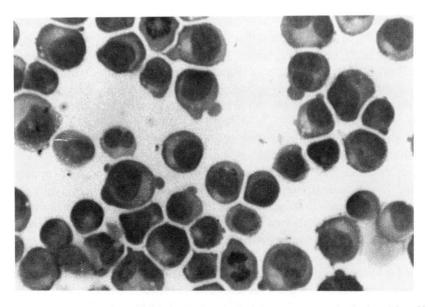

Figure 13.2 (A) K562 cells, a human myeloid cell line expressing a deregulated *bcr-abl*, are resistant to cell death via apoptosis when serum starved for 72 h.

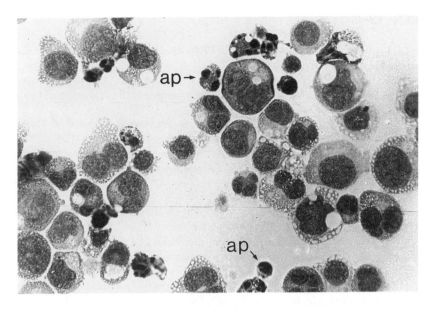

Figure 13.2 (B) When the cells are treated with an antisense corresponding to bcr-abl (10 µM) they readily undergo apoptosis upon serum starvation. Apoptotic cells (ap) are indicated in the photograph.

are. Among the genes that play roles in inducing apoptosis is c-*myc*, and while the Myc protein does not function in all forms of apoptosis, it does in several. An improved understanding of how Myc can induce or participate in the death process will give us new insights into how apoptosis occurs. Further, by examining growth factor and oncogene signals that are capable of blocking *myc*-induced death, we will gain additional perspective on the roles of such gene interactions in life, death, and transformation of a cell. Such understanding will have consequences including but beyond those of understanding cell transformation and cancer.

ACKNOWLEDGEMENTS

Some of the work described was supported by Grant AI31591 from the US National Institutes for Allergy and Infectious Disease, and by grants from The Irish Cancer Society, The Children's Leukaemia Research Project and The Health Research Board of Ireland.

REFERENCES

Arends, M. J., Morris, R. G. and Wyllie, A. H. (1990) Apoptosis: The role of the endonuclease. *American Journal of Pathology*, **136**, 593–608.

Askew, D. S., Ashmun, R. A., Simmons, B. C. and Cleveland, J. L. (1991) Constitutive c-*myc* expression in an IL-3-dependent myeloid cell line suppresses cell cycle arrest and accelerates apoptosis. *Oncogene*, **6**, 1915–1922.

Baxter, G. D., Collins, R. J., Harmon, B. V., Kumar, S., Prentice, R. L., Smith, P. J. and Lavin, M. F. (1989) Cell death in acute leukaemia. *Journal of Pathology*, **158**, 123–129.

Benhamou, L. E., Cazenave, P. A. and Sarthou, P. (1990) Anti-immunoglobulins induce death by apoptosis in WEHI-231 B lymphoma cells. *European Journal of Immunology*, **20**, 1405–1407.

Bissonnette, R. P., Escheverri, F., Mahboubi, A. and Green, D. R. (In press) Apoptotic cell death induced by c-*myc* is inhibited by *bcl*-2. *Nature*.

Bursch, W., Kleine, L. and Tenniswood, M. (1990) The biochemistry of cell death by apoptosis. *Biochemistry and Cell Biology*, **68**, 1071–1074.

Carson, D. A., Carrera, C. J., Wasson, D. B., and Yamanaka, H. (1988) Programmed cell death and adenine deoxynucleotide metabolism in human lymphocytes. *Advances in Enzyme Regulation*, **27**, 395–404.

Cohen, J. J. and Duke, R. C. (1984) Glucocorticoid activation of a calcium-dependent endonuclease in thymocyte nuclei leads to cell death. *Journal of Immunology*, **132**, 38–42.

Collins, R. J., Harmon, B. V., Souvlis, T., Pope, J. H. and Kerr, J.F.R. (1991) Effects of cycloheximide on B-chronic lymphocytic leukaemia and normal lymphocytes *in vitro*: induction of apoptosis. *British Journal of Cancer*, **64**, 518–522.

Cotter, T. G., Glynn, J. M., Echeverri, F. and Green, D. R. (1992) The induction of apoptosis by chemotherapeutic agents occurs in all phases of the cell cycle. *Anticancer Research*, **12**, 773–780.

Dieken, E. S. and Miesfeld, R. L. (1992) Transcriptional transactivation functions localized to the glucocorticoid receptor N-terminus are necessary for steroid induction of lymphocyte apoptosis. *Molecular and Cellular Biology*, **12**, 589–597.

Duke, R. C. and Cohen, J. J. (1986) IL-2 addiction: Withdrawal of growth factor activates a suicide program in dependent T cells. *Lymphokine Research*, **5**, 289–299.

Duke, R. C., Chervenak, R. and Cohen, J. J. (1983) Endogenous endonuclease-induced DNA fragmentation: An early event in cell mediated cytolysis. *Proceedings of the National Academy of Science, USA*, **80**, 6361–6365.

Eilers, M., Picard, D., Yamamoto, K. R. and Bishop, M. J. (1989) Chimaeras of Myc oncoprotein and steroid receptors cause hormone-dependent transformation of cells. *Nature*, **340**, 66–68.

Ellis, R. E., Yuan, J. Y. and Horvitz, H. R. (1991) Mechanisms and functions of cell death. *Annual Review of Cell Biology*, **7**, 663–698.

Evan, G. I., Wyllie, A. H., Gilbert, C. S., Littlewood, T. D., Land, H., Brooks, M., Waters, C. M., Penn, L. Z. and Hancock, D. C. (1992) Induction of apoptosis in fibroblasts by c-Myc protein. *Cell*, **69**, 119–128.

Fesus, L., Thomazy, V. and Falus, A. (1987) Induction and activation of tissue transglutaminase during programmed cell death. *FEBS Letters*, **224**, 104–108.

Glucksmann, A. (1951) Cell deaths in normal vertebrate ontogeny. *Biology Reviews*, **26**, 59–86.

Heine, U., Laglois, A. J. and Beard, J. W. (1966) Ultrastructural alterations in avian leukemic myeloblasts exposed to actinomycin D *in vitro*. *Cancer Research*, **26**, 1847–1858.

Hockenbery, D., Nunez, G., Milliman, C., Schreiber, R. and Korsmeyer, S. J. (1990) Bcl-2 is an inner mitochondrial membrane protein that blocks programmed cell death. *Nature*, **348**, 334–336.

Ijiri, K. and Cotten, C. S. (1987) Further studies on the response of intestinal crypt cells of different hierarchical status to eighteen different cytotoxic agents. *British Journal of Cancer,* **55**, 113–123.

Jenne, D. E. and Tschopp, J. (1989) Molecular structure and functional characterization of a human complement cytolysis inhibitor found in blood and seminal plasma: identity to sulfated glycoprotein 2, a constituent of rat testis fluid. *Proceedings of the National Academy of Science, USA,* **86**, 7123–7127.

Kerr, J. F. R., Wyllie, A. H. and Currie, A. R. (1972) Apoptosis: a basic biological phenomenon with wide ranging implications in tissue kinetics. *British Journal of Cancer,* **26**, 239–257.

Lieberman, M. W., Verbin, R. S., Landay, M., Liang, H., Farber, E., Lee, T. N. and Starr, R. (1970) A probable role for protein synthesis in intestinal epithelial cell damage induced *in vivo* by cytosine arabinoside, nitrogen mustard, or X-irradiation. *Cancer Research,* **30**, 942–951.

Lockshin, R. A. and Zakeri, Z. (1991) Programmed cell death and apoptosis. In: *Apoptosis: The Molecular Basis of Cell Death,* L. D. Tomei and F. O. Cope (eds), pp. 47–60. New York: Cold Spring Harbor Laboratory Press.

Martin, S. J., Lennon, S. V., Bonham, A. M. and Cotter, T. G. (1990) Induction of apoptosis (programmed cell death) in human leukemic HL-60 cells by inhibition of RNA and protein synthesis. *Journal of Immunology,* **145**, 1859–1867.

McConkey, D. J., Hartzell, P., Duddy, S. K., Hakansson, H. and Orrenius, S. (1988) 2,3,7,8-Tetrachlorodibenzo-p-dioxin kills immature thymocytes by Ca^{2+}-mediated endonuclease activation. *Science,* **242**, 256–259.

McConkey, D. J., Hartzell, P., Nicotera, P., and Orrenius, S. (1989) Calcium-activated DNA fragmentation kills immature thymocytes. *FASEB Journal,* **3**, 1843–1849.

McConkey, D. J., Hartzell, P. and Orrenius, S. (1990) Rapid turnover of endogenous endonuclease activity in thymocytes: Effects of inhibitors of macromolecular synthesis. *Archives of Biochemistry and Biophysics,* **278**, 284–287.

Nazareth, L. V., Harbour, D. V. and Thompson, E. B. (1991) Mapping the human glucocorticoid receptor for leukemic cell death. *Journal of Biological Chemistry,* **266**, 12 976–12 980.

Odaka, C., Kizaki, H. and Tadakuma, T. (1990) T-cell receptor-mediated DNA fragmentation and cell death in T-cell hybridomas. *Journal of Immunology,* **144**, 2096–2101.

Pallavicini, M. G., Rosette, C., Reitsma, M., Deteresa, P. S. and Gray, J. W. (1990) Relationship of c-*myc* gene copy number and gene expression: Cellular effects of elevated c-myc protein. *Journal of Cellular Physiology,* **143**, 372–380.

Pratt, R. M. and Greene, R. M. (1976) Inhibition of palatal epithelial cell death by altered protein synthesis. *Developmental Biology,* **54**, 135–145.

Robertson, A. M. G. and Thomson, J. N. (1982) Morphology of programmed cell death in the ventral nerve cord of *Caenorhabditis elegans* larvae. *Journal of Embryology and Experimental Morphology,* **67**, 89–100.

Rubin, B. Y., Smith, L. J., Hellermann, G. R., Lunn, R. M., Richardson, N. K. and Anderson, S. L. (1988) Correlation between the anticellular and DNA fragmenting activities of tumor necrosis factor. *Cancer Research,* **48**, 6006–6010.

Rustgi, A. K., Dyson, N. and Bernards, R. (1991) Amino-terminal domains of c-Myc and N-Myc proteins mediate binding to the retinoblastoma gene product. *Nature,* **352**, 541–544.

Savill, J., Dransfield, I., Hogg, N. and Haslett, C. (1990) Vitronectin receptor-mediated phagocytosis of cells undergoing apoptosis. *Nature,* **343**, 170–177.

Sawyers, C. L., Callahan, W. and Witte, O. N. (1992) Dominant negative *myc* blocks transformation by *abl* oncogenes. *Cell,* **70**, 901–910.

Sellins, K. S. and Cohen, J. J. (1991) Hyperthermia induces apoptosis in thymocytes. *Radiation Research,* **126**, 88–95.

Shi, Y., Szalay, M. G., Paskar, L., Boyer, M., Singh, B. and Green, D. R. (1990) Activation-induced cell death in T-cell hybridomas is due to apoptosis: Morphological aspects and DNA fragmentation. *Journal of Immunology*, **144**, 3326–3333.

Shibuya, H., Yoneyama, M., Ninomiya-Tsuji, J., Matsumoto, K. and Taniguchi, T. (1992) IL-2 and EGF receptors stimulate the hematopoietic cell cycle via different signalling pathways: Demonstration of a novel role for c-*myc*. *Cell*, **70**, 57–67.

Tata, J. R. (1966) Requirement for RNA and protein synthesis for induced regression of tadpole tail in organ culture. *Developmental Biology*, **13**, 77–94.

Tomei, L. D. (1991) Apoptosis: A program for death or survival? In: *Apoptosis: The Molecular Basis of Cell Death*, L. D. Tomei and F. O. Cope (eds), pp. 279–316. New York: Cold Spring Harbor Laboratory Press.

Ucker, D. S., Ashwell, J. D. and Nickas, G. (1989) Activation-driven T-cell death. I Requirements for *de novo* transcription and translation and association with genome fragmentation. *Journal of Immunology*, **43**, 3461–3469.

Vaux, D. L., Cory, S. and Adams, J. M. (1988) bcl-2 gene promotes haemopoietic cell survival and cooperates with c-*myc* to immortalize pre-B cells. *Nature*, **335**, 440–442.

Vogt, M., Lesley, J., Bogenberger, J. M., Haggblom, C., Swift, S. and Haas, M. (1987) The induction of growth factor-independence in murine myelocytes by oncogenes results in monoclonal cell lines and is correlated with cell crisis and karyotypic instability. *Oncogene Research*, **2**, 49–63.

Wurm, F. M., Gwinn, K. A. and Kingston, R. E. (1986) Inducible overproduction of the mouse c-Myc protein in mammalian cells. *Proceedings of the National Academy of Science, USA*, **83**, 5414–5418.

Wyllie, A. H. (1980) Glucocorticoid-induced thymocyte apoptosis is associated with endogenous endonuclease activation. *Nature*, **284**, 555–556.

Wyllie, A. H., Kerr, J. F. R. and Currie, A. R. (1980) Cell death: the significance of apoptosis. *International Review of Cytology*, **68**, 251–306.

Wyllie, A. H., Morris, R. G., Smith, A. L. and Dunlop, D. (1984) Chromatin cleavage in apoptosis: Association with condensed chromatin morphology and dependence on macromolecular synthesis. *Journal of Pathology*, **142**, 67–77.

Wyllie, A. H., Rose, K. A., Morris, R. G., Steel, C. M., Forster, E. and Spandidos, D. A. (1987) Rodent fibroblast tumors expressing human *myc* and *ras* genes: Growth, metastasis and endogenous oncogene expression. *British Journal of Cancer*, **56**, 251–259.

Yamada, T. and Ohyama, H. (1988) Radiation-induced interphase death of rat thymocytes is internally programmed (apoptosis). *International Journal of Radiation Biology*, **53**, 65–75.

Chapter 14

THE *BCL*-2 ONCOGENE REGULATES LYMPHOCYTE SURVIVAL AND POTENTIATES LYMPHOMAGENESIS

Andreas Strasser, Alan W. Harris and Suzanne Cory
The Walter and Eliza Hall Institute of Medical Research
PO Royal Melbourne Hospital, Victoria 3050, Australia

Programmed cell death (apoptosis) plays a prominent role in lymphocyte development and function. A recently identified regulator of survival is *bcl*-2, a gene discovered in 1984 when it was cloned from a t(14;18) chromosome translocation (Tsujimoto *et al.*, 1984; Bakhshi *et al.*, 1985; Cleary *et al.*, 1986). Most human follicular centre B cell lymphomas contain this characteristic translocation, which links the *bcl*-2 gene to the immunoglobulin heavy chain locus, thereby deregulating its expression but not altering its coding region (Tsujimoto and Croce, 1986). Translocations that introduce *bcl*-2 into the immunoglobulin κ or λ light chain locus (Adachi *et al.*, 1990) have also been identified in a small proportion of follicular centre B cell lymphoma and chronic lymphocytic leukaemia. The Bcl-2 protein associates with cytoplasmic and nuclear intracellular membranes (Chen-Levy and Cleary, 1990), including those of the mitochondria (Hockenbery *et al.*, 1990). Its hydrophobic carboxy terminus appears to be essential for function but, curiously, not for membrane association (Alnemri *et al.*, 1992).

Bcl-2 protein augments cell survival, as first realised when growth factor dependent cell lines constitutively expressing *bcl*-2 were found to survive protracted deprivation of factor (Vaux *et al.*, 1988; Nunez *et al.*, 1990; Hockenbery *et al.*, 1990). The development of transgenic mice constitutively expressing *bcl*-2 in the B (McDonnell *et al.*, 1989, 1990; Strasser *et al.*, 1990a, 1991a) and/or the T lymphoid lineage (Strasser *et al.*, 1990a, 1991b; Sentman *et al.*, 1991; Siegel *et al.*, 1992) confirmed this function in normal lymphocytes. Transgenic B and T lymphoid cells of multiple differentiation stages were found to survive abnormally well when cultured in simple medium without added lymphokines. Furthermore, while thymocytes from normal mice rapidly die by apoptosis when exposed to corticosteroids, γ-radiation, phorbol ester, calcium ionophore or anti-CD3 antibodies, thymocytes from the bcl-2 mice were partially resistant to treatment with these agents (Strasser et al., 1991b; Sentman et al., 1991; Siegel et al., 1992). As reviewed below, the effect of bcl-2 transgene expression upon T and B cell differentiation in vivo, strongly suggests that the selection processes which generate and fine-tune the immune system act by modulating Bcl-2 levels.

CONSTITUTIVE BCL-2 EXPRESSION PERTURBS B LYMPHOID HOMEOSTASIS

Transgenic mice expressing the bcl-2 transgene in the B lymphoid compartment exhibited a several-fold excess of small, non-cycling, conventional, mature B cells in all lymphoid organs (bone marrow, spleen, lymph nodes, peritoneal cavity and peripheral blood e.g. Figure 14.1) (McDonnell et al., 1989, 1990; Strasser et al., 1990a, 1991a). The number of immunoglobulin-secreting cells was also elevated, as were serum immunoglobulin levels, and immunization evoked a greatly amplified and prolonged immune response (Strasser et al., 1991a; Nunez et al., 1991). The bcl-2 transgenic B cells proliferated and differentiated normally in response to mitogens in vitro, but survived longer than control B cells (Figure 14.2). All of these properties are consistent with an increased lifespan of bcl-2 expressing mature B cells and immunoglobulin secreting plasma cells. In certain genetic backgrounds (Strasser et al., 1993b), bcl-2 mice frequently succumbed to a fatal autoimmune disease resembling human systemic lupus erythematosus (Strasser et al., 1991a). We speculate that autoreactive antibodies accumulated to pathological levels because of the longevity in these mice of immunoglobulin-secreting cells with anti-self reactivity. Perturbation of the selection mechanisms that normally operate in the bone marrow and in the germinal centres of peripheral lymphoid organs may have increased the frequency of self-reactive clones.

The inference from these results for normal B cell activation in germinal centres is that cells with useful specificities (i.e. with high affinity receptors for immunogen) may be triggered to express bcl-2 to ensure their survival and differentiation to memory B cells. Conversely, down-regulation of bcl-2 may ensure the demise of cells incapable of interacting with antigen. The available data on the regulation of Bcl-2 protein expression in germinal centre cells (Liu et al., 1991; Pezzella et al., 1990; Hockenbery et al., 1991) lends credence to this hypothesis.

Immature cells lacking immunoglobulin on the cell surface also seem to gain a survival advantage by expressing bcl-2. Young bcl-2 transgenic mice displayed increased numbers of $B220^+Ig^-$ cells in the bone marrow and spleen (unpublished observations) and these cells survived far better in vitro than control cells (Strasser et al., 1991a). Furthermore, introduction of the bcl-2 transgene into scid mice, which are incapable of functional immunoglobulin or T-cell receptor gene rearrangement

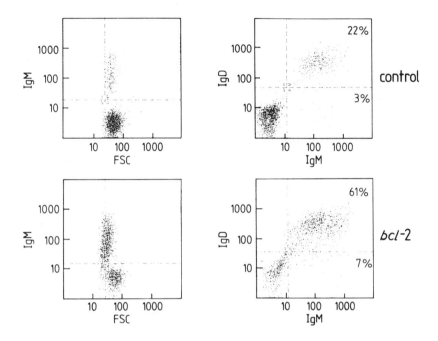

Figure 14.1 B cell excess in Eμ-bcl-2 peripheral blood. Flow cytometric analysis of nucleated peripheral blood cells stained with monoclonal antibodies against IgM and IgD. Eμ-bcl-2-22 transgenic mice have on average a 5- to 10-fold increase in peripheral blood B lymphocytes, calculated from the 3- to 5-fold increase in blood leukocytes and the 2- to 4-fold relative increase in B cells. Forward light scatter (FSC) analysis which indicates cell size shows that bcl-2 transgenic IgM$^+$ B cells are slightly smaller than those from control mice.

(Bosma et al., 1983; Schuler et al., 1986; Malynn et al., 1988), greatly increased the number of B220$^+$Ig$^-$ cells (Strasser et al., 1993c and unpublished data), presumably because of their extended survival capacity. Surprisingly, most bcl-2/scid B lineage cells, apart from lacking surface immunoglobulin, displayed a phenotype (B220^{hi}, CD23$^+$, class II MHC$^+$, BP-1$^-$, PB76$^-$) that is normally characteristic of mature, long-lived B cells. Clearly, the acquisition of these characteristics is not dependent upon the expression of cell surface immunoglobulin but may occur spontaneously when survival of immature cells is permitted by bcl-2 expression.

Developing B cells are thought to be continually subjected to positive selection for functional immunoglobulin gene rearrangement. Expression of μ polypeptide on the cell surface in association with the λ_5 and VpreB surrogate light chains (for reviews see Rolink and Melchers, 1991; Reth, 1992) has been shown to be indispensible for differentiation and survival (Kitamura et al., 1991, 1992). It is tempting to speculate that

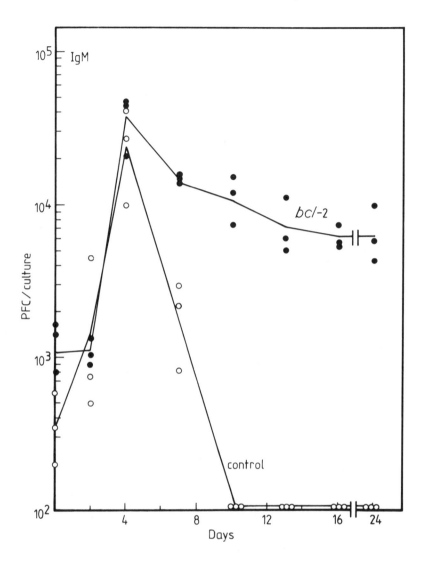

Figure 14.2 Prolonged *in vitro* LPS response of Eμ-bcl-2 spleen cells. Reverse (protein A) plaque assay for immunoglobulin secretion by spleen cells from Eμ-bcl-2-22 transgenic mice and control littermates that were stimulated *in vitro* with lipopolysaccharide (LPS). Cultures were initiated with 5×10^5 cells and stimulated by addition of LPS (20 μg/ml) on day 0. These data demonstrate extended survival *in vitro* of terminally differentiated immunoglobulin secreting cells.

engagement of these "immature" B cell antigen receptors results in the induction of Bcl-2 synthesis.

CONSEQUENCES OF CONSTITUTIVE *BCL*-2 EXPRESSION IN T LYMPHOID CELLS

T lymphocytes developing in the thymus are subjected to both positive and negative selection pressures (von Boehmer and Kisielow, 1990; Blackman et al., 1990). The first cells to be discarded are those which fail to achieve functional rearrangement of their antigen receptor genes. Within the population expressing the $\alpha\beta$ T-cell receptor heterodimer (TCR), only cells with a receptor capable of interacting with self-MHC molecules are positively selected. However, because of their potential for autoreactivity, the subset that expresses receptors which bind strongly to self-MHC complexed with self antigen must be eliminated (negative selection).

Mature post-selection T cells in the thymus medulla express high levels of Bcl-2 whereas immature cortical cells, most of which are doomed to die (Egerton et al., 1990), do not. This expression pattern (Pezzella et al., 1990; Hockenbery et al., 1991) is consistent with the view that modulation of Bcl-2 is the major effector of positive and negative selection, as proposed above for B lymphoid development. There is accumulating evidence, however, that this is too simplistic a model and that other mechanisms must also contribute. We have recently found, for example, that a *bcl*-2 transgene does not appear to confer a detectable survival advantage on immature T cells lacking productive TCR gene rearrangement. In *bcl*-2/*scid* mice, thymocytes did not accumulate *in vivo* and rapidly underwent apoptotic death *in vitro*, unlike their B lymphoid counterparts, despite expressing high levels of Bcl-2 protein (Strasser et al., 1993c and unpublished data). Furthermore, although TCRα/β-expressing cortical thymocytes were markedly advantaged by *bcl*-2 transgene expression (see Figure 14.3) and resisted apoptosis induced by glucocorticoid, γ-irradiation or anti-CD3 antibody, negative selection was not aborted (Strasser et al., 1991b; Sentman et al., 1991; Siegel et al., 1992). Overall, T-cell homeostasis was normal (Strasser et al., 1991b) and, in the periphery, T cells reactive with various Mls self-superantigens were either not detectable (Strasser et al., 1991b; Sentman et al., 1991) or very infrequent (Siegel et al., 1992).

Despite this evidence that negative selection was maintained in the face of constitutive *bcl*-2 expression, there was some indication that the process had been significantly hampered by expression of the transgene. Low levels of self Mls-reactive cells were detected in the thymus in two studies (Strasser et al., 1991b; Siegel et al., 1992), although not in another (Sentman et al., 1991). The apparent discrepancy might have reflected differences in the timing and level of *bcl*-2 expression achieved with the different transgene constructs or, more likely, differences in the avidity of the different self-superantigens for their respective T-cell receptors.

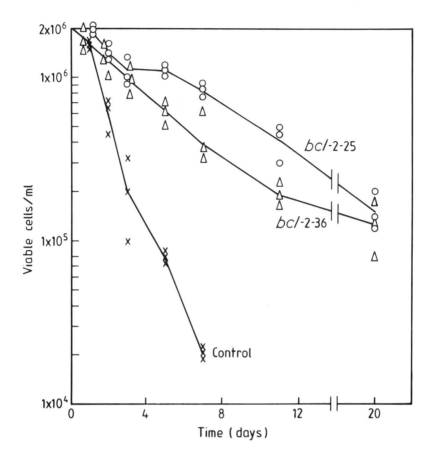

Figure 14.3 Prolonged survival of Eμ-bcl-2 thymocytes *in vitro*. Survival of thymocytes from Eμ-bcl-2-25 and -36 transgenic mice and control littermates in simple tissue culture medium. Numbers of live and dead cells were determined by trypan blue exclusion in a haemocytometer. These data demonstrate a 20- to 50-fold improved survival of bcl-2 transgenic thymocytes compared to those from control littermates.

TCR transgenic mice offer great experimental advantages for studying negative selection because most of their T cells express the same receptor. We have therefore now crossed our bcl-2 transgenic mice with mice which express TCR αβ heterodimers recognizing the male HY antigen in the context of class I MHC H-2D^b (for a review, see von Boehmer, 1990). Analysis of the doubly transgenic male progeny have confirmed that deregulated bcl-2 expression interferes significantly with clonal deletion of autoreactive thymocytes but does not totally block this process (Strasser *et al.*, 1993c and unpublished data).

Negative selection is thought to operate at two separate stages of thymocyte development in two distinct microenvironments (Speiser *et al.*, 1992). Thymocytes that bind with high affinity to self-antigens presented by self-MHC molecules on thymic epithelial cells are deleted at the CD4$^+$8$^+$ TCRα/β^{lo} stage (Speiser et al., 1992). Engagement of the TCR at this stage is probably responsible for the calcium influx that has been detected *in situ* in autoreactive thymocytes in TCR transgenic mice (Nakayama et al., 1992). Immature thymocytes are very susceptible to calcium influx and this form of apoptosis is significantly delayed in *bcl*-2 transgenic thymocytes (Strasser et al., 1991b; Siegel et al., 1992), providing a possible explanation for the accumulation of autoreactive thymocytes in *bcl*-2 transgenic and anti-HY TCR/*bcl*-2 doubly transgenic mice (Strasser et al., 1993c and unpublished data). Negative selection also operates on developing thymocytes at the CD4$^+$8$^+$TCRα/β $^{int-hi}$ stage (Speiser et al., 1992) when they are scrutinized for their binding affinity to thymic dendritic cells at the cortico-medullary junction (Hengartner et al., 1988) The prime candidate for the signal which induces apoptosis at this stage is the interaction of the T-cell FAS/APO-1 receptor (Yonehara et al., 1989; Trauth et al., 1989; Itoh et al., 1991; Oehm et al., 1992) with its putative natural ligand *gld* (Allen et al., 1990) on thymic dendritic cells. Signalling through the FAS/APO-1 molecule is known to induce a form of apoptosis that does not seem to be affected by high levels of Bcl-2 (Falk et al., 1992), providing an explanation for the fact that self-superantigen reactive T cells were either absent (Strasser et al., 1991b; Sentman et al., 1991) or rare (Siegel et al., 1992) in peripheral lymphoid organs of *bcl*-2 transgenic mice. However, since negative selection also appears to be largely intact in *lpr* mice (Sidman et al., 1992), which lack a functional FAS/APO-1 receptor (Watanabe-Fukunaga et al., 1992), down-regulation of *bcl*-2 may also play a critical role. By introducing the *bcl*-2 transgene into *lpr* mice, we are currently testing the hypothesis that constitutive *bcl*-2 expression and absence of signalling through the FAS/APO-1 receptor are both necessary to escape from clonal deletion.

THE ROLE OF *BCL*-2 IN LYMPHOMAGENESIS

By itself, *bcl*-2 is relatively innocuous as an oncogene. Long term analysis of cohorts of several independent strains of *bcl*-2 transgenic mice (Strasser et al., 1993a) have revealed very little, if any, increased propensity to develop T lymphoid tumours. However, in agreement with results obtained by Korsmeyer and colleagues (McDonnell and Korsmeyer, 1991), *bcl*-2 transgene expression was associated with a low but significant incidence (3 to 15%, depending on the line) of B lymphoid tumours. In our transgenic mice, there were principally two tumour types: plasmacytomas and novel lymphomas exhibiting a phenotype and antigen receptor gene configuration consistent with an origin very early in B lymphoid development. Tumours of mature B cells were very rare. The low frequency and long latency implies that tumour onset depended on the acquisition of somatic mutation(s) by the *bcl*-2 expressing cells. Rearrangement of the *myc* gene was common in the plasmacytomas, as reported by McDonnell and Korsmeyer (1991) for *bcl*-2 tumours designated large cell lymphomas, implying a synergistic role for *myc* and *bcl*-2 in their aetiology. Collaboration between *myc* and *bcl*-2 had been suggested earlier by the observation that stable long-term proliferating lines could be generated *in vitro* from pre-B cells expressing a *myc* transgene and a *bcl*-2 retrovirus (Vaux *et al.*, 1988). Dramatic confirmation of the synergistic potential of this oncogene combination came from doubly transgenic *myc/bcl*-2 mice which developed tumours much more rapidly than mice expressing either transgene alone (Strasser *et al.*, 1990b). Intriguingly, the tumours all lacked immunoglobulin and TCR gene rearrangements and exhibited a cell surface phenotype and gene expression pattern indicative of a primitive lymphoid stem or progenitor cell.

The recent observation that *myc* induces apoptosis of myeloid (Askew *et al.*, 1991) and fibroblastic (Evan *et al.*, 1992) cell lines under limiting growth conditions has provided a deeper appreciation of why the *myc/bcl*-2 combination is such an effective transforming agent. If, as seems likely, cytokine concentrations *in vivo* are often limiting, cells acquiring a *myc* mutation might normally drive themselves to suicide unless they had previously acquired a mutation conferring either a survival function or factor-independence. Suggestive evidence that this is indeed the case is provided by the fact that the expansion of the pre-B cell compartment induced in mice by a *myc* transgene plateaus in young animals (Langdon *et al.*, 1986). Furthermore, the preneoplastic *myc*-expressing pre-B cells were found to die much faster than normal pre-B cells in the absence of support by stromal cells (Langdon *et al.*, 1988).

In conclusion, *bcl*-2 is a novel oncogene which promotes cell survival rather than proliferation. Its primary role in human follicular lymphoma appears to be to enable the cells which have acquired a 14;18 translocation to resist apoptosis in the lymphoid germinal centre. The *bcl*-2 expressing clone can survive even under adverse circumstances and the chance accumulation of a mutation conferring a proliferative advantage converts enhanced survival capacity into a potent drive towards malignancy. Since both hypermutation of rearranged immunoglobulin variable region genes and switch recombination of the heavy chain locus take place at this stage of B lymphoid differentiation, inadvertent mistakes by the enzymes carrying out these reactions may place the clone bearing the t(14;18) translocation at increased risk of acquiring additional oncogenic mutations.

ACKNOWLEDGEMENTS

This work was supported by fellowships from the Leukemia Society of America, and the Swiss National Science Foundation to A.S. and by the National Health and Medical Research Council of Australia and the US National Cancer Institute (CA43540).

REFERENCES

Adachi, M., Tefferi, A., Greipp, P. R., Kipps, T. J. and Tsujimoto, Y. (1990) Preferential linkage of *bcl*-2 to immunoglobulin light chain gene in chronic lymphocytic leukemia. *Journal of Experimental Medicine*, **171**, 559–564.

Allen, R. D., Marshall, J. D., Roths, J. B. and Sidman, C. L. (1990) Differences defined by bone marrow transplantation suggest that *lpr* and *gld* are mutations of genes encoding an interacting pair of molecules. *Journal of Experimental Medicine*, **172**, 1367–1375.

Alnemri, E. S., Robertson, N. M., Fernandes, T. F., Croce, C. M. and Litwack, G. (1992) Overexpressed full-length human *bcl*-2 extends the survival of baculovirus-infected Sf9 insect cells. *Proceedings of the National Academy of Science, USA*, **89**, 7295–7299.

Askew, D. S., Ashman, R. A., Simmons, B. C. and Cleveland, J. L. (1991) Constitutive c-*myc* expression in an IL-3-dependent myeloid cell line suppresses cell cycle arrest and accelerates apoptosis. *Oncogene*, **6**, 1915–1922.

Bakhshi, A., Jensen, J., Goldman, P., Wright, J. J., McBride, O. W., Epstein, A. L. and Korsmeyer, S. J. (1985) Cloning the chromosomal breakpoint of t(14;18) human lymphomas: clustering around JH on chromosome 14 and near a transcriptional unit on 18. *Cell*, **41**, 899–906.

Blackman, M., Kappler, J. A., Marrack, P. (1990) The role of the T-cell receptor in positive and negative selection of developing T cells. *Science*, **248**, 1335–1340.

Bosma, G. C., Custer, R. P. and Bosma, M. J. (1983) A severe combined immunodeficiency mutation in the mouse. *Nature*, **301**, 527–530.

Chen-Levy, Z. and Cleary, M. L. (1990) Membrane topology of the Bcl-2 proto-oncogenic protein demonstrated *in vitro*. *Journal of Biological Chemistry*, **265**, 4929–4933.

Cleary, M. L., Smith, S. D. and Sklar, J. (1986) Cloning and structural analysis of cDNAs for *bcl*-2 and a hybrid *bcl*-2/immunoglobulin transcript resulting from the t(14;18) translocation. *Cell*, **47**, 19–28.

Egerton, M., Scollay, R. and Shortman, K. (1990) Kinetics of mature T-cell development in the thymus. *Proceedings of the National Academy of Science, USA*, **87**, 2579–2582.

Evan, G. I., Wyllie, A. H., Gilbert, C. S., Littlewood, T. D., Land, H., Brooks, M., Waters, C. M., Penn, L. Z. and Hancock, D. C. (1992) Induction of apoptosis in fibroblasts by c-Myc protein. *Cell*, **69**, 889–899.

Falk, M. H., Trauth, B. C., Debatin, K.-M., Klas, C., Gregory, C. D., Rickinson, A. B., Calender, A., Lenoir, G. M., Ellwart, J. W., Krammer, P. H. and Bornkamm, G. W. (1992) Expression of the APO-1 antigen in Burkitt lymphoma cell lines correlates with a shoft towards a lymphoblastoid phenotype. *Blood*, **79**, 3300–3306.

Hengartner, H., Odermatt, B., Schneider, R., Schreyer, M., Walle, G., MacDonald, H. R. and Zinkernagel, R. M. (1988) Deletion of self-reactive T cells before entry into the thymus medulla. *Nature*, **336**, 388–390.

Hockenbery, D., Nunez, G., Milliman, C., Schreiber, R. D. and Korsmeyer, S. J. (1990) Bcl-2 is an inner mitochondrial membrane protein that blocks programmed cell death. *Nature*, **348**, 334–336.

Hockenbery, D. M., Zutter, M., Hickey, W., Nahm, M. and Korsmeyer, S. (1991) Bcl-2 protein is topographically restricted in tissues characterized by apoptotic cell death. *Proceedings of the National Academy of Science, USA*, **88**, 6961–6965.

Itoh, N., Yonehara, S., Ishii, A., Yonehara, M., Mizushima, S.-I., Samashima, M., Hase, A., Seta, Y. and Nagata, S. (1991) The polypeptide encoded by the cDNA for human cell surface antigen Fas can mediate apoptosis. *Cell*, **65**, 233–243.

Kitamura, D., Kudo, A., Schaal, S., Muller, W., Melchers, F. and Rajewsky, K. (1992) A critical role of λ5 protein in B cell development. *Cell*, **69**, 823–831.

Kitamura, D., Roes, J., Kuhn, R. and Rajewsky, K. (1991) A B cell-deficient mouse by targeted disruption of the membrane exon of the immunoglobulin μ chain gene. *Nature*, **350**, 423–426.

Langdon, W. Y., Harris, A. W., and Cory, S. (1988) Growth of Eμ-myc transgenic B-lymphoid cells *in vitro* and their evolution toward autonomy. *Oncogene Research*, **3**, 271–279.

Langdon, W. Y., Harris, A. W., Cory, S., and Adams, J. M. (1986) The c-myc oncogene perturbs B lymphocyte development in E-mu-*myc* transgenic mice. *Cell*, **47**, 11–18.

Liu, Y. J., Mason, D. Y., Johnson, G. D., Abbot, S., Gregory, C. D., Hardie, D. L., Gordon, J. and MacLennan, I. C. (1991) Germinal center cells express Bcl-2 protein after activation by signals which prevent their entry into apoptosis. *European Journal of Immunology*, **21**, 1905–1910.

Malynn, B. A., Blackwell, T. K., Fulop, G. M., Rathbun, G. A., Furley, A. J., Ferrier, P., Heinke, L. B., Phillips, R. A., Yancopoulos, G. D. and Alt, F. W. (1988) The *scid* defect affects the final step of the immunoglobulin VDJ recombinase mechanism. *Cell*, **54**, 453–460.

McDonnell, T. J., Deane, N., Platt, F. M., Nunez, G., Jaeger, U., McKearn, J. P. and Korsmeyer, S. J. (1989) *bcl-2*-immunoglobulin transgenic mice demonstrate extended B cell survival and follicular lymphoproliferation. *Cell*, **57**, 79–88.

McDonnell, T. J. and Korsmeyer, S. J. (1991) Progression from lymphoid hyperplasia to high-grade malignant lymphoma in mice transgenic for the t(14; 18). *Nature*, **349**, 254–256.

McDonnell, T. J., Nunez, G., Platt, F. M., Hockenberry, D., London, L., McKearn, J. P. and Korsmeyer, S. J. (1990) Deregulated *bcl-2*-immunoglobulin transgene expands a resting but responsive immunoglobulin M and D-expressing B-cell population. *Molecular and Cellular Biology*, **10**, 1901–1907.

Nakayama, T., Ueda, Y., Yamada, H., Shores, E. W., Singer, A. and June, C. H. (1992) *In vivo* calcium elevations in thymocytes with T-cell receptors that are specific for self ligands. *Science*, **257**, 96–99.

Nunez, G., Hockenbery, D., McDonnell, T. J., Sorensen, C. M. and Korsmeyer, S. J. (1991) Bcl-2 maintains B cell memory. *Nature*, **353**, 71–73.

Nunez, G., London, L., Hockenbery, D., Alexander, M., McKearn, J. P. and Korsmeyer, S. J. (1990) Deregulated *bcl-2* gene expression selectively prolongs survival of growth factor-deprived hemopoietic cell lines. *Journal of Immunology*, **144**, 3602–3610.

Oehm, A., Behrmann, I., Falk, W., Pawlita, M., Maier, G., Klas, C., Li-Weber, M., Richards, S., Dhein, J., Trauth, B. C., Ponstingl, H. and Krammer, P. H. (1992) Purification and molecular cloning of the APO-1 cell surface antigen, a member of the tumor necrosis factor/nerve growth factor receptor superfamily. *Journal of Biological Chemistry*, **267**, 10709–10715.

Pezzella, F., Tse, A. G. D., Cordell, J. L, Pulford, K. A. F., Gatter, K. C. and Mason, D. Y. (1990) Expression of the *bcl-2* oncogene protein is not specific for the 14;18 chromosomal translation. *American Journal of Pathology*, **137**, 225–232.

Reth, M. (1992) Antigen receptors on B lymphocytes. *Annual Review of Immunology*, **10**, 97–121.

Rolink, A. and Melchers, F. (1991) Molecular an cellular origins of B lymphocyte diversity. *Cell*, **66**, 1081–1094.

Schuler, A., Phillips, R. A., Rosenberg, N., Mak, T. W., Kearney, J. F., Perry, R. P. and Bosma, M. J. (1986) Rearrangement of antigen receptor genes is defective in mice with severe combined immune deficiency. *Cell*, **46**, 963–972.

Sentman, C. L., Shutter, J. R., Hockenbery, D., Kanagawa, O. and Korsmeyer, S. J. (1991) *bcl-2* inhibits multiple forms of apoptosis but not negative selection in thymocytes. *Cell*, **67**, 879–888.

Sidman, C. L., Marshall, J. D. and von Boehmer, H. (1992) Transgenic T-cell receptor interactions in the lymphoproliferative and autoimmune syndromes of *lpr* and *gld* mutant mice. *European Journal of Immunology*, 22,

Siegel, R. M., Katsumata, M., Miyashita, T., Louie, D. C., Greene, M. I. and Reed, J. C. (1992) Inhibition of thymocyte apoptosis and negative antigenic selection in *bcl*-2 transgenic mice. *Proceedings of the National Academy of Science, USA*, 89, 7003–7007.

Speiser, D. E., Pircher, H., Ohashi, P. S., Kyburz, D., Hengartner, H. and Zinkernagel, R. M. (1992) Clonal deletion induced by either radioresistant thymic host cells or lymphohemopoietic donor cells at different stages of class I-restricted T-cell ontogeny. *Journal of Experimental Medicine*, 175, 1277–1283.

Strasser, A., Harris, A. W., Bath, M. L. and Cory, S. (1990b) Novel primitive lymphoid tumours induced in transgenic mice by cooperation between *myc* and *bcl*-2. *Nature*, 348, 331–333.

Strasser, A., Harris, A. W. and Cory, S. (1991b) *bcl*-2 transgene inhibits T-cell death and perturbs thymic self-censorship. *Cell*, 67, 889–899.

Strasser, A., Harris, A. W. and Cory, S. (1993a) Eu-*bcl*-2 transgene facilitates spontaneous transformation of early pre-B and immunoglobulin-secreting cells but not T cells. *Oncogene*, 8, 1–9.

Strasser, A., Harris, A. W. and Cory, S. (1993b) The role of *bcl*-2 in lymphoid differentiation and neoplastic transformation. *Current Topics in Microbiology and Immunology*, 182, 299–302.

Strasser, A., Harris, A. W. and Cory, S. (1993c) The role of *bcl*-2 in lymphoid differentiation and transformation. Proc. 8th International Congress of Immunology, Budapest, 1992, pp. 51–56.

Strasser, A., Harris, A. W., Vaux, D. L., Webb, E., Bath, M. L., Adams, J. M. and Cory, S. (1990a) Abnormalities of the immune system induced by dysregulated *bcl*-2 expression in transgenic mice. *Current Topics in Microbiology and Immunology*, 166, 175–181.

Strasser, A., Whittingham, S., Vaux, D. L., Bath, M. L., Adams, J.M., Cory, S., Harris, A. W. (1991a) Enforced *bcl*-2 expression in B-lymphoid cells prolongs antibody responses and elicits autoimmune disease. *Proceedings of the National Academy of Science, USA*, 88, 8661–8665.

Trauth, B. C., Klas, C., Peters, A. M. J., Matzku, S., Moller, P., Falk, W., Debatin, K.-M., Krammer, P. H. (1989) Monoclonal antibody-mediated tumor regression by induction of apoptosis. *Science*, 245, 301–305.

Tsujimoto, Y. and Croce, C. M. (1986) Analysis of the structure, transcripts, and protein products of bcl-2, the gene involved in human follicular lymphoma. *Proceedings of the National Academy of Science, USA*, 83, 5214–5218.

Tsujimoto, Y., Finger, L. R., Yunis, J., Nowell, P. C. and Croce, C. M. (1984) Cloning of the chromosome breakpoint of neoplastic B cells with the t(14;18) chromosome translocation. *Science*, 226, 1097–1099.

Vaux, D. L., Cory, S. and Adams, J. M. (1988) *bcl*-2 gene promotes haemopoietic cell survival and cooperates with c-*myc* to immortalize pre-B cells. *Nature*, 335, 440–442.

von Boehmer, H. (1990) Developmental biology of T cells in T-cell receptor transgenic mice. *Annual Review of Immunology*, 8, 531–556.

von Boehmer, H. and Kisielow, P. (1990) Self-nonself discrimination by T cells. *Science*, 248, 1369–1373.

Watanabe-Fukunaga, R., Brannan, C. I., Copeland, N. G., Jenkins, N. A. and Nagata, S. (1992) Lymphoproliferation disorder in mice explained by defects in Fas antigen that mediates apoptosis. *Nature*, 356, 314–317.

Yonehara, S., Ishii, A. and Yonehara, M. (1989) A cell-killing monoclonal antibody (anti-Fas) to a cell surface antigen co-down regulated with the receptor of tumor necrosis factor. *Journal of Experimental Medicine*, 169, 1747–1756.

Chapter 15

BCL-2 EXPRESSION IN THE PROSTATE AND ITS ASSOCIATION WITH ANDROGEN-INDEPENDENT PROSTATE CANCER

Timothy J. McDonnell[1], Patricia Troncoso[1], Shawn Brisbay[1], Leland Chung[2], Christapher Logothetis[3], Jer-Tsong Hsieh[2], Martin Cambell[1], and Shi-Ming Tu[3]

Departments of [1]Molecular Pathology, [2]Urology, and [3]Medical Oncology, The University of Texas, M.D. Anderson Cancer Center, Houston, Texas 77030, USA

Prostate cancer is now the most commonly diagnosed malignancy in American men, surpassing even carcinoma of the lung. Most patients have advanced disease at the time of diagnosis and are, therefore, not considered candidates for potentially curative surgical resection. The most effective frontline therapy for these patients currently is androgen ablation. The initial biologic consequence of this therapy is to induce programmed cell death in the responsive prostatic cancer cells (Kyprianou et al., 1990). It is unfortunate that virtually all patients will experience rapidly fatal, androgen-independent tumour recurrences for which there remains no effective therapy. Tumor recurrence is thought to represent the clonal selection of pre-existing, androgen-independent cells (Sinha et al., 1977; Kyprianou et al., 1990; Roehrborn et al., 1991; Van Weerden, 1991). Numerous investigators have established experimentally that this hormonal ablation results in the death of over 80% of the prostatic epithelial cells by the derepression of an endogenous programmed cell death cascade (Isaacs, 1984; English et al., 1985, 1989; Kyprianou and Isaacs, 1988; Kyprianou et al., 1990). Activation of this cascade has been shown to be associated with the enhanced expression of specific genes. The upregulation of genes, such as TGF-β and TRPM-2, has been shown to immediately precede the onset of programmed cell death in the prostate and has been implicated in the mediation of this process (Montpetit et al., 1986; Buttyan et al., 1989; Kyprianou and Isaacs, 1989; Kyprianou et al., 1990). The molecular basis for the emergence of androgen-independent prostate cancer is, as yet, poorly understood. It is reasonable to assume that understanding the molecular regulation of androgen-dependent programmed cell death in the prostate will provide insights into the basis for the development of androgen-independent prostate cancer as well as suggest additional approaches for therapeutic intervention.

The ability of oncogenes to influence sensitivity to programmed cell death is receiving more widespread attention. At present, bcl-2 remains the only oncogene characterized that is able to extend cellular viability by overriding normal endogenous programmed cell death mechanisms (Hockenbery et al., 1990). bcl-2 was originally identified secondary to its association with the t(14;18) translocation observed in most follicular lymphomas. A prospective study of the in vivo consequence of bcl-2 gene deregulation was initially provided in a transgenic mouse model utilizing minigene constructs which recreated the molecular features of the t(14;18) (McDonnell et al., 1989, 1990). Evidence that the enhanced viability imparted by the deregulated bcl-2 transgene is ultimately oncogenic has recently been provided (McDonnell and Korsmeyer, 1991). With the availability of anti-Bcl-2 antibodies it has recently been shown that Bcl-2 is normally expressed in CNS neurons, basal epithelial cells of the thyroid, breast, and prostate as well as epithelial cells of the intestinal crypts and the basal keratinocytes of the skin (Hockenbery et al., 1991). This interesting cytoarchitectural distribution suggests that Bcl-2 is confined to cells normally exhibiting extended viability or in the proliferative compartment of tissues whose self-renewal is characterized by the process of programmed cell death.

We undertook a detailed immunohistochemical analysis of the distribution of Bcl-2 protein within the human prostate gland. Formalin-fixed, paraffin-embedded prostate tissue was obtained from the pathology files of the Department of Pathology at the University of Texas, M.D. Anderson Cancer Center in Houston. Sections were deparaffinized and rehydrated before staining by the alkaline-phosphatase-anti-alkaline-phosphatase (APAAP) technique (Cordell et al., 1984). The primary anti-Bcl-2 antibody (Pezzella et al., 1990), MoAb 100, was generously supplied by Dr David Mason (John Radcliffe Hospital, Oxford, UK). The basal cells of the normal prostatic glandular epithelium uniformly exhibited strong positivity when stained for the presence of Bcl-2 protein (Figure 15.1). The secretory cell component of the prostatic glandular epithelium, in contrast, was virtually uniformly negative. Infiltrating lymphocytes and peripheral nerves in the prostate and periprostatic tissue also stained positively for Bcl-2 protein. Strong Bcl-2 staining was also observed in ejaculatory ducts and seminal vesicles. Interestingly, prostatic glands showing morphologic evidence of atrophy or transitional metaplasia were strongly positive (i.e. 3+).

The molecular events associated with the conversion from an androgen-dependent to an androgen-independent state are not well characterized. We examined, using immunohistochemical techniques, a total of 31 cases of prostatic carcinoma, consisting of 19 cases of androgen-dependent and 12 cases of androgen-independent

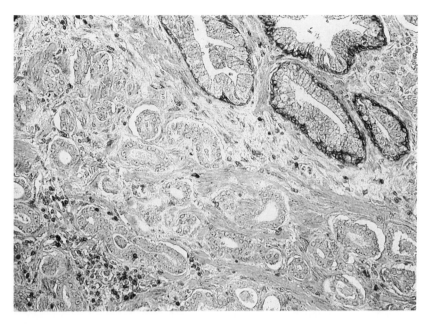

Figure 15.1 Section of prostate gland immunohisto-chemically stained for Bcl-2 protein. Normal staining of the basal epithelial cells is observed in the intact glands (upper right). With the exception of infiltrating lymphocytes, no staining for Bcl-2 is observed in the androgen-dependent adenocarcinoma (lower left). 200 × mag.

(13 samples, two samples derived from the same patient) cancer. The androgen-independent carcinomas consisted of eight adenocarcinomas (derived from seven patients) and five small cell carcinomas of the prostate. Heterogeneous expression of *bcl*-2 was observed within androgen-dependent prostate cancers with the majority of cases exhibiting undetectable levels of Bcl-2 protein (Figure 15.1). It is noteworthy, however, that three cases of androgen-dependent cancer showed diffuse 1+ to 2+ staining for Bcl-2 protein and three cases demonstrated heterogeneous staining with approximately 50% of the tumour negative and 50% showing 2+ to 3+ positivity. We are now expanding the patient population so that the potential prognostic implications of these findings may be evaluated.

In contrast, androgen-independent prostate cancers, with the exception of three bone marrow metastases obtained from black patients, possessed detectable levels of Bcl-2 protein that, characteristically, was present at high levels and diffusely distributed throughout the tumour. These data suggest that the expression of Bcl-2 protein in carcinomas of the prostate is not a primary molecular event but rather is a secondary event associated with progression in tumour recurrences following androgen ablation. Interestingly, small cell carcinomas of the prostate also expressed high levels of Bcl-2 protein comparable to that seen in androgen independent tumour recurrences (Figure 15.2). These tumours have been shown to be unresponsive to

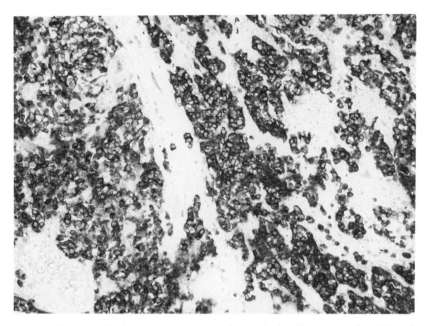

Figure 15.2 Section of prostate gland immunohisto-chemically stained for Bcl-2 showing infiltration of small cell carcinoma. Note the diffuse intense staining pattern of this androgen-independent tumour. 200 × mag.

androgen ablation (Moore et al., 1992) and exhibit a rapidly fatal clinical course. It remains to be determined whether *bcl*-2 gene deregulation in these tumours represents a primary pathogenic event.

A summary of the correlation of Bcl-2 staining with the known androgen sensitivity of the prostate cancers examined is provided in Table 15.1. The association of Bcl-2 positivity with an androgen independent phenotype was determined to be highly significant ($p<0.01$) using the Chi square test.

Table 15.1 Correlation of *bcl*-2 staining and androgen sensitivity. The results of Bcl-2 staining are related to the hormonal status of the tumour. The difference in staining pattern between androgen-dependent and androgen-independent carcinomas was statistically significant ($p<0.01$) using the chi square test.

Bcl-2 Staining	Androgen-dependent	Androgen-independent
Positive	6/19 (32%)	10/13 (77%)
Negative	13/19 (68%)	3/13 (23%)

Evidence that, in fact, *bcl*-2 expression could be directly augmented by androgen ablation was obtained using a rat experimental system. Male Sprague-Dawley rats were surgically castrated through a scrotal approach. Sham-operated animals served as controls. Animals were sacrificed at intervals of one, five, and 10 days post-castration and total cellular RNA was purified from the ventral prostate. Steady state levels of *bcl*-2 transcript were determined using S1 nuclease protection assays and a murine exon 2 *bcl*-2 probe end-labelled on the antisense strand. In contrast to other apoptosis associated genes such as TRPM-2 and TGF-b, maximum levels of *bcl*-2 mRNA were not observed until day 10, in other words, well after the peak levels of apoptosis known to occur three days following castration (Kyprianou and Isaacs, 1988). In order to determine whether this increase in steady state levels of *bcl*-2 mRNA was a consequence of depletion of serum testosterone a second cohort of rats received injections of either testosterone propionate (TP) or TP plus the antiandrogen 4-hydroxyflutamide daily for a total of 4 days following castration. The observed increase in *bcl*-2 transcripts in castrated rats was abrogated in those rats receiving injections of testosterone propionate (Figure 15.3). An increase in steady-state levels of *bcl*-2 mRNA was also observed in those animals receiving flutamide in addition to testosterone following castration. C3 and TRPM-2 probes, representing genes known to be down-regulated and up-regulated, respectively, in response to androgen withdrawal, were used for internal standards of comparison. These results are consistent with the concept that *bcl*-2 gene expression in the prostate is augmented as a direct consequence of androgen ablation.

It has previously been shown that the basal epithelial cells, in contrast to the secretory cells, are resistant to programmed cell death induced by androgen withdrawal (English *et al.*, 1989) so that our data indicates that the increase in the apparent steady-state level of *bcl*-2 mRNA is a consequence of the selection for cells resistant to apoptosis induced by androgen withdrawal and that this resistance may be a direct consequence of *bcl*-2 expression.

In summary, our findings indicate that expression of the *bcl*-2 proto-oncogene is associated with the progression of prostate cancer from androgen-dependence to androgen-independence. To our knowledge this is the first study documenting a role for *bcl*-2 in the progression of an epithelial malignancy. Follow-up experiments are in progress will determine if *bcl*-2 expression is in itself sufficient to confer an androgen-independent phenotype.

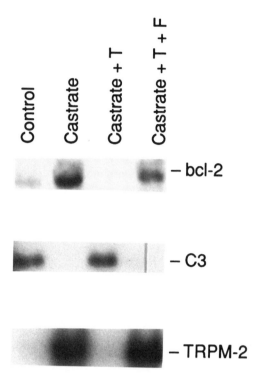

Figure 15.3 S1 nuclease protection assay showing an increase in steady-state levels of *bcl*-2 mRNA in the rat ventral prostate following castration. Animals were castrated transcrotally. Sham-operated animals served as controls. Animals were given daily injections of testosterone propionate (T, 500 mg, I. M.) or T plus the antiandrogen 4-hydroxyflutamide (F, 5 mg) for a total of four days. RNA was also probed for C3 and TRPM-2 transcripts, representing genes that are known to be negatively and positively modulated in response to androgen withdrawal, respectively.

REFERENCES

Bodley, J. W. (1990) Does diphtheria toxin have nuclease activity? *Science*, **250**, 832.

Buttyan, R., Olsson, C. A., Pintar, J., Chang, C., Bandyk, M. and Ng, P.-Y. (1989) Induction of the TRPM-2 gene in cells undergoing programmed death. *Molecular and Cellular Biology*, **9**, 3473–3481.

Chang, M. P., Baldwin, R. L., Bruce, C. and Wisnieski, B. J. (1989) Second cytotoxic pathway of diphtheria toxin suggested by nuclease activity. *Science*, **246**, 1165–1168.

Cordell, J. L., Falin, B., Erber, W. N., Gosh, A. K., Abdulaziz, Z. and Pulford, K. A. F. (1984) Immunoenzymatic labelling of monoclonal antibodies using immune complexes of alkaline phosphatase and monoclonal anti-alkaline phosphatase (APAAP complexes). *Journal of Histochemistry and Cytochemistry*, **32**, 219–229.

English, H. F., Drago, J. R. and Santen, R. J. (1985) Cellular response to androgen depletion and repletion in the rat ventral prostate: Autoradiography and morphometric analysis. *Prostate,* 7, 41–51.

English, H. F., Kyprianou, N. and Isaacs, J. T. (1989) Relationship between DNA fragmentation and apoptosis in the programmed cell death in the rat prostate following castration. *Prostate,* 15, 233–50.

Hockenbery, D., Nunez, G., Milliman, C., Schreiber, R. D. and Korsmeyer, S. J. (1990) Bcl-2 is an inner mitochondrial membrane protein that blocks programmed cell death. *Nature,* 348, 334–36.

Hockenbery, D. M., Zutter, M., Hickey, W., Nahm, M. and Korsmeyer, S. J. (1991) Bcl-2 protein is topographically restricted in tissues characterized by apoptotic cell death. *Proceedings of the National Academy of Science, USA,* 88, 6961–6965.

Isaacs, J. T. (1984) Antagonistic effect of androgen on prostatic cell death. *Prostate,* 5, 545–557.

Kyprianou, N. and Isaacs, J. T. (1988) Activation of programmed cell death in the rat ventral prostate after castration. *Endocrinology,* 122, 552–562.

Kyprianou, N. and Isaacs, J. T. (1989) Expression of transforming growth factor-b in the rat ventral prostate during castration-induced programmed cell death. *Molecular Endocrinology,* 3, 1515–1522.

Kyprianou, N., English, H.F. and Isaacs, J. T. (1990) Programmed cell death during regression of PC-82 human prostate cancer following androgen ablation. *Cancer Research,* 50, 3748–3753.

McDonnell, T. J., Deane, N., Platt, F. M., Nunez, G., Jaeger, U., McKearn, J. P. and Korsmeyer, S. J. (1989) bcl-2-immunoglobulin transgenic mice demonstrate extended B cell survival and follicular lymphoproliferation. *Cell,* 57, 79–88.

McDonnell, T. J. and Korsmeyer, S. J. (1991) Progression from lymphoid hyperplasia to high-grade malignant lymphoma in mice transgenic for the t(14;18). *Nature,* 349, 254–256.

McDonnell, T. J., Nunez, G., Platt, F. M., Hockenbery, D., London, L., McKearn, J. P. and Korsmeyer, S. J. (1990) Deregulated bcl-2-immunoglobulin transgene expands a resting but responsive immunoglobulin M and D-expressing B-cell population. *Molecular and Cellular Biology,* 10, 1901–1907.

Montpetit, M., Lawless, K. R. and Tenniswood, M. (1986) Androgen-repressed messages in the rat ventral prostate. *Prostate,* 8, 25–36.

Moore, S. R., Reinberg, Y. and Zhang, G. (1992) Small cell carcinoma of prostate: Effectiveness of hormonal versus chemotherapy. *Urology,* 34, 411–416.

Pezzella, F., Tse, A. G. D., Cordell, J. L., Pulford, K. A. F., Gatter, K. C. and Mason, D. Y. (1990) Expression of the Bcl-2 oncogene protein is not specific for the 14;18 chromosomal translocation. *American Journal of Pathology,* 137, 225–232.

Roehrborn, C. G. (1991) Oncogenes and growth factors in prostatic carcinoma. In: *Mechanisms of Progression to Hormone-independent Growth of Breast and Prostatic Cancer,* P. M. J. J. Berns, J. C. Romijn, F. H. Schroder (eds), pp. 97–122. Park Ridge: The Parthenon Publishing Group.

Sinha, A. A, Blackard, C. E. and Seal, U. S. (1977) A critical analysis of tumor morphology and hormone treatments in the untreated and estrogen-treated responsive and refractory human prostatic carcinoma. *Cancer,* 40, 2836–2850.

Van Weerden, W. M. (1991) Animal models in the study of progression of prostate and breast cancer to endocrine independency. In: *Mechanisms of Progression to Hormone-independent Growth of Breast and Prostatic Cancer,* P. M. J. J. Berns, J. C. Romijn, F. H. Schroder (eds), pp. 55–70. Park Ridge: The Parthenon-Publishing Group.

Chapter 16

THE ROLE OF TUMOUR SUPPRESSOR GENES IN APOPTOSIS

Elisheva Yonish-Rouach[1,2], Doron Ginsberg*[1]
and Moshe Oren[1]

[1]Department of Chemical Immunology,
The Weizmann Institute of Science, Rehovot 76100, Israel
[2]Institut de Recherches Scientifiques sur le Cancer, 94891 Villejuif Cedex, France
*Present address: Dana-Farber Cancer Institute, Boston, MA 02115, USA

Cancer is a multistep process, usually involving sequential changes in the cellular genotype and phenotype. One type of change which can potentially promote tumourigenesis is the abrogation of the normal cell death program (Umansky, 1982; Arends and Wyllie, 1991; Williams, 1991; Williams et al., 1992). Coupled with the acquisition of an ability to maintain unrestrained cell proliferation, illegitimate cell survival will conceivably give rise to highly malignant progeny. In fact, tumour progression is often accompanied by a reduced propensity to undergo cell death, as exemplified by the paradigms of rat hepatocellular carcinoma (Schulte-Hermann et al., 1990) and chicken bursal lymphoma (Neiman et al., 1991).

The phenotypic changes seen in tumour cells result from permanent alterations in an array of cancer-related genes. Such genes characteristically fall into either one of two categories: oncogenes and tumour suppressor genes (Bourne and Varmus, 1992, and references therein). In the broad sense, any gene whose loss of function can contribute to the development of cancer may be regarded as a tumour suppressor gene. These genes are believed to exert their effects primarily through the inhibition of cell proliferation, very much in analogy with the ability of activated oncogenes to promote neoplasia by stimulating aberrant cell proliferation. There is now extensive evidence for the importance of the antiproliferative activities of tumour suppressors (Weinberg, 1991; Massague and Weinberg, 1992). Yet, since oncogenes such as bcl-2, ras and c-myc can have pronounced effects on cell viability (this volume, Chapters 13–15), it is conceivable that tumour suppressors may also be involved in the control of apoptosis. Along this line of reasoning, if the proper activity of a protein encoded by a tumour suppressor gene is required for the induction of physiological cell death, then the inactivation of such a gene will result in illegitimate cell survival. It is thus plausible that increased cell viability, often seen during tumour progression, may result not only from the activation of oncogenes but also from the inactivation of tumour suppressor genes. Recent studies with the p53 tumour suppressor gene are consistent with this idea.

In the last several years, p53 has emerged as one the most actively investigated tumour suppressor proteins. In fact, alterations in the p53 gene, often in the form of missense mutations giving rise to full-length mutant proteins, are perhaps the most frequent specific genetic alterations in human cancer (reviewed in Hollstein et al., 1991; Rotter and Prokocimer, 1991; Oren, 1991; Tominaga et al., 1992). Hence, the ongoing expression of functional wild type (wt) p53 protein may serve as an effective barrier against the progression of neoplastic processes. Detailed investigation of the properties of p53, as well as of the differences between the wt protein and the various mutants found in tumour cells, has recently provided much new insight into the molecular basis of its action (reviewed in Levine et al., 1991; Montenarh, 1992; Oren, 1992; Vogelstein and Kinzler, 1992). It now appears that p53 is a nuclear protein capable of engaging in sequence-specific DNA binding. In addition, the protein possesses a potent transactivation domain. In fact, the binding of intact p53 to cognate DNA sequences can lead to pronounced transcriptional activation, when these DNA elements are placed within the context of a transcriptional control region. These properties suggest that p53 may be a *bona fide* transcription factor. To this date, however, the genes which serve as physiologically relevant targets for regulation by p53 still remain to be identified. Alternatively, the sequence-specific DNA binding of p53 may pertain to its putative role in the control of DNA synthesis, including the recently proposed function of p53 in orchestrating the response to DNA damage (Kastan et al., 1991,1992; Lane 1992; Oren 1992). In either case, it is noteworthy that sequence-specific DNA binding is an exclusive feature of wt p53, which is lost in practically all tumour-derived variants tested thus far (El-Deiry et al., 1992; Farmer 1992; Kern 1992). This observation strongly supports the notion that DNA binding is crucial for the tumour suppressor activity of p53.

Attempts to elucidate the biological basis for the ability of p53 to act as a tumour suppressor have mostly focused on its antiproliferative effects. Thus, it was shown that the overexpression of wt p53 in transformed cells caused a growth arrest. Typically, the affected cells accumulated in the G1 phase of the cell cycle, giving rise to the suggestion that p53 can inhibit the transition from G1 into S (reviewed in Levine et al., 1991; Montenarh, 1992; Oren, 1992; Vogelstein and Kinzler, 1992). Once again, tumour-derived p53 mutants were completely ineffective in eliciting such a growth arrest. The ability of wt p53 to block cell proliferation is likely to account, at least in part, for its tumour suppressor potential. Yet, this is not the only documented biological activity of wt p53. In fact, the induced expression of wt p53 in transformed cells has now also been shown to promote differentiation (Shaulsky et al. 1991a, 1991b) as well as apoptosis, in a manner dependent on the nature of the target cells.

Data suggesting a direct link between wt p53 activity and apoptosis were first obtained with cells of the mouse myeloid leukaemic line M1 (clone S6). These cells are completely devoid of p53, as determined at both protein and mRNA levels (Yonish-Rouach et al., 1991). The total elimination of p53 synthesis suggested the existence of a strong selective pressure against the continuous expression of the protein. It was therefore of great interest to identify processes whose execution was presumably prevented upon ablation of p53 expression. The most logical approach was to reconstitute wt p53 expression in M1 clone S6 cells, and monitor the consequences. However, like in a number of other cell systems (Hinds et al., 1989; Johnson et al., 1991), it was impossible to establish M1-derived clones constitutively producing transfected wt p53 (E.Y.-R., unpublished). This difficulty was eventually overcome through the use of a temperature-sensitive (ts) mutant of mouse p53 (Michalovitz et al., 1990). This mutant, p53val135, possesses properties indistinguishable from those of other p53 mutants as long as the cells are maintained at temperatures above 37°C. However, upon shifting the temperature down to 32.5°C, the protein acquires molecular properties and biological activities similar to those of authentic wt p53 (Michalovitz et al. 1990; Milner and Medcalf, 1990). In rodent fibroblasts, this is reflected in the ability of p53val135 to suppress oncogene-mediated neoplastic transformation, as well as to induce a reversible growth arrest (Michalovitz et al., 1990). The growth arrest appears to be caused primarily by the inhibition of transition through the late G1 phase of the cell cycle or from G1 into S (Michalovitz et al., 1990; Gannon and Lane, 1991; Martinez et al., 1991).

DNA encoding p53val135 was transfected into M1 cells, and clones stably expressing the corresponding protein were obtained at 37.5°C (Yonish-Rouach et al., 1991). Upon shift-down to 32.5°C, the protein indeed adopted the characteristics of wt p53. This was manifested, for instance, by a dramatic change in its intracellular distribution. Whereas much of the p53 protein was present in the cytoplasm at 37.5°C, it rapidly accumulated in the nucleus when the cells were transferred to 32.5°C (Figure 16.1). In this respect, the picture was very similar to that seen in fibroblasts transfected with the same ts p53 plasmids (Gannon and Lane, 1991; Ginsberg et al., 1991; Martinez et al., 1991). However, unlike these fibroblasts, the leukaemic cells did not end up in a viable growth arrested state. Rather, they underwent a rapid loss of viability. Most of the cells died within 24–36 hours after temperature down-shift, as concluded from their inability to exclude trypan blue (Yonish-Rouach et al., 1991). The process exhibited distinctive apoptotic features, such as extensive cell shrinkage and chromatin condensation (Figure 16.2a), as well as internucleosomal fragmentation of genomic DNA, giving rise to a characteristic "DNA ladder" (Figure 16.2b). In addition, cell death was at least partially retarded when cycloheximide was used to block *de novo* protein synthesis (Yonish-Rouach et al., 1992). Thus, in p53-negative leukaemic M1 cells, the induction of wt p53 activity gave rise to an apparently apoptotic response.

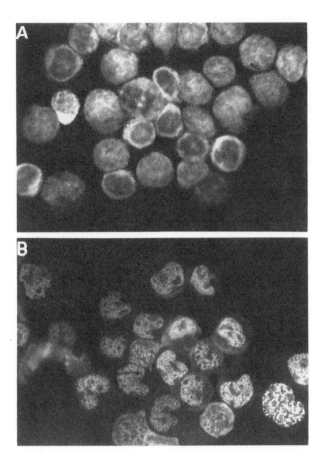

Figure 16.1 Activated p53 is translocated into the nucleus of M1-derived transfectant clones. Cells of clone LTR-6, derived by transfection of M1 myeloid leukaemia cells with the temperature-sensitive p53 mutant p53val135, were either maintained at 37.5°C (A), or transferred to 32.5°C for 4 hours before fixation (B). Cells were fixed with methanol/acetone and subjected to immunofluorescent staining with the p53-specific monoclonal antibody PAb421.

This phenomenon could either reflect a direct activity of wt p53, or an indirect consequence of its cellular effects. In particular, it could be argued that the observed cell death is actually a secondary response to a growth arrest imposed on the cells by wt p53 overexpression. This is a very plausible model, given that wt p53 has been shown to block cell cycle progression in G1 in many other systems, and that a G1 arrest precedes death when myeloid cells are deprived of survival factors (Sabourin and Hawley, 1990; Askew et al., 1991; Matsushime et al., 1991). It was therefore important to establish the relationship between cell cycle progression and apoptosis in the M1-derived p53 clones. To that end, a FACS analysis was performed on such cells

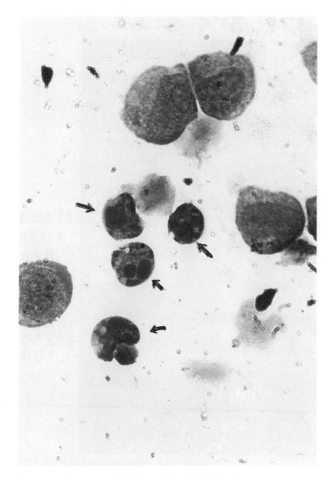

Figure 16.2A Apoptotic features of p53-mediated cell death.
Cells of clone LTR-13, derived by transfection of M1 myeloid leukaemia cells with the temperature-sensitive p53 mutant p53val135, were maintained at 32.5°C for 12 hours, and then fixed and stained with May-Grunwald-Giemsa stain. Cells exhibiting a typical apoptotic morphology are indicated by arrows.

following their shift down to 32.5°C. The results (Figure 16.3) failed to reveal any evidence for a G1 arrest. Rather, the cells appeared to proceed through the cell cycle even when a substantial fraction of them were already becoming apoptotic. The lack of a detectable G1 arrest was also established more directly, using centrifugal elutriation (Yonish-Rouach et al., 1993). These experiments demonstrated that the cells could move out of G1 and into S before losing viability. The kinetics of the process suggested that the the commitment to cell death occurred primarily in G1, whereas the actual death appeared to take place preferentially during the S phase. Thus, there does not seem to be a requirement for a prior G1 arrest in order for wt p53 to induce

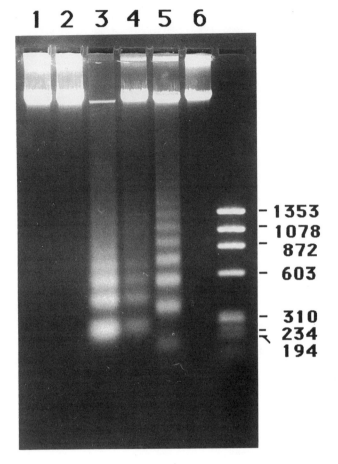

Figure 16.2B Apoptotic features of p53-mediated cell death. Cells of clones LTR-6 and SV-1, derived by transfection of M1 myeloid leukaemia cells with the temperature-sensitive p53 mutant p53val135, were either maintained at 37.5°C or shifted down to 32.5°C for 20 hours. Total cellular DNA was extracted and subjected to agarose gel electrophoresis. Lanes: 1, parental M1 cells, 32.5°C; 2, LTR-13, 37.5°C; 3, LTR-13, 32.5°C; 4. LTR-13, 32.5°C in the presence of IL-6; 5, SV-1, 32.5°C; 6, LTR-Phe, derived by transfection of M1 with the non-ts mutant p53phe132, 32.5°C. (Yonish-Rouach et al., 1991, reprinted with permission from Nature 352, 346. Copyright © 1991, Macmillan Magazines Ltd.)

apoptosis. These observations suggest that the activation of the death process in M1 cells is a direct consequence of the action of wt p53.

A distinctive feature of p53-mediated apoptosis in M1 cells is that it can be significantly inhibited by the addition of Interleukin-6 (IL-6) to the culture medium (Yonish-Rouach et al., 1991; see also Figure 16.2B, lanes 3 and 4). In M1 cells, IL-6 serves as a differentiation inducing agent (Sachs, 1990). Normal myeloid progenitors may be programmed to undergo apoptosis unless presented with appropriate signals for

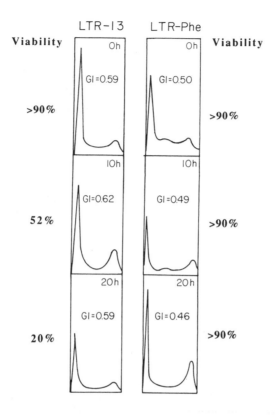

Figure 16.3 Cell cycle distribution of M1-derived transfectant clones. Cells of clone LTR-13 and clone LTR-Phe (see Figure 16.2) were shifted to 32.5°C, and samples were stained with propidium iodide and subjected to FACS analysis at the indicated times after down-shift. The fraction of cells with a G1 DNA content is indicated for each sample. Viability at each time point was established by assaying for trypan blue exclusion.

proliferation or differentiation. In this sense, agents capable of delivering such signals will act as survival factors. It is tempting to speculate that the expression of functional p53 may play a role in maintaining the dependence on such survival factors. The loss of wt p53 expression may thus have made the progenitors of the M1 line capable of survival in the absence of these factors, thereby bypassing physiological "execution" mechanisms. Consequently, the reactivation of p53 may now make these cells dependent again on survival factors, driving them into apoptosis unless such factors are provided.

While IL-6 has previously been shown to act as a potent survival factor for cells of the B lymphocytic lineage, it has not been implicated in the survival of normal myeloid progenitors. On the other hand, there is ample evidence suggesting that IL-3 can act as a survival factor for these cells (see Williams *et al.*, 1992). When applied to M1 cells expressing wt p53 activity, IL-3 could not prevent cell death (data not shown). This was not surprising, given that M1 cells do not express IL-3 receptors (Lotem and Sachs, 1989). However, when such transfected M1 cells were exposed for 3 days to IL-6, they could now be rescued from apoptosis not only by IL-6 but also by IL-3 (Yonish-Rouach et al., 1993). This most probably reflects a need for IL-6 in order to induce the expression of IL-3 receptors, as demonstrated previously for a different subclone of M1 (Lotem and Sachs, 1989). Hence, IL-3 can also serve as an effective survival factor in M1 cells subjected to wt p53 activity, provided that the cognate receptors are present.

The ability of wt p53 to promote apoptosis has also been demonstrated in a very different system (Shaw et al., 1992). Cells of line EB, derived from a human colon cancer, also fail to produce any detectable p53. As in the case of the mouse myeloid M1 line, this suggested a powerful selection against the ongoing expression of wt p53. EB cells were therefore transfected with an inducible p53 gene, this time encoding authentic human wt p53 under the transcriptional control of the metallothionein promoter. When wt p53 expression was induced by the addition of zinc ions, cultured cells detached from the monolayer within a few days. Analysis of the detached cells revealed typical apoptotic morphological features, as well as internucleosomal DNA fragmentation. Furthermore, apoptosis could be induced by wt p53 *in vivo*. Nude mice were injected with EB cells carrying the inducible p53 transgene, and tumours were allowed to develop. Upon induction of wt p53 synthesis (by injecting the mice with zinc chloride) the tumours gradually regressed, while exhibiting morphological evidence for apoptosis.

Another interesting system in which a link has now been indicated between p53 and apoptosis is human Burkitt lymphoma (BL). A hallmark of BL is the deregulation of c-*myc* expression, resulting from a chromosomal translocation event (Magrath, 1990). Like other cells overexpressing deregulated *myc* (Askew et al., 1991; Evan et al., 1992; also this volume, Chapter 13), BL cells undergo apoptosis when deprived of essential serum growth factors (Henderson et al., 1991). This response is pronounced in Epstein-Barr virus (EBV)-negative BL lines, but not in EBV-positive ones, where the presence of the viral LMP-1 protein appears to suppress apoptosis through the induction of *bcl-2* expression (Henderson et al., 1991). Interestingly, many BL lines carry p53 gene mutations (Wiman et al., 1991). This indicates that in BL, too, the sustained expression of wt p53 is disadvantageous for the progression of the disease. The functional significance of p53 gene mutations in BL was probed by introducing the ts mutant p53val135 into one such EBV-negative line (BL41), which expresses mutant p53. Three positive transfectants were tested with respect to the effect of wt p53 activation, induced by a shift-down to 32.5°C. In all three clones, wt p53 activity was found to give rise to apoptosis even in the presence of high serum concentrations; apoptosis was defined on the basis of both cell morphology and DNA fragmentation (K. Wiman, personal communication). Similar to the case of transfected M1 cells, apoptosis first became apparent within 10 hours after transfer to 32°C, and cell death was practically complete within 2–3 days. In addition to extending the connection between p53 and the control of cell death, these facts strengthen the notion that coordinate expression of activated p53 and deregulated c-*myc* is often incompatible with cell survival.

The results of the above transfection experiments, which suggest a cause-and-effect relationship between wt p53 activation and apoptosis, are also consistent with observations made in cells induced to undergo apoptosis upon abrogation of a physiological hormonal stimulus. In a mouse model system, abrupt termination of lactation results in involution of the mammary glands. The process has distinct apoptotic features, including the induction of SGP-2, a gene associated with apoptosis in other cell types (Strange et al., 1992). It was found that during involution there is a marked and relatively rapid increase in p53 mRNA levels (Strange et al., 1992). The p53 transcript levels stay elevated for a number of days, while gradual cell loss is in progress.

A similar situation was also found in hormone-depleted rat prostate cells. Following the castration of male rats, glandular epithelial cells in the prostate undergo extensive apoptosis (Colombel et al., 1992). Within the first 24 hours after castration, there is a several-fold increase in the steady state levels of p53 mRNA and protein, peaking approximately a week later. Despite the enhanced p53 expression, the cells actually proceed through the cell cycle; like in M1 transfectants (see above), replicative DNA synthesis practically continues until the onset of extensive chromosomal DNA fragmentation (Colombel et al., 1992). Interestingly, c-myc mRNA also increases in abundance, albeit with somewhat slower kinetics. Moreover, when androgens are readministered to long-term castrated rats, thereby inducing prostate epithelial proliferation rather than apoptosis, p53 expression is drastically reduced (R. Buttyan, personal communication). These observations are in line with the possibility that p53 may participate in orchestrating the apoptotic response *in vivo* under physiological conditions.

That wt p53 may be required in order to promote cell death, at least under certain circumstances, is also suggested by studies involving REF52 cells (Hicks et al., 1991). In this fibroblastic cell line, overexpression of activated mutant *ras* eventually leads to cell death. The lethal effect of activated *ras* can be abrogated by SV40 large T antigen, as well as by mutants of p53. Both these types of proteins can interfere with the normal functions of wt p53, as reflected for instance in their ability to block sequence-specific DNA binding by wt p53 (Farmer et al., 1992; Kern et al., 1992; Shaulian et al., 1992). Hence, it is plausible that in this particular cell system, too, wt p53 may contribute to the induction of apoptosis elicited by inappropriate levels of deregulated *ras* activity. As discussed earlier, overexpression of p53 in transformed cells often leads only to a growth arrest, with no effect on cell viability. What, then, determines whether wt p53 activation will turn on the death program in a particular cell population? One possibility is that the outcome may depend on the overall balance of signals operating on these cells. It is conceivable that apoptosis is activated whenever a cell is subjected to mutually incompatible growth-regulatory signals. This may occur, for instance, when cells harboring a deregulated oncogene are placed in a growth restrictive environment. Whereas normal cells will withdraw from the cell cycle, those carrying the activated oncogene may not be able to do so. The persistence of these deregulated cells will be most undesirable for the organism, and a possible function of the apoptotic pathway may therefore be to eliminate such cells (Umansky, 1982). Support for this notion comes from studies with deregulated c-*myc* expression. In two very different systems, namely myeloid cells and fibroblasts, it has now been shown that deregulated c-*myc* will drive the cells into rapid apoptosis under conditions which otherwise would have led to a growth arrest (Askew *et al.*, 1991; Evan *et al.*, 1992). In the case of the myeloid cells, this is achieved through the withdrawal of the survival factor IL-3, while in fibroblasts the same effect is seen upon serum deprivation.

One could speculate that M1 cells are also constitutively subjected to positive growth signals, responsible for their malignant phenotype. The introduction of activated wt p53 may bring in a potent antiproliferative signal, highly incompatible with these strong constitutive positive signals. The resultant signal imbalance may then, in turn, trigger an apoptotic response. Although this conjecture still has to be formally proven, two recent observations suggest that it may indeed be correct. First, the onset of cell death is significantly delayed when M1 cells are deprived of serum in parallel with the activation of wt p53 (Yonish-Rouach *et al.*, 1993). Furthermore, the protective action of IL-6 on M1 transfectants expressing wt p53 activity (see above) is accompanied by a very efficient G0/G1 arrest (Levy *et al.*, in press); the presence of IL-6 alone is insufficient for the induction of such an efficient arrest. It is tempting to speculate that serum withdrawal, as well as exposure to IL-6, complement the contribution of wt p53. Consequently, the balance of signals is shifted such that wt p53 activation is now more compatible with cell survival.

The issue of the factors determining whether wt p53 activation will cause apoptosis or a growth arrest or no phenotype at all can also be approached from a somewhat different angle. Recent studies have given rise to the attractive idea that wt p53 may be involved in protecting the cell from permanent DNA damage (Kastan *et al.*, 1991; Kuerbitz *et al.*, 1992; Lane, 1992). The suggested mechanism is that when the genomic DNA is damaged, p53 imposes a G1 arrest on the cell. This allows the damage to be repaired before the cell undergoes S phase replicative DNA synthesis, which would otherwise propagate and perpetuate the damage. In such a scenario, p53 may be either directly involved in sensing the DNA damage, or more probably in responding to it. In either case, it is conceivable that in the absence of DNA damage, cells will behave similarly whether or not they contain active wt p53. Such a contention is consistent with the finding that p53 "knock-out" mice, which make no p53 at all, develop apparently normally (Donehower *et al.*, 1992). One could therefore speculate that the consequences of p53 activation in a given cell will be determined by the status of its DNA, as well as by the efficacy of its DNA repair system (Figure 16.4). In a normal cell, which expresses normal basal levels of wt p53, the DNA is expected to be intact. Such cells should therefore not constitute a relevant biological target for wt p53, and the introduction of excessive wt p53 will not be predicted to lead to major phenotypic alterations. The same will presumably also hold true for transformed cells, as long as they still express active wt p53. Indeed, overexpression of wt p53 does not affect grossly cells which have still retained endogenous wt p53 synthesis (Eliyahu *et al.*, 1989; Baker *et al.*, 1990; Casey *et al.*, 1991). On the other hand, in transformed cells which either make only mutant p53 or fail to make p53 at all, one would expect

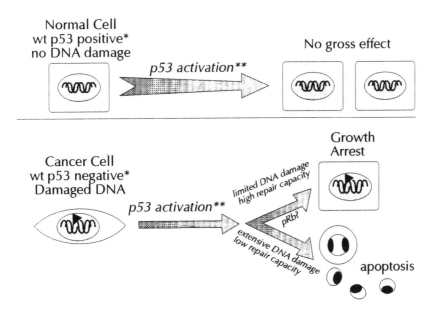

Figure 16.4 Possible relationship between p53, DNA damage and apoptosis. See text for details. *wild type p53 positivity implies the presence of physiological basal levels of wt p53 activity; negativity implies the lack of functional wt p53. **p53 activation may be achieved either by an absolute increase in the cellular levels of wt p53 or by specific biochemical activation of pre-existing protein. The figure also addresses the possibility that the presence of functional pRb may, under certain circumstances, be required for a proper G1 growth arrest. Under those circumstances, the loss of pRb function may deprive the cell of the growth arrest option and leave apoptosis as the only available response to DNA damage. Based in part on a model proposed by Lane (1992).

permanent DNA damage to have accumulated. When wt p53 activity is now restored to these cells, this is predicted to lead to a vigorous response. If the DNA lesions are limited and can be successfully handled by the cellular DNA repair mechanisms, the cell will probably undergo a G1 arrest. Alternatively, if the DNA damage is extensive, or if the cells do not harbor an effective DNA repair system, the apoptotic option may be chosen (Figure 16.4). This model is highly speculative, and may represent a gross oversimplification. Yet, it makes some testable predictions.

While there is a growing number of indications that wt p53 may play a role in the activation of the cell death pathway, no comparable observations have been reported for any other tumour suppressor. In order to gain some clues as to possible involvement of such genes in the control of cell viability, one may turn to DNA tumour viruses for help. DNA tumour viruses have been instrumental in setting the ground for some of the major recent breakthroughs in tumour suppressor research. These viruses target specific proteins in the infected cells and subvert their function in a manner that serves the purpose of successful viral replication. One aspect of this mechanism is the abrogation of the anti-proliferative effects of known and putative tumour suppressor proteins, such as p53, pRb, p107 and p300 (Doerfler, 1992). The purpose of such inactivation is presumably to permit uninterrupted entry into the S phase of the cell cycle, which is necessary for efficient viral DNA synthesis. In addition, however, it is also possible that some of the interactions between viral gene products and cellular proteins may impinge on the cell death pathway. It is conceivable that the gross distortion of intracellular homeostasis, brought about by the virus, may normally mark the cell for rapid apoptosis. If this occurs too fast, the viral life cycle may not be completed effectively. The ability to slow down the onset of apoptosis may thus be of great advantage to such viruses.

The notion that viruses may target the apoptotic machinery is supported by several lines of evidence. First, the oncogenic Epstein-Barr virus encodes the LMP-1 protein, which appears to induce the overexpression of *bcl*-2, thereby making the infected cells refractory to apoptosis (see above). In addition, the insect baculoviruses produce a protein which inhibits the apoptotic death of infected cells (Clem et al., 1991). One could therefore speculate that the inactivation of tumour suppressor proteins by DNA tumour viruses could also serve a similar purpose. However, while this may indeed be true for p53, this model seems to be very inappropriate where pRb and p300 are concerned. New insight into this issue has been gained through studies addressing the mechanism of transformation by adenoviruses. Indeed, these viruses also express a protein which suppresses apoptosis (White et al., 1992). However, this protein does not appear to inactivate any known tumour suppressor. Rather, in the process of lytic infection by the virus, it counteracts the apoptotic effects of the adenovirus E1A gene products. The latter, which play a major role in adenovirus-mediated transformation, actually activate apoptosis rather than suppressing it (White et al., 1991). Most interestingly, the domain of E1A which is responsible for the induction of cell death has been mapped to its N-terminal part, in a segment encompassing conserved region 1 (White et al., 1991). As conserved region 1 is essential for the ability of the E1A polypeptides to bind and presumably inactivate pRb and perhaps additional growth-restrictive gene products (Whyte et al., 1989), it would appear that the presence of these tumour suppressor proteins in a functional state may actually serve as a barrier to apoptosis.

How can one rationalize the seemingly paradoxical suggestion that certain tumour suppressor proteins may prevent apoptosis rather than facilitate it? The answer may have to do with the signal imbalance model discussed above. One of the predictions of this model is that apoptosis will occur whenever a cell can not undergo a proper growth arrest despite the presence of pertinent signals. Just as activated c-*myc* may induce apoptosis by depriving the cell of the growth arrest option, so could the inactivation of pRb have very similar consequences (see Figure 16.4). In fact, such conjecture will be consistent with the many reports suggesting an antagonism between the actions of Myc and pRb (e.g. Pietenpol et al., 1990; Rustgi, 1991). In more molecular terms, this could for instance imply that excess c-Myc protein may turn on particular genes whose expression plays a role in the apoptotic process, whereas pRb may suppress the same genes. In any event, it should be kept in mind that any model addressing the role of pRb and related tumour suppressor proteins in apoptosis is bound to be highly speculative and probably very inaccurate, owing to the paucity of pertinent data. It is conceivable, for instance, that when one looks into a larger variety of experimental systems, situations will be found where pRb actually contributes to cell death rather than seemingly preventing it. Furthermore, even in the case of the adenovirus E1A proteins, it still remains to be formally proven that the activity of pRb is indeed relevant to the apoptotic effects of the viral E1A proteins.

In conclusion, there are very good theoretical reasons to expect tumour suppressor genes to play a role in the control of cell death. Yet, the actual evidence in support of this notion is rather preliminary and inconclusive. However, in view of the extensive effort that is presently being invested in learning more about the biology and biochemistry of tumour suppressors, this situation is likely to change in the very near future.

REFERENCES

Arends, M. J. and Wyllie, A. H. (1991) Apoptosis: mechanisms and role in pathology. *International Review of Experimental Pathology*, 32, 223–256.

Askew, D. S., Ashmun, R. A., Simmons, B. C. and Cleveland, J. L. (1991) Constitutive c-myc expression in an IL-3-dependent myeloid cell line suppresses cell cycle arrest and accelerates apoptosis. *Oncogene*, 6, 1915–1922.

Baker, S. J., Markowitz, S., Fearon, E. R., Willson, J. K. V. and Vogelstein, B. (1990) Suppression of human colorectal carcinoma cell growth by wild-type p53. *Science*, 249, 912–915.

Bourne, H. R. and Varmus, H. E. (1992) Oncogenes and cell proliferation. *Current Opinion in Genetics and Development*, 2, 1–3.

Casey, G., Lo-Hsueh, M., Lopez, M., Vogelstein, B. and Stanbridge, E. (1991) Growth suppression of human breast cancer cells by the introduction of a wild-type p53 gene. *Oncogene*, 6, 1791–1798.

Clem, R. J., Fechheimer, M. and Miller, L. K. (1991) Prevention of apoptosis by a baculovirus gene during infection of insect cells. *Science*, 254, 1388–1390.

Colombel, M., Olson, C. A., Ng, P. Y. and Buttyan, R. (1992) Hormone-regulated apoptosis results from re-entry of differentiated prostate cells into a defective cell cycle. *Cancer Research*, 52, 4314–4319.

Doerfler, W. (1992) *Malignant transformation by DNA viruses*. Weinheim: Verlag Chemie Heidelberg.

Donehower, L. A., Harvey, M., Slagle, B. L., McArthur, M. J., Montgomery, C. A., Buttel, J. S. and Bradley, A. (1992) p53-deficient mice are developmentally normal but susceptible to spontaneous tumours. *Nature*, 356, 215–221.

El-Deiry, W. S., Kern, S. E., Pietenpol, J. A., Kinzler, K. W. and Vogelstein, B. (1992) Definition of a consensus binding site for p53. *Nature Genetics*, 1, 45–49.

Eliyahu, D., Michalovitz, D., Eliyahu, S., Pinhasi-Kimhi, O. and Oren, M. (1989) Wild-type p53 can inhibit oncogene-mediated focus formation. *Proceedings of the National Academy of Science, USA*, 86, 8763–8767.

Evan, G. E., Wyllie, A. H., Gilbert, C. S., Littlewood, T. D., Land, H., Brooks, M., Waters, C. M., Penn, L. Z. and Hancock, D. C. (1992) Induction of apoptosis in fibroblasts by c-Myc protein. *Cell*, 69, 119–128.

Farmer, G., Bargonetti, J., Zhu, H., Friedman, P., Prywes, R. and Prives, C. (1992) Wild-type p53 activates transcription *in vitro*. *Nature*, 358, 83–86.

Gannon, J. V. and Lane, D. P. (1991) Protein synthesis required to anchor a mutant p53 protein which is temperature-sensitive for nuclear transport. *Nature*, 349, 802–806.

Ginsberg, D., Michalovitz, D., Ginsberg, D. and Oren, M. (1991) Induction of growth arrest by temperature-sensitive p53 mutant is correlated with increased nuclear localization and decreased stability of the protein. *Molecular and Cellular Biology*, 11, 582–585.

Henderson, S., Rowe, M., Gregory, C., Croom-Carter, D., Wang, F., Longnecker, R., Kieff, E. and Rickinson, A. (1991) Induction of *bcl*-2 expression by Epstein-Barr virus latent membrane protein 1 protects infected B cells from programmed cell death. *Cell*, 65, 1107–1115.

Hicks, G. G., Egan, S. E., Greenberg, A. H. and Mowat, M. (1991) Mutant p53 tumour suppressor alleles release *ras*-induced cell cycle growth arrest. *Molecular and Cellular Biology*, 11, 1344–1352.

Hinds, P., Finlay, C., and Levine, A. J. (1989) Mutation is required to activate the p53 gene for cooperation with the *ras* oncogene and transformation. *Journal of Virology*, 63, 739–746.

Hollstein, M., Sidransky, D., Vogelstein, B. and Harris, C. C. (1991) p53 mutations in human cancers. *Science*, 253, 49–91.

Johnson, P., Gray, D., Mowat, M. and Benchimol, S. (1991) Expression of wild type p53 is not compatible with continued growth of p53-negative tumour cells. *Molecular and Cellular Biology*, 11, 1–11.

Kastan, M. B., Onyekwere, O., Sidransky, D., Vogelstein, B. and Craig, R. W. (1991) Participation of p53 protein in the cellular response to DNA damage. *Cancer Research*, 51, 6304–6311.

Kern, S., Pietenpol, J. A., Thiagalingam, S., Seymour, A., Kinzler, K. W. and Vogelstein, B. (1992) Oncogenic forms of p53 inhibit p53-regulated gene expression. *Science*, 256, 827–830.

Kuerbitz, S. J., Plunkett B. S., Walsh V. W. and Kastan M. B. (1992) Wild type p53 is a cell cycle checkpoint determinant following irradiation. *Proceedings of the National Academy of Science, USA*, 89, 7491–7495.

Lane, D. P. (1992) p53, guardian of the genome. *Nature*, 358, 15–16.

Levine, A. J., Momand, J. and Finlay, C. (1991) The p53 tumour suppressor gene. *Nature*, 351, 453–456.

Levy, N., Yonish-Rouach, E., Oren, M. and Kimchi, A. (In press). Complementation by wild type p53 of interleukin-6 effects on M1 cells: induction of cell cycle exit and relationship to c-myc suppression. *Molecular and Cellular Biology*.

Lotem, J. and Sachs, L. (1989) Induction of dependence on haematopoietic proteins for viability and receptor upregulation in differentiating myeloid leukaemic cells. *Blood*, 74, 579–585.

Magrath, I. (1990) The pathogenesis of Burkitt's lymphoma. *Advances in Cancer Research*, 55, 133–169.

Martinez, J., Georgoff, I., Martinez, J. and Levine, A. J. (1991) Cellular localization and cell cycle regulation by a temperature-sensitive p53 protein. *Genes Development*, 5, 151–159.

Massague, J. and Weiberg, R. A. (1992) Negative regulators of growth. *Current Opinion in Genetics and Development*, 2, 28–32.

Matsushime, H., Roussel, M. F., Ashmun, R. A. and Sherr, C. J. (1991) Colony-stimulating factor 1 regulates novel cyclins during the G1 phase of the cell cycle. *Cell*, 65, 701–713.

Michalovitz, D., Halevy, O. and Oren, M. (1990) Conditional inhibition of transformation and of cell proliferation by a temperature-sensitive mutant of p53. *Cell*, 62, 671–680.

Milner, J. and Medcalf, E. A. (1990) Temperature dependent switching between wild type and mutant forms of p53-val135. *Journal of Molecular Biology*, 216, 481–484.

Montenarh, M. (1992) Biochemical, immunological, and functional aspects of the growth-suppressor/oncoprotein p53. *Critical Reviews in Oncogenesis,* **3**, 233–256.

Neiman, P. E., Thomas, S. J. and Loring, G. (1991) Induction of apoptosis during normal and neoplastic B-cell development in the bursa of Fabricius. *Proceedings of the National Academy of Science, USA,* **88**, 5857–5861.

Oren, M. (1991) Role of p53 in neoplasia. In: *Biochemical and molecular aspects of selected cancers,* T. G. Pretlow and T. P. Pretlow (eds.), pp. 373–391. New York: Academic Press.

Oren, M. (In press) p53 — the ultimate tumour suppressor gene? *FASEB Journal.*

Pietenpol, J. A., Stein, R. W., Moran, E., Yaciuk, P., Schlegel, R., Lyons, R. M., Pittelkow, M. R., Munger, K., Howley, P. M. and Moses, H. L. (1990) TGF-β1 inhibition of c-*myc* transcription and growth in keratinocytes is abrogated by viral transforming proteins with pRb binding domains. *Cell,* **61**, 777–785.

Rotter, V. and Prokocimer, M. (1991) p53 and human malignancies. *Advances in Cancer Research,* **57**, 257–272.

Rustgi, A. K., Dyson, N. and Bernards, R. (1991) Amino-terminal domains of c-*myc* and N-*myc* proteins mediate binding to the retinoblastoma gene product. *Nature,* **352**, 541–544.

Sabourin, L. A. and Hawley, R. G. (1990) Suppression of programmed death and G1 arrest in B-cell hybridomas by Interleukin-6 is not accompanied by altered expression of immediate early response genes. *Journal of Cellular Physiology,* **145**, 564–574.

Sachs, L. (1990) The control of growth and differentiation in normal and leukemic blood cells. *Cancer,* **65**, 2196–2206.

Schulte-Hermann, R., Timmermann-Trosiener, I., Barthel, G. and Bursch, W. (1990) DNA synthesis, apoptosis, and phenotypic expression as determinants of growth of altered foci in rat liver during phenobarbital promotion. *Cancer Research,* **50**, 5127–5135.

Shaulian, E., Zauberman, A., Ginsberg, D. and Oren, M. (In press) Identification of a minimal transforming domain of p53: negative dominance through abrogation of sequence-specific DNA binding. *Molecular and Cellular Biology.*

Shaulsky, G., Goldfinger, N., Peled, A. and Rotter, V. (1991a) Involvement of wild-type p53 in pre-B cell differentiation *in vitro. Proceedings of the National Academy of Science, USA,* **88**, 8982–8986.

Shaulsky, G., Goldfinger, N. and Rotter, V. (1991b) Alterations in tumor development *in vivo* mediated by expression of wild type and mutant p53 proteins. *Cancer Research,* **51**, 5232–5237.

Shaw, P., Bovey, R., Tardy, S., Sahli, R., Sordat, B. and Costa, J. (1992) Induction of apoptosis by wild type p53 in a human colon tumor-derived cell line. *Proceedings of the National Academy of Science, USA,* **89**, 4495–4499.

Strange, R., Li, F., Saurer, S., Burkhardt, A. and Friis, R. R.(1992) Apoptotic cell death and tissue remodeling during mouse mammary gland involution. *Development,* **115**, 49–58.

Tominaga, O., Hamelin, R., Remvikos, Y., Salmon, R. J. and Thomas, G. (1992) p53 from basic research to clinical applications. *Critical Reviews in Oncogenesis,* **3**, 257–282.

Umansky, S. (1982) The genetic program of cell death. Hypothesis and some applications: transformation, carcinogenesis, ageing. *Journal of Theoretical Biology,* **97**, 591–602.

Vogelstein, B. and Kinzler, K. W. (1992) p53 function and dysfunction. *Cell,* **70**, 523–526.

Weinberg, R. A. (1991) Tumour suppressor genes. *Science,* **254**, 1138–1146.

White, E., Cipriani, R., Sabbatini, P. and Denton, A. (1991) Adenovirus E1b 19-kilodalton protein overcomes the cytotoxicity of E1A proteins. *Journal of Virology,* **65**, 2968–2978.

White, E., Sabbatini, P., Debbas, M., Wold, W. S., Kusher, D. I. and Gooding, L. R. (1992) The 19-kilodalton adenovirus E1b transforming protein inhibits programmed cell death and prevents cytolysis by tumor necrosis factor α. *Molecular and Cellular Biology,* **12**, 2570–2580.

Whyte, P., Williamson, N. M. and Harlow, E. (1989) Cellular targets for transformation by the adenovirus E1A proteins. *Cell*, **56**, 67–75.

Williams, G. T (1991) Programmed cell death: apoptosis and oncogenesis. *Cell*, **65**, 1097–1098.

Williams, G. T., Smith, C. A., McCarthy, N. J. and Grimes, E. A. (In press) Apoptosis: the final control point in cell biology. *Trends in Cell Biology*.

Wiman K. G., Magnusson K. P., Ramqvist G. and Klein G. (1991) Mutant p53 detected in a majority of Burkitt lymphoma cell lines by monoclonal antibody PAb240. *Oncogene*, **6**, 1633–1640.

Yonish-Rouach, E., Grunwald, D., Wilder, S., Kimchi, A., May, E., Lawrence, J.J., May, P. and Oren, M. (1993) p53-mediated cell death: relationship to cell cycle control. *Molecular and Cellular Biology*, **13**, 1415–1423.

Yonish-Rouach, E., Resnitzky, D., Lotem, J., Sachs, L., Kimchi, A. and Oren, M. (1991) Wild-type p53 induces apoptosis of myeloid leukaemic cells that is inhibited by interleukin-6. *Nature*, **352**, 345–347.

Chapter 17

ALTERED GENE EXPRESSION DURING APOPTOSIS

Sherilyn D. Goldstone[1] and Martin F. Lavin[2]

[1]Department of Pathology, University of Sydney, NSW 2006, Australia
[2]Queensland Cancer Fund Research Unit, Queensland Institute of Medical Research, The Bancroft Centre, 300 Herston Road, Brisbane, Queensland 4029, Australia

In all higher organisms, programmed cell death, or apoptosis, is required during development, differentiation, and for the maintenance of tissue homeostasis. Defects in the signalling pathway which would normally result in apoptosis allow abnormal proliferation, thus contributing to both developmental abnormalities and neoplasia. Considerable evidence exists to suggest a role for altered gene expression in the induction of apoptosis. Distinct gene loci involved in the regulation of apoptosis during development have been identified in the nematode *Caenorhabditis elegans* (Ellis and Horvitz, 1986). Complementation studies between glucocorticoid-sensitive and glucocorticoid-resistant derivatives of a human T-cell leukaemia line, which both have intact glucocorticoid receptor function and differ only in the lytic response to dexamethasone, indicate the existence of a genetic locus involved in apoptosis in human cells (Yuh and Thompson, 1987). Similar results have been obtained in complementation studies between glucocorticoid-sensitive and glucocorticoid-resistant murine cell lines (Gasson and Bourgeois, 1983; Gasson et al., 1983). Azacytidine treatment of these resistant cell lines results in reversion to sensitivity, suggesting a role for methylation of the "lethality" locus in the expression of this phenotype. Further evidence for the existence of a locus involved in apoptosis comes from the observation that glucocorticoid sensitivity and susceptibility to cytotoxic T lymphocyte killing in a murine thymoma line share a common element, in that a mutant derivative of the same cell line, which retained intact glucocorticoid receptors and wild type cell surface antigens, was resistant to both glucocorticoid-induced and CTL-mediated apoptosis. A single spontaneous reversion event was able to restore sensitivity to both glucocorticoid-induced and CTL-mediated cell death, indicating that a single locus was involved in the induction of apoptosis by these two stimuli (Ucker, 1987).

Inhibition of both RNA and protein synthesis prevents the onset of apoptosis in a variety of model systems, including both glucocorticoid- and γ-radiation-induced death of mouse thymocytes (Wyllie et al., 1984; Sellins and Cohen, 1987), activation-induced death of T-cell hybridomas (Shi et al., 1990) and insect metamorphosis (Schwartz et al., 1990). Although the requirement for macromolecular synthesis during apoptosis is well documented, there have been few reports of either altered gene expression or alteration in cellular proteins occurring during this process.

A variety of strategies have been employed by a number of groups, including ourselves, in the effort to identify so-called "death genes". The candidate gene approach has implicated the oncogenes bcl-2 and c-myc and the tumour suppressor p53 in the onset of apoptosis; either as "survival genes", in the case of bcl-2 and p53, or in the case of c-myc, as a "switching factor", as discussed in Chapters 13–16. Subtractive and/or differential screening of cDNA libraries is an approach which has been widely used in order to isolate clones representing mRNAs with altered expression in a particular mRNA population. We and others have successfully used this approach for the isolation of genes showing altered expression during apoptosis in a variety of model systems. We previously reported the isolation of a cDNA clone, representing a 14 kDa β-galactoside binding protein, overexpressed 10-fold during glucocorticoid-induced apoptosis in the human T-cell leukaemia line CEM-C7 (Goldstone and Lavin, 1991). Schwartz et al. (1990) have described the isolation of cDNA clones representing four genes that are abundantly overexpressed in the intersegmental muscles of the hawkmoth Manduca sexta during the involution of these muscles which occurs during metamorphosis. Hoshikawa et al. (1988) have reported the isolation of eight cDNA clones representing genes whose expression was decreased in response to dexamethasone treatment of thymocytes isolated from adrenalectomized rats, while Harrigan et al. (1989) have isolated 11 cDNA clones overexpressed during apoptosis induced in a murine thymoma cell line by treatment with either cyclic AMP or dexamethasone. Preliminary sequence analysis of four of these 11 cDNAs showed that one clone was homologous to the mouse VL30 retrovirus-like element, another was homologous to the mouse chondroitin sulfate proteoglycan core protein, while the remaining two clones had no significant homology to any previously described gene. Buttyan et al. (1989) have isolated a cDNA clone from rat regressing ventral prostate tissue that is expressed at high levels in a variety of rodent models of apoptosis, including the regression of prostate tissue after castration, renal atrophy, involution of the interdigital web during embryogenesis, and the regression of bladder tumours following chemotherapy. Sequence analysis of this cDNA showed significant homology to the clusterin gene, expressed constitutively by Sertoli cells in the rat testis. Most recently, Owens et al. (1991) have reported the isolation of two novel cDNAs associated with apoptosis induced in murine thymocytes by treatment with either dexamethasone or γ-radiation. On the basis of computer modelling, one of these cDNAs is thought to encode an integral membrane protein, and the other a zinc finger protein. Although the expression of all of the genes outlined above was found to be altered during apoptosis, direct involvement of any of these genes in the process of apoptosis remains to be proven.

The human leukaemic cell line CEM-C7 may be induced to undergo apoptosis by either treatment with glucocorticoids or exposure to γ-radiation. Treatment of this cell line with 10^{-6} M dexamethasone results in an irreversible G1 phase cell cycle arrest together with a decline in viability, first becoming apparent 18–24 hours post-treatment. By 48 hours post-treatment 80% of cells are no longer viable (Harmon et al., 1979). This cell cycle arrest and subsequent cell death has been shown to be accompanied by the morphological changes characteristic of apoptosis, namely reduction in cell volume, chromatin condensation, loss of microvilli, membrane blebbing, and ultimately the degradation of the DNA into the "ladder" pattern seen as a result of specific internucleosomal cleavage by a Ca^{2+}/Mg^{2+}-dependent endonuclease activated during the process of apoptosis (Duvall and Wyllie, 1986; Cohen and Duke, 1984). CEM-C7 cells are also sensitive to the induction of apoptosis by treatment with ionizing radiation; exposure to a 20 Gy dose of γ-radiation results in DNA fragmentation, first becoming obvious at 4 hours post-irradiation, with cell death occurring between 6–8 hours post-irradiation (Baxter et al., 1989).

The initiation of the programme of altered gene expression seen during glucocorticoid-induced apoptosis is mediated by the action of the activated glucocorticoid receptor as a transcription factor involved in the regulation of expression of glucocorticoid-responsive genes (Beato, 1989). The molecular mechanisms involved in the regulation of gene expression following exposure to ionizing radiation are less well understood. Exposure to ionizing radiation results in direct DNA damage and altered DNA/nucleoprotein conformation. It has been suggested that this conformational change elicits a signal, mediated by protein kinases, resulting in the activation of certain transcription factors which directly modify the expression of responsive genes (Weichselbaum et al., 1991). While both inducing stimuli result in altered gene expression, it would be expected that the expression of only a percentage of these genes would be altered in association with cell death, rather than the inducing stimulus itself. Therefore, in our study, a double selection approach was used in order to isolate cDNA clones representing mRNAs showing altered expression in association with the induction of apoptosis. By isolating clones representing mRNAs whose expression were similarly altered during the induction of apoptosis by either glucocorticoid treatment or exposure to γ-radiation, we hoped to identify genes whose expression were specifically altered in association with cell death.

The initial selection process involved the differential screening of a cDNA library constructed from mRNA extracted from CEM-C7 cells 18 hours post-treatment with dexamethasone, in order to isolate clones showing altered expression in association with glucocorticoid-induced apoptosis. The second selection step involved using the clones isolated in the initial screening process as probes for expression analysis of mRNA isolated from CEM-C7 cells over the timecourse of induction of apoptosis by both glucocorticoid treatment and exposure to γ-radiation. Using this strategy, we have identified five clones showing altered expression during glucocorticoid-induced apoptosis, two of which show a similar expression pattern during apoptosis induced by γ-radiation.

CLONE 32

The cDNA insert for this clone was approximately 5 kb in length, and detected a single mRNA species of >9.5 kb, whose expression was increased at 24 hours post-treatment with dexamethasone (Figure 17.1A). Repeat experiments did not reproducibly detect the same level of overexpression, however, some degree of overexpression was present in each case, varying from two-fold to ten-fold. Nuclear run-off transcription assays demonstrated that this mRNA was continuously transcribed over the timecourse of induction of apoptosis with no apparent increase in the rate of transcription at any point within the timecourse, suggesting stabilization of the mRNA as a mechanism for the increased expression.

In order to distinguish between overexpression occurring as a result of glucocorticoid treatment, and overexpression more intimately involved in the induction of apoptosis, the levels of the mRNA represented by clone 32 were examined in mRNA samples isolated from CEM-C7 cells induced to undergo apoptosis by treatment with γ-radiation. In this case, the timecourse of induction of apoptosis is much shorter, with DNA fragmentation becoming obvious by 4 hours post-treatment, and death occurring between 6–8 hours post-treatment. An approximately two-fold increase in expression was seen at 2 hours post-treatment, in general accord with the degree of overexpression seen in response to dexamethasone treatment (Figure 17.1B).

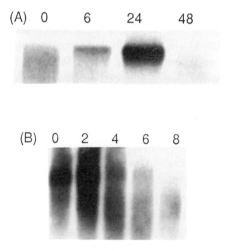

Figure 17.1 Expression of the mRNA represented by clone 32 during apoptosis induced in CEM-C7 cells by treatment with either dexamethasone (A), or γ-radiation (B). The indicated times are given in hours post-treatment.

Sequence analysis and comparison with the GENBANK and EMBL DNA databases detected no significant homology between clone 32 and any other sequence present in these databases, with the exception of an Alu repeat located within the central region of the clone. Clone 32 thus is likely to represent a novel gene.

CLONE 310

The cDNA for this clone was 200 bp in length, and represented an mRNA of approximately 400 bp, whose expression was decreased five-fold at six hours post-treatment with dexamethasone. In untreated cells, the expression of this mRNA was reduced later in the timecourse, with suppression of approximately three-fold first becoming evident at the 24 hour timepoint (Figure 17.2A). When apoptosis was induced by exposure to γ-radiation, the expression of the mRNA represented by clone 310 was also reduced, however, in this case, untreated cells showed no reduction in expression (Figure 17.2B). As the timecourse in this case is considerably shorter, at 8 hours versus 48 hours, suppression would not necessarily be expected to occur.

Sequence analysis of this cDNA and comparison with both the GENBANK and EMBL databases showed no significant homology between the sequence of clone 310 and any other sequence present in the database, thus suggesting that this cDNA also represents a previously undescribed gene.

CLONES 38 and 61

The cDNA insert for clone 38 was 100 bp in size, and the RNA represented by this clone was <200 bp in size. In response to glucocorticoid treatment, this RNA was

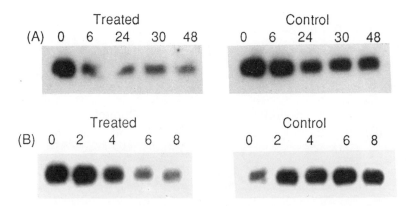

Figure 17.2 Expression of the mRNA represented by clone 310 during apoptosis induced in CEM-C7 cells by treatment with either dexamethasone (A), or γ-radiation (B). The indicated times are given in hours post-treatment.

overexpressed four-fold relative to the 0 hour-timepoint at 6 hours post-treatment, with expression decreasing at all later timepoints (Figure 17.3A). A parallel experiment using mRNA isolated from untreated cells showed a similar degree of overexpression of this RNA at 6 hours post-treatment, with decreasing expression at later timepoints, suggesting that the initial overexpression seen in both treated and untreated cells may be due to the serum stimulus received at the time of plating. Interestingly, although the decline in expression of this RNA after the 6 hour-timepoint was more pronounced in the treated cells than the control population, by 48 hours post-treatment there was an approximately two-fold greater amount of this RNA in the treated cells as compared the equivalent timepoint in the control cells. A similar expression pattern was seen in mRNA isolated from CEM-C7 cells induced to undergo apoptosis by treatment with γ-radiation; initial overexpression was seen at 2 hours post-irradiation, followed by a gradual decrease in expression up to 8 hours post-irradiation (Figure 17.3B). Again, expression levels were reduced at the 4- and 6 hour-timepoints in treated cells compared to the equivalent timepoints in control cells, with increased expression in the treated cells at the 8 hour-timepoint compared to the control, in general agreement with the pattern of expression seen in response to dexamethasone treatment. Sequence analysis and comparison with the GENBANK database showed that this clone was identical to part of the human mitochondrial D-loop region.

The cDNA insert for clone 61 was 120 bp in size, and represented a 1.6 kb mRNA whose expression was increased approximately 4-fold at 30 and 48 hours post-treatment with dexamethasone (Figure 17.4A). In untreated cells, expression of this mRNA was found to decrease over the corresponding timecourse. This mRNA was also overexpressed during apoptosis induced by treatment with γ-radiation, with maximum expression seen at 2 hours post-irradiation, and declining thereafter (Figure

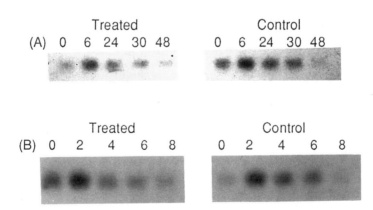

Figure 17.3 Expression of the mitochondrial primer RNA represented by clone 38 during apoptosis induced in CEM-C7 cells by treatment with either dexamethasone (A), or γ-radiation (B). The indicated times are given in hours post-treatment.

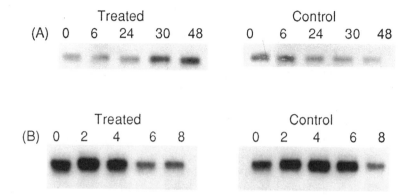

Figure 17.4 Expression of the mitochondrial cytochrome oxidase subunit 1 gene represented by clone 61 during apoptosis induced in CEM-C7 cells by treatment with either dexamethasone (A), or γ-radiation (B). The indicated times are given in hours post-treatment.

17.4B). However, in this case overexpression was also seen in control cells, becoming evident by the 2 hour timepoint, maximal at the 4 hour timepoint, and declining thereafter. In the equivalent experiment, using dexamethasone as the inducing stimulus, this early induction of overexpression was not seen, presumably as by 8 hours post-treatment expression had returned to basal levels.

Sequence analysis of this clone and comparison with the GENBANK database showed that this clone represented the mitochondrial cytochrome oxidase subunit I gene. Mitochondria have a separate, autonomously replicating genome, encoding the 12S and 16S mitochondrial rRNAs, 22 mitochondrial tRNA genes, and 13 key subunits of five enzymes involved in oxidative phosphorylation (Anderson *et al.*, 1981; Attardi and Schatz, 1988). Although the replication of the mitochondrial genome is autonomous, the regulation of mitochondrial gene expression is entirely dependent upon nuclear-encoded factors, as the mitochondrial genome does not encode any protein involved in the regulation of either mitochondrial DNA or RNA synthesis.

Both mitochondrial DNA replication and transcription are regulated by protein/DNA interactions within the displacement (D)-loop region of the mitochondrial genome. Transcription of the mitochondrial genome is initiated bidirectionally from two promoter sites within the D-loop region, which direct the synthesis of the polycistronic primary transcripts produced from each strand of the mitochondrial genome. Replication of the mitochondrial genome also begins within the D-loop region and is intimately connected with transcription, in that transcription from one of the two promoter sequences results in a primary transcript which is cleaved by an endoribonuclease at specific sites to yield the primer RNAs required for initiation of DNA synthesis, as well as the RNAs encoded by this strand of the mitochondrial genome. (Chang and Clayton, 1987; Clayton, 1991). Clone 38 represents a 100 bp region of the D-loop, and thus represents a primer RNA involved in the initiation of mitochondrial DNA replication.

Certain stimuli which result in cellular proliferation or differentiation are known to affect the mitochondrial contribution to cellular respiratory activity, through a combination of gene dosage effect, by modulation of the copy number of the mitochondrial genome per cell, and altered transcriptional activity of the mitochondrial genome (Nagley, 1991). Increases in mitochondrial gene expression, resulting from increased transcription of the mitochondrial genome and/or increased stability of mitochondrial transcripts have been reported after transformation of rat fibroblasts with the *myc* oncogene, adenovirus E1A, or polyoma virus large T antigen (Glaichenhaus *et al.*, 1986). Altered expression of the mitochondrial genome has also been reported as a result of serum stimulation of quiescent hepatoma H-35 cells (Kadowaki and Kitagawa, 1988), while glucocorticoids, estrogens, and thyroid hormone exert similar effects on mitochondrial gene expression in a variety of cell lines (Kadowaki and Kitagawa, 1988; Van Itallie and Dannies, 1988; Van Itallie, 1990). This altered gene expression is thought to be a specific secondary effect of hormone treatment, and results in increased cellular respiratory activity.

The relatively late overexpression of clone 61 in response to dexamethasone treatment is consistent with the hypothesis that this is a specific secondary effect of dexamethasone treatment, in agreement with previously published observations (Kadowaki and Kitagawa, 1988; Van Itallie, 1990). Given that the mitochodrial genome is transcribed in a polycistronic manner, and that the expression of clone 38 was decreased during dexamethasone-induced apoptosis, it would seem that the increased expression of clone 61 seen in response to dexamethasone treatment results from selective stabilization of this mRNA in response to glucocorticoid treatment. In support of this argument, during apoptosis induced by γ-radiation, expression levels of both clone 38 and clone 61 were decreased.

CLONE 41

The cDNA insert for this clone was 200 bp in size, and detected a 500 b mRNA species. The expression of this mRNA increased over the timecourse of glucocorticoid-induced apoptosis, reaching a maximum of ten-fold induction by 48 hours post-treatment, while expression remained constant in untreated cells over a similar timecourse (Figure 17.5A). When apoptosis was induced by γ-radiation, the expression of this mRNA was decreased over the 8 hour-timecourse, with decreased expression evident by 2 hours post-irradiation and continuing thereafter (Figure 17.5B).

Sequence analysis and comparison with the GENBANK and EMBL databases identified this clone as representing the human HL-14 gene, which encodes a 14 kDa β-galactoside binding protein (Gitt and Barondes, 1986). Proteins of this class have been isolated from a number of vertebrate species, and all share similar carbohydrate binding properties, as well as cross-reacting immunological determinants (Drickamer,

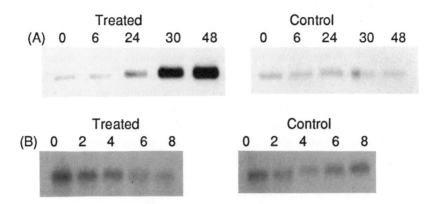

Figure 17.5 Expression of the HL-14 gene represented by clone 41 during apoptosis induced in CEM-C7 cells by treatment with either dexamethasone (A), or γ-radiation (B). The indicated times are given in hours post-treatment.

1988). There is considerable sequence homology between isolates from different vertebrate species, as would be expected on the basis of their similar properties (Ohyama et al., 1986; Paroutaud, 1987; Clerch et al., 1988; Raz et al., 1988; Couraud et al., 1989). The appearance and cellular distribution of these proteins is developmentally regulated, and coordinated with the expression of complementary structures in many organs and tissues. The specific carbohydrate binding properties of this class of proteins has led to the suggestion that they function as lectins, mediating cellular recognition events which require the binding of specific glycoconjugates (Barondes, 1984). However, recently an unexpected physiological role for the 14 kDa β-galactoside binding proteins has been identified, with the demonstration that the murine protein acts as a negative regulator of proliferation, exerting this effect by controlling the cell cycle at two points, exit from G_0 phase and passage through G_2 phase into mitosis. This antiproliferative effect is autocrine in nature, independent of carbohydrate binding activity, and appears to be a receptor mediated phenomenon (Wells and Malluci, 1991).

It is interesting to note that glucocorticoids induce overexpression of the HL-14 gene, concurrent with the onset of cell cycle arrest seen in CEM-C7 cells in response to glucocorticoid treatment. As expression of the HL-14 mRNA was downregulated in response to γ-radiation, it would seem that the increased expression of this mRNA seen during dexamethasone-induced apoptosis is a glucocorticoid-specific response, and not necessarily related to apoptosis *per se*. However, this does not preclude an involvement for the HL-14 gene in the cell cycle arrest seen after glucocorticoid treatment. In support of this argument, overexpression of the HL-14 gene is seen during glucocorticoid-induced terminal differentiation of chick embryo epidermal cells to squamous keratinized epidermis, a process associated with the cessation of cell division (Oda et al., 1989).

In summary, we have used a double selection approach to identify genes showing altered expression during the induction of apoptosis in a human leukaemic T-cell line by two different stimuli, glucocorticoid treatment and γ-radiation. The initial step involved the use of differential hybridization to identify clones showing altered expression during glucocorticoid-induced apoptosis, and the second step, screening these clones by examining their expression patterns during apoptosis induced by γ-radiation. We hoped to identify clones showing similar expression patterns during apoptosis induced by both stimuli, thus eliminating clones whose expression was altered in response to either inducing stimulus alone, rather than in association with cell death. The expression of the mRNAs represented by two of the five clones originally isolated, clones 61 (mitochondrial cytochrome oxidase subunit I) and 41 (b-galactoside binding protein), were increased during apoptosis induced by treatment with dexamethasone. However, the expression of both of these mRNAs was down-regulated during apoptosis induced by γ-radiation. Thus it would appear that the over-expression of both of these mRNAs in response to dexamethasone treatment is a glucocorticoid-induced phenomenon and not necessarily related to apoptosis *per se*. As the programme of altered gene expression seen during glucocorticoid-induced apoptosis is initiated by the action of the activated glucocorticoid receptor as a transcription factor involved in the regulation of glucocorticoid-sensitive genes, it was expected that only a percentage of the clones identified in the initial screening step would show a similar expression pattern during apoptosis induced by γ-radiation. This was indeed found to be the case, as of the five cDNAs isolated upon initial screening, only three were found to represent mRNAs whose expression were similarly altered during apoptosis induced by both stimuli. One of these, clone 38, represented a mitochondrial primer RNA, and interestingly, the two others, clones 32 and 310, are likely to represent novel genes, as no homology was seen between the sequence of either of these clones and any sequence present in the GENBANK and EMBL databases. These two cDNAs may therefore represent previously uncharacterized genes whose expression is altered in association with the onset of apoptosis.

In the effort to elucidate the molecular mechanisms which regulate the onset of apoptosis, much attention has been focussed on the search for so-called "death genes". There is considerable evidence to suggest that altered gene expression is required for the induction of apoptosis, and a number of genes have been demonstrated to show altered expression during the induction of apoptosis in a variety of model systems. However, to date, with the exception of the nematode *ced-3* and *ced-4* genes, none have been shown to be directly involved in this process. Considered in the broader light of recent advances in our knowledge of the molecular mechanisms involved in the regulation of cellular proliferation, it is tempting to hypothesise that the induction of apoptosis does not depend on the expression of any one gene. Rather, the induction of apoptosis may result from a switching off of the proliferative pathway, initiated by the action of the inducing stimulus in rendering the cell unable to pass the next cell cycle checkpoint due to lack of the appropriate balancing stimulus. Thus, the onset of apoptosis would be determined by the balance between positive and negative stimuli acting on the target cell at any one time. It follows therefore that numerous genes, each of which would function as crucial regulators within the proliferative pathway, must be considered as candidate "death genes". The answer to the question "division or death?" is ultimately one of balance, and our understanding of the molecular mechanisms which determine the outcome of this question will come therefore from a study of the relationship between proliferation, differentiation, and death, rather than for a search for a limited number of "death genes".

REFERENCES

Anderson, S., Bankier, A. T., Barell, B. G., de Bruijn, M. H. L., Coulson, A. R., Drouin, J., Eperon, I. C., Nierlich, D. P., Roe, B. A., Sanger, F., Schrier, P. H., Smith, A. J. H., Staden, R. and Young, I. G. (1981) Sequence and organization of the human mitochondrial genome. *Nature*, 332, 166–171.

Attardi, G. and Schatz, G. (1988) Biogenesis of mitochondria. *Annual Review of Cell Biology*, 4, 289–333.

Barondes, S. H. (1988) Bifunctional properties of lectins: Lectins redefined. *Trends in Biochemical Science*, 13, 480–482.

Baxter, G. D., Smith, P. J. and Lavin, M. F. (1989) Molecular changes associated with induction of cell death in a human T-cell leukaemia line: putative nucleases identified as histones. *Biochemical and Biophysical Research Communications*, 162, 30–37.

Beato, M. (1989) Gene regulation by steroid hormones. *Cell*, 56, 335–344.

Buttyan, R., Olsson, C. A., Pintar, J., Chang, C., Bandyk, M., Ng, P.-Y. and Sawczuk, I. S. (1989) Induction of the TRPM-2 gene in cells undergoing programmed cell death. *Molecular and Cellular Biology*, 9, 3473–3481.

Chang, D. D. and Clayton, D. A. (1987) A mammalian mitochondrial RNA processing enzyme contains nucleus-encoded RNA. *Science*, 235, 1178–1184.

Clayton, D. A. (1991) Nuclear gadgets in mitochondrial DNA replication and transcription. *Trends in Biochemical Science*, 16, 107–111.

Clerch, L. B., Whitney, P., Hass, M., Brew, K., Miller, T., Werner, R. and Massaro, D. (1988) Sequence of a full-length cDNA for rat lung β-galactoside-binding protein: Primary and secondary structure of the lectin. *Biochemistry*, 27, 692–699.

Cohen, J. J., and Duke, R. C. (1984) Glucocorticoid activation of a calcium-dependent endonuclease in thymocyte nuclei leads to cell death. *Journal of Immunology*, 132, 38–42.

Couraud, P.-O., Casetini-Borocz, D., Bringman, T. S., Griffith, J., McGrogan, M. and Nedwin, G. E. (1989) *Journal of Biological Chemistry*, 264, 1310–1316.

Drickamer, K. (1988) Two distinct classes of carbohydrate-recognition domains in animal lectins. *Journal of Biological Chemistry*, 263, 9557–9560.

Duvall, E. and Wyllie, A. H. (1986) Death and the cell. *Immunology Today*, 7, 115–119.

Ellis, H. M. and Horvitz, H. R. (1986) Genetic control of programmed cell death in the nematode *C. elegans*. *Cell*, 44, 817–829.

Gasson, J. C. and Bourgeois, S. (1983) A new determinant of glucocorticoid sensitivity in lymphoid cell lines. *Journal of Cell Biology*, 96, 409–415.

Gasson, J. C., Ryden, T., and Bourgeois, S. (1983) Role of *de novo* DNA methylation in the glucocorticoid resistance of a T-lymphoid cell line. *Nature*, 302, 621–623.

Gitt, M. A. and Barondes, S.H. (1986) Evidence that a human soluble β-galactoside-binding lectin is encoded by a family of genes. *Proceedings of the National Academy of Science, USA,* **83**, 7603–7607.

Glaichenhaus, N., Leopold, P., Cuzin, F. (1986) Increased levels of mitochondrial gene expression in rat fibroblast cells immortalized or transformed by viral and cellular oncogenes. *The EMBO Journal,* **5**, 1261–1265.

Goldstone, S. D. and Lavin, M. F. (1991) Isolation of a cDNA clone, encoding a human galactoside binding protein, overexpressed during glucocorticoid-induced cell death. *Biochemical and Biophysical Research Communications,* **178**, 746–750.

Harmon, J. M., Norman, M. R., Fowlkes, B. J. and Thompson, E. B. (1979) Dexamethasone induces irreversible G1 arrest and death of a human lymphoid cell line. *Journal of Cellular Physiology,* **98**, 267–278.

Harrigan, M. T., Baughman, G., Campbell, N. F. and Bourgeois, S. (1989) Isolation and characterization of glucocorticoid- and cyclic AMP-induced genes in T lymphocytes. *Molecular and Cellular Biology,* **9**, 3438–3446.

Hoshikawa, Y., Izawa, M. and Ichii, S. (1988) Cloning of glucocorticoid-responsive mRNA in the rat thymus. *Endocrinologia Japonica,* **35**, 429–437.

Kadowaki, T. and Kitagawa, Y. (1988) Enhanced transcription of mitochondrial genes after growth stimulation and glucocorticoid treatment of Reuber hepatoma H-35. *FEBS Letters,* **233**, 52–58.

Nagley, P. (1991) Coordination of gene expression in the formation of mammalian mitochondria. *Trends in Genetics,* **7**, 1–4.

Oda, Y., Ohyama, Y., Obinata, A., Endo, H. and Kasai, K.-I. (1989) Endogenous β-galactoside-binding lectin expression is suppressed in retinol-induced mucous metaplasia of chick embryonic epidermis. *Experimental Cell Research,* **182**, 33–43.

Ohyama, Y., Hirabayashi, J., Oda, Y., Ohno, S., Kawasaki, H., Suzuki, K., Kasai, K.-I. (Year) Nucleotide sequence of 14 kDa β-galactoside binding lectin mRNA. *Biochemical and Biophysical Research Communications,* **134**, 51–56.

Owens, G. P., Hahn, W. E. and Cohen, J. J.(1991) Identification of mRNAs associated with programmed cell death in immature thymocytes. *Molecular and Cellular Biology,* **11**, 4177–4188.

Paroutaud, P., Levi, G., Teichberg, V. I. and Strosberg, A. D. (1987) Extensive amino acid sequence homologies between animal lectins. *Proceedings of the National Academy of Science, USA,* **84**, 6345–6348.

Raz, A., Carmi, P. and Pazerini, G. (1988) Expression of two different endogenous galactoside-binding lectins sharing sequence homology. *Cancer Research,* **48**, 645–649.

Schwartz, L. M., Kocz, L. and Kay, B. K. (1990) Gene activation is required for developmentally programmed cell death. *Proceedings of the National Academy of Science, USA,* **87**, 6594–6598.

Sellins, K. S. and Cohen, J. J. (1987) Gene induction by radiation leads to DNA fragmentation in lymphocytes. *Journal of Immunology,* **139**, 3199–3206.

Shi, Y.-F., Szalay, M. G., Paskar, L., Boyer, M., Singh, B. and Green, D. R. (1990) Activation-induced cell death in T-cell hybridomas is due to apoptosis. *Journal of Immunology,* **144**, 3326–3333.

Ucker, D. S. (1987) Cytotoxic T lymphocytes and glucocorticoids activate an endogenous suicide process in target cells. *Nature,* **327**, 62–65.

Van Itallie, C. M. (1990) Thyroid hormone and dexamethasone increase the level of a messenger ribonucleic acid for a mitochondrially encoded subunit but not for a nuclear encoded subunit of cytochrome c oxidase. *Endocrinology,* **127**, 55–62.

Van Itallie, C. M. and Dannies, P. S. (1988) Estrogen induces accumulation of the mitochondrial ribonucleic acid for subunit II of cytochrome oxidase in pituitary tumour cells. *Moecular Endocrinology*, **2**, 332–337.

Weichselbaum, D. R., Hallahan, D. E., Sukhatme, V., Dritschilo, A., Sherman, M. L. and Kufe, D. W. (1991) Biological consequences of gene regulation after ionizing radiation exposure. *J. National Cancer Institute*, **83**, 480–484.

Wells, V. and Mallucci, L. (1991) Identification of an autocrine negative growth factor: Mouse β-galactoside-binding protein is a cytostatic factor and cell growth regulator. *Cell*, **64**, 91–97.

Wyllie, A. H., Morris, R. G., Smith, A. L. and Dunlop, D. (1984) Chromatin cleavage in apoptosis: association with condensed chromatin morphology and dependence on macromolecular synthesis. *Journal of Pathology*, **142**, 67–77.

Yuh, Y.-S. and Thompson, E. B. (1987) Complementation between glucocorticoid receptor and lymphocytolysis in somatic cell hybrids of two glucocorticoid-resistant human leukaemic clonal cell lines. *Somatic Cellular and Molecular Genetics*, **13**, 33–46.

APOPTOSIS IN THE IMMUNE SYSTEM

Chapter 18

THE ROLE OF APOPTOSIS IN THE REGULATION OF THE IMMUNE RESPONSE

J. J. T. Owen, I. M. Allan, N. C. Moore and E. J. Jenkinson

Department of Anatomy, Medical School,
University of Birmingham, Birmingham, UK

INTRODUCTION

The susceptibility of thymocytes to apoptosis, especially in response to treatment with glucocorticoids or irradiation, has been known for a number of years (Wyllie et al., 1980). However, it has been appreciated only recently that apoptosis of thymocytes plays a crucial role in shaping the repertoire of T-cell receptor (TCR) specificities during T-cell development in the thymus (Smith et al., 1989). Moreover, mature T cells can also be induced to undergo apoptosis *in vitro* (Russell et al., 1991) and this process probably plays an important role in the regulation of normal T-cell responses to antigens *in vivo*. Of considerable interest are recent studies which have suggested that an enhanced susceptibility to apoptosis might underlie the loss of T cells found following HIV infection (Groux et al., 1992). Thus apoptosis might have a part to play in immunopathology.

Apoptosis is equally important in regulating and refining B lymphocyte responses. Thus apoptosis in newly formed B lymphocytes with receptors which recognise self antigens probably plays a part in preventing subsequent auto-antibody production (Nemazee and Burki, 1989). Apoptosis is also known to play a part in the selection of the antibody repertoire at a later stage following hypermutation of Ig genes in B cells located in germinal centres (Liu et al., 1992).

APOPTOSIS IN IMMATURE THYMOCYTES SHAPES THE REPERTOIRE OF TCR SPECIFICITIES

On entering the thymus from the bloodstream, lymphoid stem cells are induced to undergo a programme of proliferation, TCR gene rearrangement and differentiation resulting in the production of large numbers of CD4$^+$8$^+$ thymocytes expressing a diverse repertoire of TCR specificities. Cell kinetic studies suggest that most of these immature thymocytes have a very limited life span of 3–4 days (Shortman et al., 1990). The massive level of cell death within the thymus, reflects the stringent selection of the TCR repertoire. Thus only lymphocytes recognising self-MHC molecules are signalled to mature into CD4$^+$ (helper) or CD8$^+$ (cytotoxic) T cells (Teh et al., 1988). This process is known as positive selection. However, potentially autoreactive cells recognising self peptide/MHC complexes, although present within the CD4$^+$8$^+$ immature thymocyte population, are not allowed to proceed to full maturation and hence are absent from the peripheral T-cell pool (Kappler et al., 1987). This process is known as negative selection.

Direct evidence that apoptosis of immature thymocytes is the mechanism involved in negative selection was obtained using organ cultures of embryonic thymus (Smith et al., 1989). Day 14 mouse embryo thymus lobes generate TCR$^+$ immature CD4$^+$8$^+$ T cells over a 7-day organ culture period. Engaging the TCR of the these cells by means of antibodies results in the production of cells with a characteristic apoptotic appearance and DNA which is fragmented into "ladder" bands. Furthermore, the superantigen Staphylococcal enterotoxin B (SEB), when added to thymic organ cultures, induces apoptosis in Vβ8$^+$ immature T cells which specifically recognise this superantigen (Jenkinson et al., 1990).

DNA fragmentation is characteristic of apoptosis in thymocytes (Wyllie, 1980) although it may not accompany the process in all cell types. With a view to quantitating apoptosis on a cell to cell basis, a number of investigators have used flow cytometry studies to study DNA fragmentation and chromatin condensation accompanying the process. In particular, decreased uptake of intercalating DNA binding dyes has been reported in apoptotic cells. Thus cells undergoing apoptosis can be pinpointed as a population of low fluorescence (Ao peak) distinct from the Go/G1 cell cycle region of the fluorescence distribution (Telford et al., 1992).

In recent studies, we have used the intercalating DNA binding dye 7-amino-actinomycin D (7AAD) to identify apoptotic cells in thymus organ cultures. In view of the spectral characteristics of this dye, it is possible to measure the surface phenotype of apoptotic cells using antibodies conjugated with fluorescein (FITC) or phycoerythrin (PE). Figure 18.1, shows results obtained from thymic organ cultures treated overnight with antibodies to the TCR. A distinct apoptotic population can be identified by flow cytometry and analysis of the surface phenotype of these cells with antibodies to CD8 and CD4, conjugated with FITC and PE respectively, shows a reduced level of CD4 and CD8 expression associated with the apoptotic process. These results not only demonstrate the potent induction of apoptosis in immature thymocytes by anti-TCR

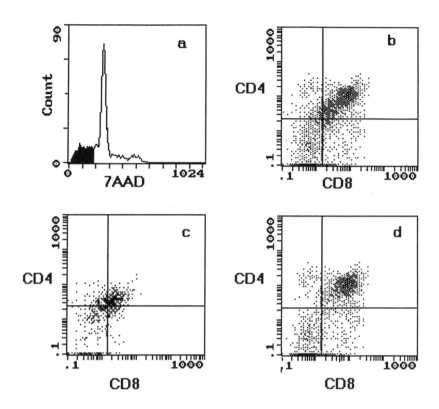

Figure 18.1 14 day mouse embryo thymus lobes, cultured for 8 days were treated overnight with antibodies to the constant region of the TCR β chain (Kubo et al., 1989). Lobes were teased and thymocytes labelled with anti-CD8 FITC and anti CD4 PE antibodies (Becton Dickinson). Cells were then fixed in 80% ethanol and stained with 7-AAD at a final concentration of 5 μg/ml. (a) shows the flow cytometric cell cycle histogram of thymocytes stained with 7-AAD. The apoptotic peak (shaded) contains 29.9% of cells (note comparable thymic organ cultures not treated with antibodies to the TCR have an apoptotic peak containing only 4.8% of cells). (b) shows a dot plot of overall CD4/CD8 expression of all cells (apoptotic and normal). (c) shows a dot plot of CD4/CD8 expression in the apoptotic population. (d) shows a dot plot of CD4/CD8 expression of normal cells (the non-shaded region of (a)).

antibodies but they confirm the reduced expression of CD4 and CD8 which has been used as an indirect measure of apoptosis in other studies (Swat et al., 1991). In further studies, we have found that apoptotic cells retain TCR expression and so it should be possible to assess the deletion of cells expressing defined TCR specificities in response to antigens added to thymus organ cultures.

SIGNALLING PATHWAYS IN THYMOCYTE APOPTOSIS

In order to identify signalling pathways employed during negative selection of the TCR repertoire, we have examined the effects of anti-TCR antibodies on the generation of inositol phosphates through hydrolysis of inositol lipids in cultured thymus lobes. The results show an accumulation of inositol phosphates indicating receptor stimulated phosphatidylinositol 4-5-bisphosphonate hydrolysis (Conroy et al., 1991). In turn, these events are known to result in elevation of intracellular Ca^{2+} levels which forms part of the signalling cascade in various cell types. Elevation of Ca^{2+} levels is likely to be important in induction of apoptosis because treatment of thymocytes with Ca^{2+} ionophores leads to apoptosis (Wyllie et al., 1984; McConkey et al., 1989). Initiation of phosphoinositol hydrolysis and Ca^{2+} elevation in apoptosis parallels the known role of these signalling events in activation of mature T cells into proliferation (Desai et al., 1990). Further similarities with signalling pathways in mature T cells are found in recent evidence that apoptosis can be blocked by cyclosporin (Zacharchuk et al., 1990) and that activation of the proto-oncogene c-*myc* is important (Shi et al., 1992) in apoptosis.

Thus these results suggest similarities between activation pathways producing apoptosis in immature T cells and proliferation in mature T cells raising the possibility that the major difference in the cellular response to these signals is inherent to the two different developmental stages. Perhaps the most significant piece of information with regard to this notion is the fact that the proto-oncogene *bcl-2* has the important functional property of inhibiting apoptosis (Sentman et al., 1991; Strasser et al., 1991). Moreover using an anti-human Bcl-2 monoclonal antibody, Hockenbury et al. (1991) have shown that *bcl-2* expression is more concentrated in the mature T cells of the thymic medulla than in the cortex where the vast majority of cells are of the immature $CD4^+8^+$ type. Thus the susceptibility of immature thymocytes to apoptosis might reside in their low Bcl-2 expression and this idea is supported by recent studies using transgene techniques which have shown that when *bcl-2* expression is targeted to immature thymocytes these cells become resistant to apoptosis induction (Sentman et al., 1991; Strasser et al., 1991).

We have studied *bcl-2α* mRNA expression during ontogeny in selected thymocyte populations using semi-quantitative RT-PCR (Dallman and Porter, 1992). Our results show that immature thymocytes lacking TCR expression and predominantly $CD4^+8^+$ in phenotype, express little *bcl-2α* mRNA (Figure 18.2). However, $CD3^+$ thymocytes do show *bcl-2α* mRNA expression suggesting that somewhere between the $CD4^+8^+$

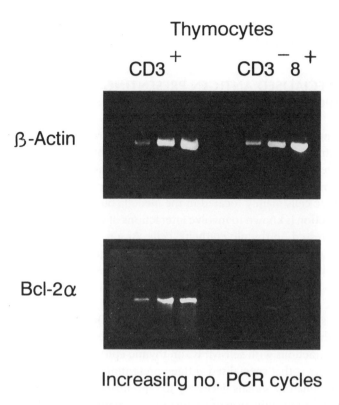

Figure 18.2 Thymocyte subpopulations were obtained from newborn mouse thymocytes depleted of macrophages and $\gamma\partial^+$ T cells, using magnetic beads (Dynal) coated with a mixture of F4/80 and 5C6 antibodies (gift from S. Gordon) and a pan C∂ antibody (clone 3A10, a gift from S. Tonagawa). The more mature CD3$^+$ T cells were selected on anti CD3 (clone KT3) coated Dynal beads, and the more immature CD4$^+$8$^+$ CD3$^-$ cells were selected from CD3 depleted thymocytes populations using Dynal beads coated with anti CD8 antibody (clone YTS 169.4). (Note most immature CD8$^+$ cells also express CD4). Cytosolic RNA, extracted from 10^5 cells using NP-40, was DNase 1 treated prior to RT-PCR. All samples were analysed for β actin and bcl-2α mRNA expression. PCR fragments were visualised using gel electrophoresis and ethidium bromide staining. The figure clearly shows that bcl-2α mRNA maps to the more mature CD3$^+$ cells.

TCR$^-$ stage of development and the TCR$^+$ stage of development, bcl-2 expression is upregulated. Whether this upregulation is characteristic of all CD3$^+$ thymocytes or whether it occurs only after cells have been positively selected remains to be established. In addition, we have found that thymocytes isolated from day 18 thymus lobes express little bcl-2α mRNA but very immature predominantly CD4$^-$CD8$^-$ thymocytes, isolated from day 14 lobes do express abundant bcl-2α mRNA (data not shown). Overall these results support the notion that as CD4$^+$8$^+$ thymocytes enter the phase when they may be selected by self-antigens expressed in the thymus they are susceptible to negative selection signals associated with their low expression of bcl-2.

THE ROLE OF SPECIALISED ANTIGEN PRESENTING CELLS IN THYMOCYTE APOPTOSIS

The processes of negative and positive selection involve interactions between thymocytes and thymic stromal cells. The thymic stroma consists of two major populations of cells which express both class I and class II MHC antigens (Jenkinson and Owen, 1990). These are epithelial cells which are predominantly located in the cortex of the thymus and dendritic cells mainly located in the medulla but also extending into the cortex. Positive selection is known to involve interactions of $CD4^+8^+$ thymocytes with MHC antigens expressed on thymic epithelial cells. As a result of positive selection, thymocytes which will be useful in immune responses i.e. capable of recognising self-MHC complexes which in immune responses will be associated with foreign peptides, will be selected for maturation (von Boehmer et al., 1989). The nature of the signals involved in positive selection are not known. In particular, it is unclear how $CD4^+8^+$ thymocytes can be signalled via interactions through the TCR to diverge along either of two separate pathways. Thus interactions with self-peptide/MHC complexes leads to apoptosis but interactions with self-MHC on thymic epithelial cells leads to maturation. One possibility is that specialised antigen presenting cells (probably dendritic cells) provide co-stimulatory signals necessary for apoptosis in a comparable manner to their requirement during activation of proliferation in mature T cells (Steinman and Young, 1991). Thymic epithelial cells may lack the ability to provide these co-stimulatory signals but instead activate a signalling pathway leading to maturation.

In attempts to investigate this hypothesis, we have carried out experiments aimed at studying interactions between $CD4^+8^+$ thymocytes and highly purified thymic epithelial cell populations depleted of dendritic cells (Jenkinson et al., 1992). Under these circumstances, we find that we can obtain efficient positive selection of thymocytes but inefficient induction of apoptosis when thymocytes expressing $V\beta8^+$ TCR's are engaged by the superantigen SEB. Addition of dendritic cells to these cultures restores their capacity to induce apoptosis within the $V\beta8$ population. Hence these experiments provide general support for the hypothesis outlined above.

APOPTOSIS IN PERIPHERAL IMMUNE RESPONSES

A number of recent studies have shown that it is possible to induce apoptosis in mature T cells by various strategies. Cross-linking CD4 molecules renders T cells susceptible to apoptosis by subsequent antibody cross-linking of the $a\beta$ TCR (Newell et al., 1990). Engagement of class I MHC molecules on cytolytic T-cell precursors together with cross-linking of the TCR also induces apoptosis (Sambhara and Miller, 1991). Activation of T cells by antigen makes cells susceptible to apoptosis by cross linking of the

TCR several days later (Russell et al., 1991). It is not possible to be certain how these in vitro strategies relate to the control of immune responses in vivo. However, they could be involved in inducing tolerance in autoreactive cells that might have escaped deletion in the thymus. Moreover, apoptosis of T cells previously stimulated by antigen might provide a mechanism for regulating T-cell responses, especially for switching off responses after protective immunity has been achieved.

Of particular interest is the possibility that perturbations in susceptibility to apoptosis might underlie the immunopathology of some diseases. Notable in this respect are recent studies which have shown that $CD4^+$ T cells of asymptomatic HIV infected individuals undergo apoptosis when stimulated in vitro (Groux et al., 1992; Meyaard et al., 1992). These results raised the possibility that heightened susceptibility to apoptosis induced by HIV infection might be responsible for loss of $CD4^+$ helper cells during AIDS.

Apoptosis in B lymphocytes is also known to play a crucial role in refining immune responses following hypermutation of Ig genes in B cells of germinal centres (Liu et al., 1992). There is also some evidence for the purging of autoreactive B cells shortly after their generation in bone marrow by induction of apoptosis paralleling similar events within the thymus (Nemazee and Burki, 1989).

SUMMARY AND CONCLUSIONS

There can now be little doubt that apoptosis plays a very important role in regulating immune responses. Clearly there is an urgent need to define the signalling pathways which lead to apoptosis as opposed to cell proliferation. Knowledge of these pathways might provide clues to defining the perturbations which occur in apoptosis in diseased states. Fortunately following a considerable period of relative neglect, the current heightened interest in apoptosis in the immune system is likely to lead to rapid progress in these areas.

ACKNOWLEDGEMENTS

We thank Christine Hepherd for help with the production of the manuscript, Alan Murdoch and Deidre McLoughlin for technical assistance and Alison Orchard for the photography. The work was supported by the Medical Research Council. I.M. Allan is a Beit Memorial Research Fellow.

REFERENCES

Conroy, L. A., Jenkinson, E. J., Owen, J. J. T. and Michell, R. H. (1991) The role of inositol lipid hydrolysis in the selection of immature thymocytes. *Biochemical Society Transactions*, 19, 905.

Dallman, M. J. and Porter, A. C. G. (1992) Semi-quantitative PCR for the analysis of gene expression. In: *PCR A Practical Approach*, M. J. McPherson, P. Quirke and S. R. Taylor (eds), pp. 215–224. Oxford University Press.

Desai, D. M., Newton, M. E., Kadlecek, T. and Weiss, A. (1990) Stimulation of the phosphatidylinositol pathway can induce T-cell activation. *Nature*, 348, 66–69.

Groux, H., Torpier, G., Monte, D., Monton, Y., Capron, A. and Ameison, J. C. (1992) Activation-induced death by apoptosis in CD4[+] T cells from HIV-infected asymptomatic individuals. *Journal of Experimental Medicine*, 175, 331–340.

Hockenbury, D., Zutter, M., Hickey, W., Nahm, M. and Korsmeyer, S. J. (1991) Bcl-2 protein is topographically restricted in tissues characterized by apoptotic cell death. *Proceedings of the National Academy of Science, USA.*, 88, 6961–6965.

Jenkinson, E. J., Anderson, S. A. and Owen, J. J. T. (1992) Studies on T-cell maturation on defined stromal cell populations *in vitro*. *Journal of Experimental Medicine*, 176, 845–854.

Jenkinson, E. J., Kingston, R. and Owen, J. J. T. (1990) Newly generated thymocytes are not refractory to deletion when the aβ component of the T-cell receptor is engaged by the superantigen staphylococcal enterotoxin B. *European Journal of Immunology*, 20, 2517–2520.

Jenkinson, E. J. and Owen, J. J. T. (1990) T-cell differentiation in thymus organ cultures. *Seminars in Immunology*, 2, 51–58.

Kapper, J. W., Roehm, N. and Marrack, P. (1987) T-cell tolerance by clonal elimination in the thymus. *Cell*, 49, 272–280.

Kubo, R. T., Boin, W., Kappler, J. W., Marrack, P. and Pigeon, M. (1989) Characterization of a monoclonal antibody which detects all murine aβ T-cell receptors. *Journal of Immunology*, 142, 2736–2742.

Liu, Y. J., Johnson, S. D., Gordon, J. and MacLennan, I. C. M. (1992) Germinal centres in T-cell dependent antibody responses. *Immunology Today*, 13, No.1, 17–21.

McConkey, D. J., Hartzell, P., Amador-Perez, J. F., Orrenius, S. and Jonal, M. (1989) Calcium-dependent killing of immature thymocytes by stimulation via the CD3/T-cell receptor complex. *Journal of Immunology*, 143, 1801–1806.

Meyaard, L., Otto, S. A., Jonkes, R. R., Mijuster, M. J., Keet, R. P. M., and Miedema, F. (1992) Programmed death of T cells in HIV-1 infection. *Science*, 257, 217–219.

Nemazee, D. A. and Burki, K. (1989) Clonal deletion of B lymphocytes in a transgenic mouse bearing anti-MHC class I antibody genes. *Nature*, 337, 562–566.

Newell, M. K., Haughn, L. J., Maroun, C. R. and Julius, M. H. (1990) Death of mature T cells by separate ligation of CD4 and the T-cell receptor for antigen. *Nature*, 347, 286–289.

Russell, J. H., White, C. L., Loh, D. Y. and Meleady-Rey, P. (1991) Receptor-stimulated death pathway is opened by antigen in mature T cells. *Proceedings of the National Academy of Science, USA.*, 88, 2151–2155.

Sambhara, S. R. and Miller, R. S. (1991) Programmed cell death of T cells signalled by the T-cell receptor and the a3 domain of class I MHC. *Science*, 252, 1424–1427.

Sentman, C. L., Shutter, T. R., Hockenburg, D., Kanagawa, O. and Korsmeyer, S. J. (1991) Bcl-2 inhibits multiple forms of apoptosis but not negative selection in thymocytes. *Cell*, 67, 879–888.

Shi, Y., Glynn, J. M., Guilbert, L. J., Cotter, T. G., Bissonnette, R. P. and Green, D. R. (1992) Role for c-*myc* in activation-induced apoptotic cell death in T-cell hybridomas. *Science*, 257, 212–214.

Shortman, K., Egerton, M., Spangrude, G. J., and Scollay, R. (1990) The generation and fate of thymocytes. *Seminars in Immunology*, 2, 3–12.

Smith, C. A., Williams, G. T., Kingston, R., Jenkinson, E. J. and Owen, J. J. T. (1989) Antibodies to the CD3/T-cell receptor complex induce death by apoptosis in immature T cells in thymic cultures. *Nature*, 337, 181–183.

Steinman, R. M. and Young, J. W. (1991) Signals arising from antigen-presenting cells. *Current Opinion in Immunology*, 3, 301–372.

Strasser, A., Harris, A. W. and Cory, S. (1991) bcl-2 transgene inhibits T-cell death and perturbs thymic self-censorship. *Cell*, 67, 889–899.

Swat, W., Iguatowitz, L., von Boehmer, H. and Kisielow, P. (1991) Clonal deletion of immature $CD4^+8^+$ thymocytes in suspension culture by extrathymic antigen-presenting cells. *Nature*, 351, 150–153.

Teh, H. S., Kisielow, P., Scott, B., Kishi, H., Vematsu, Y., Bluthman, H. and von Boehmer, H. (1988) Thymic major histocompatibility complex antigens and the $\alpha\beta$ T-cell receptor determine the CD4/CD8 phenotype of T cells. *Nature*, 335, 229–233.

Telford, W. G., King, L. W. and Fraker, P. J. (1992) Comparative evaluation of several DNA binding dyes in the detection of apoptosis-associated chromatin degradation by flow cytometry. *Cytometry*, 13, 137–143.

Wyllie, A. H. (1980) Glucocorticoid-induced thymocyte apoptosis is associated with endogenous endonuclease activation. *Nature*, 284, 555–556.

Wyllie, A. H., Kerr, J. F. R. and Currie, A. R. (1980) Cell death: The significance of apoptosis. *International Review of Cytology*, 68, 251–306.

Wyllie, A. H., Morris, R. G., Smith, A. L. and Dunlop, D. (1984) Chromatin cleavage in apoptosis: association with condensed chromatin morphology and dependence on macromolecular synthesis. *Journal of Pathology*, 142, 67–77.

Von Boehmer, H., Teh, H. S. and Kisielow, P. (1989) The thymus selects the useful, neglects the useless and destroys the harmful. *Immunology Today*, 10, 57–61.

Zacharchuk, C. M., Mercep, M., Chakrabozti, P. K., Simons, S. S. and Ashwell, J. D. (1990) Programmed T lymphocyte death: cell activation and steroid-induced pathways are mutually antagonistic. *Journal of Immunology*, 145, 4037–4045.

Chapter 19

THE APO-1 SYSTEM
At the Crossroad of Lymphoproliferation, Autoimmunity and Apoptosis

Peter H. Krammer

*German Cancer Research Center, Tumorimmunology Program,
Im Neuenheimer Feld 280, D-6900 Heidelberg, Germany*

In an attempt to downregulate tumour growth by engaging cell surface molecules involved in control of cell proliferation, we raised monoclonal antibodies in mice against the human B lymphoblastoid cell line SKW6.4 (Trauth et al., 1989). Instead of testing antibodies in hybridoma supernatants solely by staining of the cells used for immunization, we tested the antibodies by their effect on growth of SKW6.4 cells. Among several thousand hybridoma antibodies, one—anti-APO-1—was found that completely abrogated SKW6.4 growth already at minute quantities of as little as a few nanograms of immunoglobulin. The monoclonal antibody anti-APO-1 was shown to react with a cell surface molecule which we named APO-1. The name APO-1 was chosen to indicate the ability of APO-1 to mediate signals inducing programmed cell death, apoptosis, when triggered by anti-APO-1. Anti-APO-1-induced apoptosis in SKW6.4 cells occurred only minutes after incubation of the cells with anti-APO-1 antibodies and showed all the characteristic features of apoptosis such as the morphology of apoptosis in light and electron microscopy and polynucleosomal DNA fragmentation (Kohler et al., 1990). Depending on the type of APO-1 positive cells and the concentration of anti-APO-1 antibodies used the kinetics of apoptosis induced varied widely and ranged from minutes to several hours.

To understand its structural features we purified APO-1 from cultured SKW6.4 cells. An homogenous APO-1 preparation was obtained by a combination of SKW6.4 cell membrane solubilization, anti-APO-1 affinity chromatography, and reverse-phase HPLC (Oehm et al., 1992). APO-1 was shown to be a molecule of a pI of approximately 5.7 and an apparent molecular weight of 48 kDa, 8 kDa of which are accounted for by glycosylation. We sequenced APO-1 peptides and used PCR technology to clone the APO-1 cDNA from a cDNA bank of SKW6.4 cells. The APO-1 cDNA predicted a protein of 335 amino acids with a leader peptide of 16 amino acids, an intracellular part of 145 amino acids, a single transmembrane part of 19 amino acids, and an extracellular part of 155 amino acids with 3 cysteine-rich domains and two glycosylation sites. The cysteine-rich domains of APO-1 showed significant homology with members of the NGF-receptor/TNF-receptor superfamily and sequence identity with the recently cloned Fas antigen (Itoh *et al.*, 1991). The other members of the NGF/TNF-receptor superfamily are: the NGF receptor, TNF receptor 1 and 2, CD27, CD30, CD40, OX40, 4-1BB, and the Shope fibroma virus T2 antigen (Figure 19.1). Taken together, the structural features of APO-1 and its relationship to proteins of the NGF-/TNF-receptor superfamily suggest that APO-1 is an orphan receptor for an as yet uncharacterized ligand.

Alignment of amino acids of the intracellular part of APO-1 with those of CD40 and TNF receptors 1 and 2 showed a threonine residue that was previously shown to be important for the intact signalling capacity of CD40 (Oehm *et al.*, 1990). This finding indicated that the intracellular part of APO-1 might be important for the transmission of apoptotic signals. Presently, further insight into the APO-1 signalling mechanisms are still missing. Initial recent evidence, however, shows that anti-APO-1-mediated apoptosis requires extensive crosslinking of APO-1 molecules on the cell surface (Dhein *et al.*, 1992).

APO-1 was found to be expressed on normal activated T and B cells and on various malignant T- and B-cell tumours (Trauth *et al.*, 1989). Particularly, ATL cell lines were found to express APO-1 and to be exquisitely sensitive to anti-APO-1-mediated apoptosis (Debatin *et al.*, 1990). Furthermore, APO-1 was found on Burkitt lymphoma (group III) and LCL cell lines. Only LCL cell lines, however, were sensitive to anti-APO-1-mediated apoptosis (Falk *et al.*, 1992). The important conclusion derived from these data is that two requirements need to be fulfilled for cells to be sensitive towards APO-1-mediated apoptosis: 1) expression of APO-1, and 2) an intact apoptosis signalling pathway. So far unpublished data suggest a non-lineage restricted distribution of APO-1 on some but not all normal tissues and on lymphoid and non-lymphoid tumours (Leithauser *et al.*, unpublished data). Since both the above described requirements, however, need to be fulfilled for successful anti-APO-1-mediated apoptosis, a prediction of sensitivity of normal tissues and tumours by mere characterization of APO-1 antigen expression alone is not possible. Delineation of apoptotic sensitivity in the APO-1 system, therefore, requires functional tests of apoptosis in each case.

This statement is particularly important in cases in which anti-APO-1-mediated elimination of cells by apoptosis was invoked to play a significant role. The first case is the *lpr* syndrome. In *lpr/lpr* mice lymphoproliferation of $CD4^-$, $CD8^-$, $CD3^+$, $B220^+$, T cells, particularly in the lymph nodes, is observed together with autoimmune phenomena such as autoantibodies, similar to those seen in human lupus erythematosus. Recent evidence suggested that expression of Fas/APO-1 in the mutant mice is

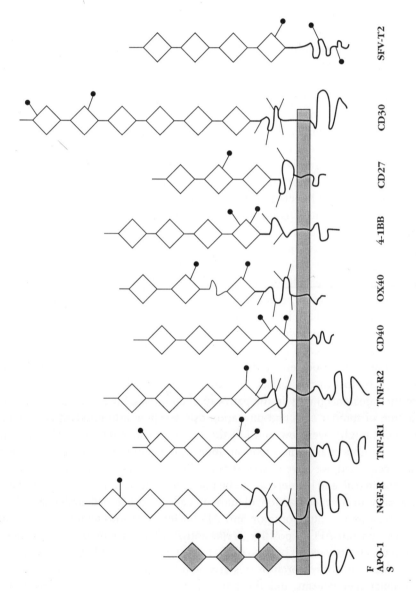

Figure 19.1 The TNF/NGF-receptor superfamily.

impaired (Watanabe-Fukunaga et al., 1992; Krammer and Debatin, 1992) and might lead to failure of apoptosis in T cells. These data were further interpreted as the most likely cause of a failure of T-cell selection in the thymus and hence accumulation of aberrant T cells in the periphery. The fact, however, that APO-1 is not expressed to a very high degree on most thymic T cells, but rather on thymic epithelial cells (Leithauser et al., unpublished data) makes this interpretation rather unlikely. It might therefore be more revealing to test for imbalanced peripheral T- and B-cell elimination and tolerance in lpr mutant mice. Such an imbalance might involve the APO-1/Fas antigen and provide further insight into the pathophysiology of autoimmune diseases.

A second case in which the APO-1 system might be helpful is the investigation of the role of apoptosis in tumour growth and regression. Apoptosis occurs normally in tumours albeit at largely different rates. Since apoptosis coexists with proliferation, tumour growth might be envisaged as an imbalance between these two phenomena with proliferation winning over apoptosis. If one could find means to specifically increase apoptosis in tumours and leave vital normal tissue intact, one might greatly improve chances of tumour regression. The APO-1 system might help to study this situation. We have transplanted APO-1-positive, apoptosis-sensitive B lymphoblastoid tumours into nu/nu or scid mice. Mice carrying large tumours of approximately 2 cm in diameter were injected with anti-APO-1 and tumour sizes were measured following this injection. In comparison to mice injected with control antibodies, dramatic anti-tumour effects were seen in the anti-APO-1 injected mice. Within days after anti-APO-1 injection, tumours began to regress and were macroscopically lost completely after about one week only. We found that tumour regression was dependent on antibody half-life, concentration, and penetration into the tumour. Most importantly, however, complement-dependent lysis and antibody-dependent cytotoxicity (ADCC) could be excluded and apoptosis was the main if not the only mechanism of induced tumour regression (Dhein et al., 1392). These data reinforce the importance of apoptosis as a future means of anti-tumour therapy.

The potent anti-tumour effects by induction of apoptosis might be quenched in several ways: by loss of cell surface molecules triggering this process, or by an alteration of the signal machinery in the pathway of apoptosis. Such changes might occur in a fraction of tumour cells and turn apoptosis-sensitive into resistant cells. Finally, resistant cells might overgrow and kill the tumour host. Again, the APO-1 system might help to understand this situation. Indeed, depending on the model system of malignant cells used, we have observed anti-APO-1 resistant tumour cells *in vitro* and *in vivo* (Debatin et al. and Daniel et al., to be published). Such resistance occurred either spontaneously in a small fraction of tumour cells or developed under anti-APO-1 antibody pressure in culture or upon anti-APO-1 injection into tumour-bearing mice. Furthermore, an anti-APO-1 positive, *in vitro* anti-APO-1 resistant tumour transplanted into *scid* mice, also proved to be anti-APO-1 resistant *in vivo* and killed the animals.

Future directions of research in this area will have to be aimed at understanding the molecular mechanisms and the genes involved that determine sensitivity or resistance to apoptosis. This understanding might ultimately provide us with means to advantageously manipulate this process and influence the clinical course of, for example, autoimmunity and cancer.

REFERENCES

Debatin, K.-M., Goldmann, C. K., Bamford, R., Waldmann, T. A. and Krammer, P. H. (1990) Monoclonal antibody mediated apoptosis in adult T-cell leukaemia. *Lancet*, **335**, 497–500.

Dhein, J., Daniel, P. T., Trauth, B. C., Oehm, A., Moller, P. and Krammer, P. H. (1992) Induction of apoptosis by monoclonal antibody anti-APO-1 class switch variants is dependent on crosslinking of APO-1 cell surface antigens. *Journal of Immunology*, **149**, 3166–3173.

Falk, M. H., Trauth, B. C., Debatin, K.-M., Klas, C., Gregory, C. D., Rickinson, A. B., Calender, A., Lenoir, G. M., Ellwarth, J. W., Krammer, P. H. and Bornkamm, G. W. (1992) Expression of the APO-1 antigen in Burkitt lymphoma cell lines correlates with a shift towards a lymphoblastoid phenotype. *Blood*, **79**, 3300–3306.

Itoh, N., Yonehara, S., Ishii, A., Yonehara, M., Mizushima, S.-I., Sameshima, M., Hase, A., Seto, Y. and Nagata. S. (1992) The polypeptide encoded by the cDNA for human cell surface antigen Fas can mediate apoptosis. *Cell*, **66**, 233–243.

Kohler, H.-R., Dhein, J., Alberti, G. and Krammer, P. H. (1990) Ultrastructural analysis of apoptosis induced by the monoclonal antibody anti-APO-1 on a lymphoblastoid B cell line. *Ultrastructral Pathology*, **14**, 513–518.

Krammer, P. H. and Debatin, K.-M. (1992) When apoptosis fails. *Current Biology*, **2** (7), 383–385.

Mallett, S. and Barclay, A. N. (1991) A new superfamily of cell surface proteins related to the nerve growth factor receptor. *Immunology Today*, **12**, 220–223.

Oehm, A., Behrmann, I., Falk, W., Li-Weber, M., Maier, G., Klas, C., Richards, S., Dhein, J., Daniel, P. T., Knipping, E., Trauth, B. C., Ponstingl, H. and Krammer, P. H. (1992) Molecular cloning and expression of the APO-1 cell surface antigen, a new member of the TNF receptor superfamily. *Journal of Biological Chemistry*, **267**, 10 709–10 715.

Trauth, B. C., Klas, C., Peters, A. M. J., Matzku, S., Moller, P., Falk, W., Debatin, K.-M. and Krammer, P. H. (1989) Monoclonal antibody-mediated tumor regression by induction of apoptosis. *Science*, **245**, 301–305.

Watanabe-Fukunaga, R., Brannan, C. I., Copeland, N. G., Jenkins, N. A. and Nagata, S. (1992) Lymphoproliferation disorder in mice explained by defects in Fas antigen that mediates apoptosis. *Nature*, **356**, 314–317.

Chapter 20

APOPTOSIS IN THE TARGET ORGAN OF AN AUTOIMMUNE DISEASE

Michael P. Pender

Neuroimmunology Research Unit, Department of Medicine, Clinical Sciences Building, Royal Brisbane Hospital, Brisbane 4029, Australia

INTRODUCTION

An autoimmune disease is one where tissues are damaged as a result of an immune attack targeted to self-antigens expressed within those tissues. Autoimmune diseases may be organ-specific or non-organ-specific. In organ-specific autoimmune diseases, such as Hashimoto's thyroiditis or type I diabetes mellitus, the tissue damage is restricted to the particular target organ whose epithelial cells (parenchymal cells) express the targeted antigen(s). In non-organ-specific autoimmune diseases, such as systemic lupus erythematosus, multiple organs are damaged by an immune response directed at antigens common to different organs.

Experimental autoimmune encephalomyelitis (EAE) is an organ-specific T-cell mediated demyelinating disease of the central nervous system (CNS) and is widely studied as a model of the human demyelinating disease, multiple sclerosis. EAE is also a prototype for T-cell mediated autoimmunity in general. It is induced by inoculation with myelin antigens, such as myelin basic protein (MBP), together with adjuvants, or by the passive transfer of T lymphocytes sensitized to myelin antigens. Myelin in the normal CNS is formed by compaction of the plasma membranes of oligodendrocytes. The mechanism of demyelination in EAE is unknown although it is widely held that the primary target of the immune attack is the myelin sheath rather than the oligodendrocyte, and that this myelin damage is caused by macrophages recruited by infiltrating autoreactive T cells.

Apoptosis is a mechanism of cytotoxic T-cell induced target cell death *in vitro* (Golstein *et al.*, 1991; Ju, 1991). Therefore the presence of apoptosis of parenchymal cells within the target organ of an autoimmune disease may be an indication of primary destruction of those cells by cytotoxic T cells. We examined the spinal cords of Lewis rats with EAE to determine whether oligodendrocyte apoptosis was present. At light and electron microscopy we identified apoptotic cells within the grey and white matter of the spinal cord (Pender *et al.*, 1991) (Figure 20.1). We concluded that some of these apoptotic cells were most likely oligodendrocytes. However, we also concluded that other apoptotic cells were most likely T lymphocytes. Using the technique of pre-embedding immunocytochemistry we subsequently showed that half of the apoptotic cells within the spinal cord in acute EAE are T lymphocytes, predominantly β T cells (Pender *et al.*, 1992). In the present chapter I will discuss the possible significance of these findings of oligodendrocyte apoptosis and T-cell apoptosis for the pathogenesis and immunoregulation of this autoimmune disease.

OLIGODENDROCYTE APOPTOSIS IN EAE

We hypothesized that oligodendrocyte apoptosis is occurring in EAE as a result of specific T-cell cytotoxicity and that this contributes to the CNS demyelination (Pender *et al.*, 1991). It would be predicted that oligodendrocyte apoptosis would be followed by phagocytosis of the oligodendrocyte and the multiple myelin sheaths supported by that oligodendrocyte. Hence, demyelination carried out by macro-

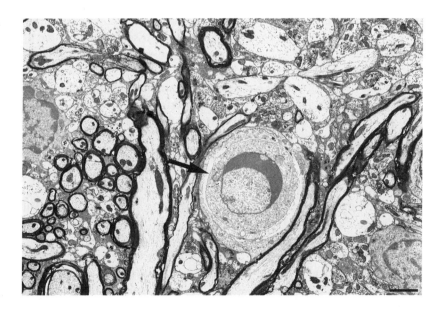

Figure 20.1 Electron micrograph of the spinal cord of a rat with acute MBP-EAE 11 days after inoculation. An apoptotic cell with a characteristic crescent of marginated compacted chromatin in its nucleus (*arrow*) lies adjacent to myelinated fibres at the grey-white matter junction. Bar = 2 μm.

phages invading and ingesting the myelin sheath could be initiated by specific cytotoxic T cells inducing oligodendrocyte apoptosis. A difficulty with this hypothesis is that EAE is mediated by CD4$^+$ T cells, which recognize antigen in the context of class II major histocompatibility complex (MHC) molecules (Pettinelli and McFarlin, 1981; Zamvil et al., 1985), but that such class II MHC molecules are not expressed by oligodendrocytes (Turnley et al., 1991). However, MBP-specific or oligodendrocyte-specific T cells and MBP-specific T-cell hybridoma cells kill oligodendrocytes in vitro (Kawai and Zweiman, 1988, 1990; Jewtoukoff et al., 1989; Kawai et al., 1991). We have observed normal β T cells in intimate contact with apoptotic cells in vivo in EAE (Pender et al., 1992). While we have not yet determined whether these particular apoptotic cells are oligodendrocytes, we expect that T-cell mediated oligodendrocyte cytotoxicity would involve such contact. Possible mechanisms for oligodendrocyte-directed cell-mediated cytotoxicity include: (1) cytotoxicity mediated by CD4$^+$ T cells in the context of oligodendrocyte expression of class II MHC antigen below the threshold for immunocytochemical detection; (2) cytotoxicity mediated by CD8$^+$ T cells in the context of the expression of class I MHC antigen which can be induced on oligodendrocytes (Turnley et al., 1991); (3) non-MHC-restricted T-cell cytotoxicity (Jewtoukoff et al., 1989); (4) cytotoxicity mediated by natural killer cells; (5) bystander cytotoxicity (Erb et al., 1990).

APOPTOSIS OF PARENCHYMAL CELLS IN THE TARGET ORGAN IN OTHER AUTOIMMUNE DISEASES

Kerr et al. (1979) have proposed that the "piecemeal necrosis" that occurs in the liver in human chronic active hepatitis, a putative autoimmune disease, is due to apoptosis of hepatocytes as a result of interaction with T lymphocytes. Apoptosis of hepatocytes in association with lymphocytic infiltration has also been reported to occur in transgenic mice expressing foreign class I MHC antigen in the liver, after irradiation and reconstitution with spleen cells (Morahan et al., 1989).

APOPTOSIS OF T LYMPHOCYTES IN THE CNS IN EAE

Possible Mechanisms

As mentioned in the introduction we have recently shown that apoptosis of β T lymphocytes occurs in the spinal cord of Lewis rats with acute EAE induced by inoculation with MBP and adjuvants (MBP-EAE) (Pender et al., 1992). We used the technique of pre-embedding immunolabelling which allows sufficient preservation of the ultrastructure to permit recognition of apoptotic changes while at the same time preserving surface antigens so that the identity of the apoptotic cells can be determined by immunocytochemistry. We found that 5% of the β T lymphocytes in the spinal cord were apoptotic 2–5 days after the onset of neurological signs.

At present we do not know what proportion of the apoptotic T cells in the spinal cord in EAE are MBP-specific. As only a minority of the infiltrating cells in the CNS in EAE are MBP-specific (Smith and Waksman, 1969; Sedgwick et al., 1987), it is possible that all of the apoptotic T cells that we observe are MBP-specific. Selective elimination by apoptosis within the spinal cord may contribute to the low yield of MBP-specific cells in lymphocytes extracted from the spinal cord of rats with acute MBP-EAE (Cohen et al., 1987).

Apoptotic elimination of autoreactive T cells in the target organ of this autoimmune disease may account for the subsidence of inflammation in the CNS and the self-limited nature of this disease in the Lewis rat.

Several mechanisms could be responsible for apoptosis of β T lymphocytes in the target organ of this autoimmune disease. Firstly, T cells may undergo apoptosis in the CNS as a result of an intrinsic programme to self-destruct after fulfilling their effector function. This seems unlikely, as cytotoxic T cells can recycle after killing their targets (Martz, 1976). Secondly, apoptosis of encephalitogenic T cells may occur as a result of cytotoxicity by T cells specifically targeted against the encephalitogenic T cells (cytolytic T-T-cell interactions) (Sun et al., 1988). Thirdly, the endogenous corticosterone released during the course of EAE (Levine et al., 1980; MacPhee et al., 1989) may cause the T-cell apoptosis, as glucocorticoids induce apoptosis of lymphoid cells (La Pushin and de Harven, 1971; Wyllie, 1980; Zubiaga et al., 1992). Lastly, T-cell apoptosis in the CNS may represent activation-induced cell death. This process results in clonal deletion of autoreactive immature T cells in the thymus during normal development (Smith et al., 1989; Murphy et al., 1990); it also eliminates mature T cells *in vitro* (Newell et al., 1990; Liu and Janeway, 1990; Russell et al., 1991; Lenardo, 1991) and *in vivo*, in the lattercase providing a mechanism of extrathymic (peripheral) tolerance to foreign or self-antigen (Jones et al., 1990; Webb et al., 1990; Kawabe and Ochi, 1991; Rocha and von Boehmer, 1991; Carlow et al., 1992).

T-cell activation by occupancy of the antigen-specific receptor can result in proliferation, anergy or apoptosis, the outcome being determined by the presence and timing of other signals such as the co-stimulatory signal (Lafferty et al., 1983; Mueller et al., 1989) and the cytokines, interleukin-2 (IL-2) (De Silva et al., 1991; Lenardo, 1991) and interferon-γ (IFN-γ) (Liu and Janeway, 1990). In the early induction phase of acute MBP-EAE, activation of MBP-specific T cells in lymphoid organs leads to proliferation; however, we hypothesize that, in the later effector phase of the disease, activation of MBP-specific T cells within the CNS may lead to apoptosis, because of the failure of non-specialized antigen-presenting cells (APCs) such as astrocytes to produce the co-stimulatory signal or because of the different availability of cytokines (Pender et al., 1992). This hypothesis is supported by our finding that the proportion of T cells dying by apoptosis in the CNS in EAE is much higher than that observed in normal or antigen-primed lymph nodes, which suggests that the environment in the CNS may be an important factor contributing to the induction of T-cell apoptosis in this autoimmune disease (Pender et al., 1992). Further evidence for the importance of the target organ environment in determining T-cell apoptosis is provided by the observation that the frequency of T-cell apoptosis is greater in the CNS than in the lymph nodes or inoculation site in EAE (H. Lassmann, personal communication).

Activation-induced T-cell Apoptosis in the Target Organ as a Possible Mechanism of Extrathymic Self-Tolerance

Previously it had been suggested that induced expression of class II antigens on parenchymal cells such as astrocytes may allow presentation of self antigens to potentially autoreactive T cells and induce autoimmune disease (Bottazzo et al., 1983; Fontana et al., 1984). More recently, however, it has been suggested that the expression of class II MHC on parenchymal cells may serve as an extrathymic mechanism for maintaining self tolerance (Markmann et al., 1988) and that astrocytes downregulate T-cell proliferation (Matsumoto et al., 1992). Ohmori et al. (1992) have recently shown that there is little T-cell proliferation within the CNS in EAE and attributed this to the downregulatory effects of glial cells expressing class II MHC antigen. However, we would suggest that the downregulation may take the form of activation-induced T-cell apoptosis rather than T-cell anergy, which was proposed by Ohmori et al. (1992). Thus interaction of autoreactive T cells with class II MHC-expressing parenchymal nonspecialized APCs may lead to activation-induced T-cell apoptosis in the target organ and be a protective mechanism against the perpetuation of autoimmunity. As a corollary to this hypothesis, individuals with parenchymal cells capable of producing the co-stimulatory signal may be susceptible to self-perpetuating chronic autoimmune disease in contrast to the self-limited monophasic disease that characterizes acute EAE in the Lewis rat.

When considering the concept of activation-induced T-cell apoptosis in the CNS or other target organ, the term "activation" needs to be clarified. Autoreactive T cells that have recently become fully activated effector cells following interaction with specialized APCs in the lymph nodes would be expected to be able to cross the intact blood-brain barrier (Wekerle et al., 1986) and produce target organ damage after recognition of the relevant peptide in the context of MHC antigen but without the need for a co-stimulatory signal. On the other hand, autoreactive T cells that are still progressing through the cell cycle following initial priming in the lymph node may be susceptible to activation-induced apoptosis in the target organ after interaction with nonspecialized APCs that fail to deliver the co-stimulatory signal (see Russell et al., 1991). Furthermore, it is not known what happens to a fully activated T cell if that T cell does not meet its target within a limited period of time.

Activation-induced apoptosis of autoreactive T cells in the target organ may also explain why normal animals, which harbour potentially encephalitogenic T cells (Schluesener and Wekerle, 1985), do not spontaneously develop EAE. Inoculation with myelin antigens and adjuvants is necessary to induce the production of enough fully activated autoreactive effector T cells in the lymph nodes to cause target organ damage.

Lewis rats that have recovered from acute EAE induced by inoculation with MBP are tolerant to MBP, as evidenced by resistance to re-induction of EAE by active immunization (Willenborg, 1979). An acute elimination of MBP-specific T cells within the CNS would not be expected to produce longstanding tolerance, since precursors in the lymphoid organs (Willenborg, 1979) could expand to produce new effector T cells; however, an ongoing low level of T-cell apoptosis in the CNS could contribute to the tolerant state that develops after an attack of acute EAE.

Relationship Between T-cell Apoptosis and T-cell Anergy in Extrathymic Self Tolerance

T-cell anergy has been proposed as an extrathymic mechanism for the maintenance of self tolerance (Schwartz, 1989). T-cell anergy can be induced in T-cell clones by T-cell receptor occupancy (signal 1) in the absence of co-stimulation (signal 2) (Mueller et al., 1989). It may be reversed by stimulation with IL-2 (Beverly et al., 1992) and may represent defects in the early stages of T-cell intracellular signalling (LaSalle et al., 1992). It is apparent that the same conditions that lead to activation-induced T-cell apoptosis (lack of co-stimulatory signal) lead to T-cell anergy. This raises the possibility that T-cell anergy represents not an antigen-induced change in the function of certain T cells, as generally assumed, but rather the survival of a subpopulation of T cells after clonal deletion by activation-induced apoptosis. Indeed relatively few T helper 1 cells (less than 20%) survive anergizing treatment *in vitro* (cited as a personal communication with M. Jenkins by Ucker et al.(1992)). A subpopulation of T cells may survive activation-induced apoptosis as a result of a low avidity of the T-cell receptor or as a result of defective intracellular signalling through the T-cell receptor. Such surviving T cells would have the characteristics of "anergic" T cells—failure to proliferate on specific antigenic stimulation. Thus many instances of T-cell anergy may simply represent the survival of defective T cells that, because of their signalling defect, can not be eliminated by activation-induced apoptosis.

Possible Effect of Cyclosporin A on Activation-induced T-cell Apoptosis in the Target Organ

Cyclosporin A is an immunosuppressive agent that acts by binding to an endogenous intracellular protein, cyclophilin. The cyclosporin A/cyclophilin complex inhibits the protein phosphatase, calcineurin, which is required in the intracellular signalling pathway between T-cell receptor ligation and the transcription of early genes (Schreiber and Crabtree, 1992). Cyclosporin A inhibits activation-induced cell death of thymocytes, T-cell hybridomas and mature T cells (Shi et al., 1989; Liu and Janeway, 1990). Interestingly, low dose cyclosporin A converts acute EAE into a chronic relapsing disease in the Lewis rat (Polman et al., 1988; Pender et al., 1990). One possible explanation for this effect of cyclosporin A is that it may inhibit activation-induced T-cell apoptosis within the CNS thus allowing perpetuation of the autoimmune attack.

Alternative explanations for the perpetuating effect of cyclosporin A on EAE include inhibition of activation-induced apoptosis of autoreactive thymocytes which may be produced in the adult.

APOPTOSIS OF T CELLS IN THE TARGET ORGAN OF OTHER AUTOIMMUNE DISEASES

Activation-induced apoptosis of autoreactive T cells within the target organ as a result of interaction with parenchymal non-specialized APCs may be a general phenomenon in self-limited autoimmune diseases and not restricted to autoimmune CNS disease. Fields and Loh (1992) have recently demonstrated pancreatic acinar autoimmune disease and the peripheral elimination of autoreactive T cells in doubly transgenic mice expressing a foreign class I MHC on pancreatic acinar cells. The site and mechanism of peripheral T-cell elimination were not determined, but activation-induced apoptosis of T cells within the pancreatic acinae could be the explanation for this extrathymic induction of tolerance. Honma et al. (1989) have reported apoptosis of infiltrating lymphocytes within the connective tissue of erythema nodosum-like lesions of corticosteroid-treated patients with Behçet's syndrome, a systemic inflammatory disease of unknown but possibly autoimmune aetiology.

CONCLUSION

Apoptosis occurs in the target organ of the T-cell mediated autoimmune disease, EAE. A minority of the apoptotic cells are likely to be parenchymal cells of the CNS, namely oligodendrocytes, while the majority are infiltrating T lymphocytes. We hypothesize that oligodendrocyte apoptosis is occurring as a result of specific T-cell cytotoxicity and that this contributes to the demyelination. There are several possible mechanisms for the T-cell apoptosis but the most likely are corticosteroid-induced apoptosis or activation-induced apoptosis as a result of interaction with non-specialized APCs. It is hypothesized that T-cell apoptosis in the target organ may be a protective mechanism that also occurs in other self-limited, T-cell mediated autoimmune diseases and that it may be a general mechanism for maintaining extrathymic tolerance.

ACKNOWLEDGEMENTS

I am grateful to Drs P. A. McCombe and Z. Tabi for helpful discussion and for comments on the manuscript. This work was supported by the National Health and Medical Research Council of Australia and by the National Multiple Sclerosis Society of Australia.

REFERENCES

Beverly, B., Kang, S-M., Lenardo, M. J. and Schwartz, R. H. (1992) Reversal of *in vitro* T-cell clonal anergy by IL-2 stimulation. *International Immunology*, 4, 661–671.

Bottazzo, G. F., Pujol-Borrell, R., Hanafusa, T. and Feldmann, M. (1983) Role of aberrant HLA-DR expression and antigen presentation in induction of endocrine autoimmunity. *Lancet*, (ii), 1115–1119.

Carlow, D. A., Teh, S. J., van Oers, N. S. C., Miller, R. G., and Teh, H-S. (1992) Peripheral tolerance through clonal deletion of mature $CD4^-CD8^+$ T cells. *International Immunology*, 4, 599–610.

Cohen, J. A., Essayan, D. M., Zweiman, B. and Lisak, R. P. (1987) Limiting dilution analysis of the frequency of antigen-reactive lymphocytes isolated from the central nervous system of Lewis rats with experimental allergic encephalomyelitis. *Cell Immunology*, 108, 203–213.

De Silva, D. R., Urdahl, K. B. and Jenkins, M. K. (1991) Clonal anergy is induced *in vitro* by T-cell receptor occupancy in the absence of proliferation. *Journal of Immunology*, 147, 3261–3267.

Erb, P., Grogg, D., Troxler, M., Kennedy, M. and Fluri, M. (1990) $CD4^+$ T-cell mediated killing of MHC class II-positive antigen-presenting cells. I. Characterization of target cell recognition by *in vivo* or *in vitro* activated $CD4^+$ killer T cells. *Journal of Immunology*, 144, 790–795.

Fields, L. E. and Loh, D. Y. (1992) Organ injury associated with extrathymic induction of immune tolerance in doubly transgenic mice. *Proceedings of the National Academy of Science, USA*, 89, 5730–5734.

Fontana, A., Fierz, W. and Wekerle, H. (1984) Astrocytes present myelin basic protein to encephalitogenic T-cell lines. *Nature*, 307, 273–276.

Golstein, P., Ojcius, D. M. and Young, J. D-E. (1991) Cell death mechanisms and the immune system. *Immunology Reviews*, 121, 29–65.

Honma, T., Bang, D., Saito, T., Nakagawa, S. and Ueki, H. (1989) Assessment of apoptosis of infiltrating lymphocytes in erythema nodosum-like lesions of corticosteroid-treated patients with Behçet's syndrome. *Journal of Submicroscopic Cytology and Pathology*, 21, 691–701.

Jewtoukoff, V., Lebar, R. and Bach, M-A. (1989) Oligodendrocyte-specific autoreactive T cells using an α/β T-cell receptor kill their target without self restriction. *Proceedings of the National Academy of Science, USA*, 86, 2824–2828.

Jones, L. A., Chin, L. T., Longo, D. L. and Kruisbeek, A. M. (1990) Peripheral clonal elimination of functional T cells. *Science*, 250, 1726–1729.

Ju, S-T. (1991) Distinct pathways of CD4 and CD8 cells induce rapid target DNA fragmentation. *Journal of Immunology*, 146, 812–818.

Kawabe, Y. and Ochi, A. (1991) Programmed cell death and extrathymic reduction of $V\beta8^+$ $CD4^+$ T cells in mice tolerant to *Staphylococcus aureus* enterotoxin B. *Nature*, 349, 245–248.

Kawai, K. and Zweiman, B. (1988) Cytotoxic effect of myelin basic protein-reactive T cells on cultured oligodendrocytes. *Journal of Neuroimmunology*, 19, 159–165.

Kawai, K. and Zweiman, B. (1990) Characteristics of *in vitro* cytotoxic effects of myelin basic protein-reactive T-cell lines on syngeneic oligodendrocytes. *Journal of Neuroimmunology*, 26, 57–67.

Kawai, K., Heber-Katz and E., Zweiman, B. (1991) Cytotoxic effects of myelin basic protein-reactive T-cell hybridoma cells on oligodendrocytes. *Journal of Neuroimmunology*, 32, 75–81.

Kerr, J. F. R., Cooksley, W. G. E., Searle, J., Halliday, J. W., Halliday, W. J., Holder, L., Roberts, I., Burnett, W. and Powell, L. W. (1979) The nature of piecemeal necrosis in chronic active hepatitis. *Lancet*, (ii), 827–828.

Lafferty, K. J., Prowse, S. J., Simeonovic, C. J. and Warren, H.S. (1983) Immunobiology of tissue transplantation: a return to the passenger leukocyte concept. *Annual Review of Immunology*, 1, 143–173.

La Pushin, R. W. and de Harven, E. (1971) A study of gluco-corticosteroid-induced pyknosis in the thymus and lymph node of the adrenalectomized rat. *Journal of Cellular Biology*, 50, 583–597.

LaSalle, J. M., Tolentino, P. J., Freeman, G. J., Nadler, L. M. and Hafler, D. A. (1992) Early signaling defects in human T cells anergized by T-cell presentation of autoantigen. *Journal of Experimental Medicine*, 176, 177–186.

Lenardo, M. J. (1991) Interleukin-2 programs mouse β T lymphocytes for apoptosis. *Nature*, 353, 858–861.

Levine, S., Sowinski, R. and Steinetz, B. (1980) Effects of experimental allergic encephalomyelitis on thymus and adrenal: relation to remission and relapse. *Proceedings of the Society of Experimental and Biological Medicine*, 165, 218–224.

Liu, Y. and Janeway, C. A. (1990) Interferon plays a critical role in induced cell death of effector T cell: a possible third mechanism of self-tolerance. *Journal of Experimental Medicine*, 172, 1735–1739.

MacPhee, I. A. M., Antoni, F. A. and Mason, D. W. (1989) Spontaneous recovery of rats from experimental allergic encephalomyelitis is dependent on regulation of the immune system by endogenous adrenal corticosteroids. *Journal of Experimental Medicine*, 169, 431–445.

Markmann, J., Lo, D., Naji, A., Palmiter, R. D., Brinster, R. L. and Heber-Katz, E. (1988) Antigen presenting function of class II MHC expressing pancreatic beta cells. *Nature*, 336, 476–479.

Martz, E. (1976) Multiple target cell killing by the cytolytic T lymphocyte and the mechanism of cytotoxicity. *Transplantation*, 21, 5–11.

Matsumoto, Y., Ohmori, K. and Fujiwara, M. (1992) Immune regulation by brain cells in the central nervous system: microglia but not astrocytes present myelin basic protein to encephalitogenic T cells under *in vivo*-mimicking conditions. *Immunology*, 76, 209–216.

Morahan, G., Brennan, F. E., Bhathal, P. S., Allison, J., Cox, K. O. and Miller, J. F. A. P. (1989) Expression in transgenic mice of class I histocompatibility antigens controlled by the metallothionein promoter. *Proceedings of the National Academy of Science, USA*, 86, 3782–3786.

Mueller, D. L., Jenkins, M. K. and Schwartz, R. H. (1989) Clonal expansion versus functional clonal inactivation: a co-stimulatory signalling pathway determines the outcome of T-cell antigen receptor occupancy. *Annual Review of Immunology*, 7, 445–480.

Murphy, K. M., Heimberger, A. B. and Loh, D. Y. (1990) Induction by antigen of intrathymic apoptosis of $CD4^+CD8^+TCR^{lo}$ thymocytes *in vivo*. *Science*, 250, 1720–1723.

Newell, M. K., Haughn, L. J., Maroun, C. R. and Julius, M. H. (1990) Death of mature T cells by separate ligation of CD4 and the T-cell receptor for antigen. *Nature*, 347, 286–289.

Ohmori, K., Hong, Y., Fujiwara, M. and Matsumoto, Y. (1992) *In situ* demonstration of proliferating cells in the rat central nervous system during experimental autoimmune encephalomyelitis: evidence suggesting that most infiltrating T cells do not proliferate in the target organ. *Laboratory Investigation*, 66, 54–62.

Pender, M. P., McCombe, P. A., Yoong, G. and Nguyen, K. B. (1992) Apoptosis of $\alpha\beta$ T lymphocytes in the nervous system in experimental autoimmune encephalomyelitis: its possible implications for recovery and acquired tolerance. *Journal of Autoimmunity*, 5, 401–410.

Pender, M. P., Nguyen, K. B., McCombe, P. A., and Kerr, J. F. R. (1991) Apoptosis in the nervous system in experimental allergic encephalomyelitis. *Journal of Neurological Science*, 104, 81–87.

Pender, M. P., Stanley, G. P., Yoong, G. and Nguyen, K. B. (1990) The neuropathology of chronic relapsing experimental allergic encephalomyelitis induced in the Lewis rat by inoculation with whole spinal cord and treatment with cyclosporin A. *Acta Neuropathologica*, 80, 172–183.

Pettinelli, C. B., McFarlin, D. E. (1981) Adoptive transfer of experimental allergic encephalomyelitis in SJL/J mice after *in vitro* activation of lymph node cells by myelin basic protein: requirement for Lyt $1^+ 2^-$ T lymphocytes. *Journal of Immunology*, **127**, 1420–1423.

Polman, C. H., Matthaei, I., de Groot, C. J. A., Koetsier, J. C., Sminia, T. and Dijkstra, C. D. (1988) Low-dose cyclosporin A induces relapsing remitting experimental allergic encephalomyelitis in the Lewis rat. *Journal of Neuroimmunology*, **17**, 209–216.

Rocha, B. and von Boehmer, H. (1991) Peripheral selection of the T-cell repertoire. *Science*, **251**, 1225–1228.

Russell, J. H., White, C. L., Loh, D. Y. and Meleedy-Rey, P. (1991) Receptor-stimulated death pathway is opened by antigen in mature T cells. *Proceedings of the National Academy of Science, USA*, **88**, 2151–2155.

Schluesener, H. J. and Wekerle, H. (1985) Autoaggressive T lymphocyte lines recognizing the encephalitogenic region of myelin basic protein: *in vitro* selection from unprimed rat T lymphocyte populations. *Journal of Immunology*, **135**, 3128–3133.

Schreiber, S. L. and Crabtree, G. R. (1992) The mechanism of action of cyclosporin A and FK506. *Immunology Today*, **13**, 136–142.

Schwartz, R. H. (1989) Acquisition of immunologic self-tolerance. *Cell*, **57**, 1073–1081.

Sedgwick, J., Brostoff, S. and Mason, D. (1987) Experimental allergic encephalomyelitis in the absence of a classical delayed-type hypersensitivity reaction. Severe paralytic disease correlates with the presence of interleukin-2 receptor-positive cells infiltrating the central nervous system. *Journal of Experimental Medicine*, **165**: 1058–1075.

Shi, Y., Sahai, B. M. and Green, D. R. (1989) Cyclosporin A inhibits activation-induced cell death in T-cell hybridomas and thymocytes. *Nature*, **339**, 625–626.

Smith, C. A., Williams, G. T., Kingston, R., Jenkinson, E. J. and Owen, J. J. T. (1989) Antibodies to CD3/T-cell receptor complex induce death by apoptosis in immature T cells in thymic cultures. *Nature*, **337**, 181–184.

Smith, S. B. and Waksman, B. H. (1969) Passive transfer and labelling studies on the cell infiltrate in experimental allergic encephalomyelitis. *Journal of Pathology*, **99**, 237–244.

Sun, D., Qin, Y., Chluba, J., Epplen, J. T. and Wekerle, H. (1988) Suppression of experimentally induced autoimmune encephalomyelitis by cytolytic T-T-cell interactions. *Nature*, **332**, 843–845.

Turnley, A. M., Miller, J. F. A. P. and Bartlett, P.F . (1991) Regulation of MHC molecules on MBP positive oligodendrocytes in mice by IFN-γ and TNF-α. *Neuroscience Letters*, **123**, 45–48.

Ucker, D. S., Meyers, J. and Obermiller, P. S. (1992) Activation-driven T-cell death. II. Quantitative differences alone distinguish stimuli triggering nontransformed T-cell proliferation or death. *Journal of Immunology*, **149**, 1583–1592.

Webb, S., Morris, C. and Sprent, J. (1990) Extrathymic tolerance of mature T cells: clonal elimination as a consequence of immunity. *Cell*, **63**, 1249–1256.

Wekerle, H., Linington, C., Lassmann, H. and Meyermann, R. (1986) Cellular immune reactivity within the CNS. *Trends in Neurosciences*, **9**, 271–277.

Willenborg, D. O. (1979) Experimental allergic encephalomyelitis in the Lewis rat: studies on the mechanism of recovery from disease and acquired resistance to reinduction. *Journal of Immunology*, **123**, 1145–1150.

Wyllie, A. H. (1980) Glucocorticoid-induced thymocyte apoptosis is associated with endogenous endonuclease activation. *Nature*, **284**, 555–556.

Zamvil, S., Nelson, P., Trotter, J., Mitchell, D., Knobler, R., Fritz, R. and Steinman, L. (1985) T-cell clones specific for myelin basic protein induce chronic relapsing paralysis and demyelination. *Nature*, **317**, 355–358.

Zubiaga, A. M., Munoz, E. and Huber, B. T. (1992) IL-4 and IL-2 selectively rescue T-cell subsets from glucocorticoid-induced apoptosis. *Journal of Immunology*, **149**, 107–112.

Chapter 21

PEPTIDE EPITOPE-INDUCED APOPTOSIS OF HUMAN CYTOTOXIC T LYMPHOCYTES

A. Suhrbier and S. R. Burrows

Queensland Institute of Medical Research, The Bancroft Centre,
300 Herston Road, Brisbane, Queensland 4029, Australia

Attempts to identify the mechanism(s) of cytotoxic T-cell (CTL)-mediated lysis have been filled with controversy and contradiction. The reasons appear to be that there are multiple lytic mediators used differentially by different CTL and different target cells have different susceptibilities and responses to the different mediators. A necrotic mediator, the granule protein perforin, has been well characterised. Considerable speculation remains as to the nature of the CTL killing mechanism(s) responsible for rapid target cell apoptosis. Two phenomena, peptide mediated CTL-CTL killing and bystander lysis, cast some light on the characteristics of the apoptotic mediator.

PERFORIN-MEDIATED LYSIS RESULTS IN TARGET CELL NECROSIS

CTL appear to have at least two lytic mechanisms, one which involves granule release, is calcium dependent and involves perforin and another which is perforin independent, does not require exocytosis and induces rapid target cell apoptosis (Tren et al., 1987; Young and Liu, 1988a and b; Berke et al., 1991; Lancki et al., 1991). Perforin, which forms pores in the target cell membrane (Tschopp and Nabholz, 1990), causes necrotic death of nucleated cells (Zychlinsky et al., 1991) and, unlike most CTL, appears incapable of inducing apoptosis (Duke et al., 1989). Co-localising with perforin in the lytic granules of CTL are a series of serine proteases known as granzymes (Peters et al., 1991). Although incapable of causing cell death by themselves they are suggested to be important for efficient functioning of perforin and/or to release the CTL from its target to allow it to kill another target (Hudig et al., 1991).

That different CTL make contrasting use of the two mediators was elegantly demonstrated by the differential ability of a series of $CD4^+$ CTL clones to lyse RBC and nucleated target cells (Lancki et al., 1991). A mast cell tumour line transfected with perforin gene was capable of lysing RBC but was very poor at lysing nucleated target cells (Shiver and Henkart, 1991). These differences might simply reflect differences in the quantity of functional perforin secreted by these effector cells. RBC are much more sensitive to perforin (and complement) than nucleated cells. However, an important observation made by Lancki et al. (1991) was that some CTL clones were inefficient at lysing RBC but were fully capable of lysing nucleated target cells in a 3.5 hour chromium release assay.

CTL-MEDIATED APOPTOSIS

The characteristics of target cell apoptosis induced by CTLs include rapid (as early as 5 min) fragmentation of target cell DNA, subsequent nuclear condensation and margination of chromatin. Three to five hours after delivery of the lethal hit, the target cell lyses, a process conventionally measured using the chromium release assay (Goldstein et al., 1991; Taylor and Cohen, 1992).

Tumour necrosis factor, lymphotoxin and interferon gamma (Schmid et al., 1986; Ju et al., 1990) are candidates for such apoptotic agents, although when these compounds are added in the absence of CTL the speed of cell death is slow compared to direct target cell lysis (Goldstein et al., 1991). In addition, certain target cells capable of being lysed by CTL are resistant to very high concentrations of these lymphokines and many fully functional CTL do not express detectable amounts of them. A novel CTL cytotoxin, related to TNF and LT, partially enriched in granules from CTL, also induces slow (>4 h) DNA fragmentation in a variety of tumour cells (Liu et al., 1987). It has been suggested that lymphotoxins, entering target cells with the help of perforin, might be considerably more active and induce much faster DNA fragmentation (Schmid et al., 1986; Liu et al., 1987). A series of granule proteins capable of inducing DNA fragmentation of permeabilised cells have also been described. TIA-1 (Tian et al., 1991), a 15 kDa protein capable of inducing DNA fragmentation of digitonin-permeabilised thymocytes; fragmentin (Shi et al., 1991), a natural killer cell protein related to granzyme B, which induced membrane damage and DNA fragmentation in the presence (but not absence) of non-lytic amounts of perforin; granzyme A, which induces DNA release from detergent-permeabilised cells or from intact cells in conjunction with perforin (Hayes et al., 1989). All these mediators require prior membrane damage for DNAase activity, indicating an absolute requirement for at least a small amount of perforin in the apoptotic death of the target cell.

CTL-MEDIATED LYSIS IS UNIDIRECTIONAL

CTL-mediated lysis is classically described as unidirectional i.e. the CTL does not die following lethal hit delivery to the target cell. If the lytic mediator is soluble and has no specificity for the target cell this observation demands resistance of the CTL to its own secreted lytic agents. A model by Peters *et al.* (1990) argues that the excreted granules do have specificity for the target. CTL appear to resist perforin mediated lysis (Liu *et al.*, 1989) by expressing surface membrane proteins, which inhibit perforin binding activity (Jiang *et al.*, 1990). Single CTL clones have been reported to be immune from self-lysis (Luciani *et al.*, 1986; Vitiello *et al.*, 1989; Robbins *et al.*, 1991) although they seem susceptible to lysis by some other CTL. A hierarchy of susceptibility has been suggested (Skinner and Marbrook, 1987) i.e. CTL A can kill CTL B which can kill CTL C, but B and C cannot kill A and C cannot kill B. One might speculate that A secretes enough perforin to overcome B and C's, but not its own contingent of resistance proteins; B secretes enough to overcome C's, but not its own or A's contingent etc.

PEPTIDE EPITOPE-INDUCED APOPTOSIS OF HUMAN CTL

The application of synthetic peptide technology has demonstrated that $CD8^+$ CTL recognise nine amino acid peptides bound to class I MHC (Figure 21.1). Since the CTL themselves express class I MHC, the addition of the synthetic peptide epitope, for which the CTL clone is specific, to the clone results in the CTL being recognised by themselves (Walden and Eisen, 1990) or their siblings as potential targets. In contrast to two reports, which describe no CTL lysis under these conditions (Luciani *et al.*, 1986; Robbins *et al.*, 1991), cognate peptide mediated lysis of mature primed CTL clones (both $CD8^+$ and $CD4^+$) has been described by several groups (Ottenhoff and Mutis, 1990; Pemberton *et al.*, 1990; Walden and Eisen, 1990; Berke, 1991; Lenardo, 1991). The critical observation made by Moss *et al.* (1991) was that cognate epitope induced death was apoptotic.

Several mechanisms induce apoptosis of T cells. Both immature (Odaka *et al.*, 1990; Shi *et al.*, 1990) and mature unprimed (naive) T cells (Liu and Janeway, 1990; Newell *et al.*, 1990; Russel *et al.*, 1991) appear susceptible to antigen-induced cell death (AICD); an apoptotic suicide process triggered after T cells recognize antigen on an antigen presenting cell. This phenomenon is postulated to be a mechanism for both thymic (immature T cells) and extrathymic (mature naive T cells) tolerance, although it has been reported to occur in primed CTL, dependent on prior exposure of the CTL to IL-2 (Lenardo, 1991). Another postulated mechanism for extrathymic tolerance is veto, whereby mature naive T-cells apoptose following recognition of antigen presented by a cell expressing CD8 (Sambhara and Miller, 1991). As peptide epitope-induced apoptosis occurs in clonal $CD8^+$ populations (Moss *et al.*, 1991) effector CTL may be vetoed by peptide-sensitised $CD8^+$ targets. An effector mechanism of peptide

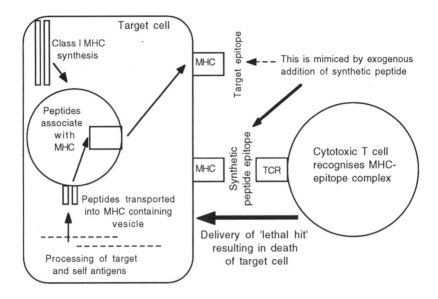

Figure 21.1 Exogenous addition of synthetic peptide epitope mimics natural processing and presentation. Epitopes recognised by CD8+ CTL are peptides between 8 and 10 amino acids long. CTL activation requires 200-300 MHC/epitope/TCR complexes and accessory molecule interactions e.g. CD8, LFA3. Target cell death is usually, but need not be, apoptotic. MHC-Major histocompatability complex; TCR-T-cell receptor.

epitope-induced death of primed CTL clones has been postulated (for murine CTL) to involve single-cell self-recognition and destruction (Walden and Eisen, 1990).

We have demonstrated that the above suicidal effector mechanisms do not play a significant role during peptide epitope-induced apoptosis of primed CTL. Fratricidal CTL-CTL killing was found to be the dominant mechanism with CTL behaving like conventional targets for CTL-mediated lysis. In other words within a clonal population of CTL, one CTL can behave like a conventional target for CTL-mediated apoptosis by one of its brothers (Suhrbier et al., 1993). This observation argues against peptide-mediated apoptosis being a model for, or having any role in, peripheral tolerance or the down regulation of an immune response. Peptide-mediated apoptosis is a manifestation of CTL-mediated target cell apoptosis, all that is different is that sibling CTL are the target. One might postulate that the inability to demonstrate PHA or peptide-mediated CTL-CTL killing in some clones may reflect the exclusive use of perforin by these clones.

Microscopic observations of doublets of CTL clones in the presence of cognate peptide showed that only one CTL in a doublet underwent the chromatin margination characteristic of apoptosis (Suhrbier et al., 1993) (Figure 21.2). Thus effector CTL did not die while delivering the apoptotic lethal hit to fellow target CTL, thus unidirectional lysis was maintained in this system. However, if pelleted together at the bottom

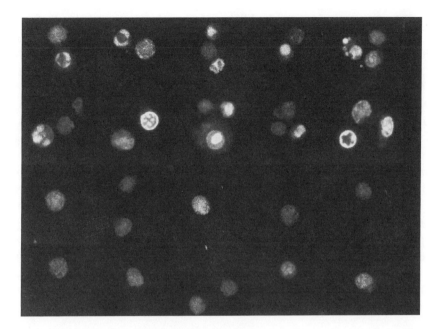

Figure 21.2 Confocal laser micrographs of ten CTL doublets or triplets (top ten of the twenty micrograph composite) and single CTL (bottom ten micrographs of the figure) after 4 h in the presence of cognate peptide and 0.25 µg/ml acridine orange. Apoptosis, resulting in characteristic chromatin margination (Smith et al., 1991), is only observed in one of a pair of CTL. Single cells do not kill themselves. (Suhrbier et al., 1993, reprinted with permission from the *Journal of Immunology*, © Copyright 1993, American Association of Immunologists.)

of a well all the CTL died after overnight incubation with cognate peptide (Burrows et al., 1992). Thus effector CTL must at some point become susceptible to lysis by other CTL and are therefore not innately resistant to the apoptotic mediator(s). Thus a paradox exists; cloned CTL are not refractory to the apoptotic mediator(s) of brother CTL, but are resistant to their own apoptotic mediator when killing peptide-sensitized CTL. Might CTL only be resistant to their own apoptotic mediator(s) at the time when they are engaged in delivery of their own lethal apoptotic hit and thereafter revert to a susceptible state?

BYSTANDER APOPTOSIS BY HUMAN CYTOTOXIC T LYMPHOCYTES

CTL-mediated killing has classically been associated with a high level of specificity for the target cell (Kupfer and Singer, 1989; Shiver and Henkart, 1991). However, when CTL clones and specific target cells were co-pelleted with innocent bystander cells, for which the CTL is not specific, these bystander cells are also lysed (along with the specific targets) (Lanzavecchia et al., 1986; Gromkowski et al., 1988; Barker et al., 1989; Duke, 1989; Fleischer, 1989; Erb et al., 1990). The bystander lysis induced by human

CD8$^+$ CTL clones was found to be HLA unrestricted and particularly efficient when CTL were co-pelleted with allogeneic bystander cells in the presence of cognate synthetic peptide (i.e. the peptide for which the CTL was specific) (Burrows et al., in press). We assume that the peptide-mediated CTL-CTL killing gave rise to bystander lysis (Burrows et al., 1993). Bystander cells were found to die by apoptosis indicating that the CTL apoptotic mediator(s), which induced specific target cell apoptosis, was responsible for bystander apoptosis. The mediator(s) of bystander apoptosis was short lived, acted concurrently with granule exocytosis but did not involve TNF, death of the specific target cell, nor could lytic activity be transferred in supernatants. Most importantly RBC could not be bystander-lysed. Even CTL rosetted with sheep RBC could kill each other (CTL-CTL killing) but could not efficiently bystander-lyse the sRBC. This suggests that bystander apoptosis may not require any contribution from perforin-mediated membrane damage (Burrows et al., in press).

Curiously CTL were not resistant to the bystander apoptosis induced by another allogeneic CTL clone in the presence of synthetic peptide cognate for the allogeneic CTL (Burrows et al., in press). A similar paradox arises; peptide mediated CTL-CTL killing is unidirectional, so the effector is resistant to its own apoptotic mediator yet CTL remain sensitive to the same mediator when it is presented by other CTL.

SELF-PROTECTION FROM THE APOPTOTIC MEDIATOR(S)

To ask whether CTL might be transiently resistant to the apoptotic mediator(s) while they were themselves in the process of delivering a lethal hit to target cells, and thereafter revert to being sensitive, we set up two experiments. (i) To see whether CTL are resistant to direct CTL-mediated apoptosis when the former CTL was killing its target, CTL clone A was co-pelleted with target or non-target cells plus an allospecific CTL clone, which was specific for CTL A. (ii) To see whether CTL were resistant to bystander apoptosis when the CTL was killing its target, CTL clone A was co-pelleted with target or non-target cells plus an allogeneic CTL clone (CTL B) in the presence of epitope cognate for CTL B. In both cases CTL A was lysed just as efficiently, directly (a) or as bystanders (b), when they were, or were not, actively killing target cells (Figure 21.3).

The inability to demonstrate reduced killing of CTL when CTL are actively engaged in target cell lysis argues against a form of transient resistance (or self-protection) whereby the CTL is resistant only during the delivery of its own lethal hit. Unidirectional lysis in the absence of self-protection cannot easily be envisaged if the lytic mediator is soluble and has no specificity for the target cell. One would have to argue that the part of the effector CTL membrane in close proximity to the target or bystander cell was resistant to the CTL mediator whilst the rest of the CTL's plasma membrane remained susceptible. These data are, therefore, consistent with the fast apoptotic CTL-mediated killing mechanism(s) involving a unidirectional membrane-mediated event not requiring any self-protection (Berke, 1991).

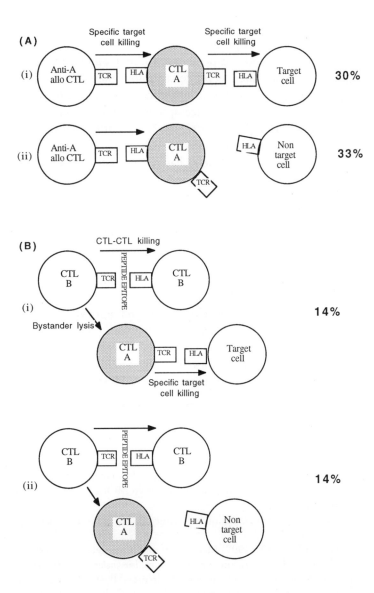

Figure 21.3 CTL are not protected from lethal apoptotic hit received (A) directly or (B) as a bystander when they are delivering the same lethal hit to target cells. The level of CTL A (shaded) % lysis was identical when CTL A was (i) or was not (ii) killing a target. Cells were co-pelleted in a standard 5 h chromium release assay. (Data from Suhrbier et al., 1993 and Burrows et al., in press).

THE APOPTOTIC MEDIATOR(S)

Several features of the CTL apoptotic mediator can be described. The apoptotic mediator(s) operated in bystander experiments without significant RBC membrane lysis suggesting that CTL-mediated target-cell apoptosis may require no contribution from perforin. Some xenogeneic cells were also susceptible to bystander lysis (Burrows et al., in press) and human CTL can induce apoptosis in mouse cells and vice versa (Gromkowski et al., 1986; Howell and Martz, 1987), indicating that this mediator(s) is highly conserved. Animals might be reluctant to make antibodies to such self-proteins, which might explain why this molecule(s) has not yet been identified. One might speculate that a receptor on the CTL is transiently activated/expressed on the cell surface, interacts with a ligand on the target cell and transmits a unidirectional apoptotic signal via this interaction. (CTL would also express the ligand). A precedent for such short-lived membrane proteins has been described in other systems (Werfel et al.,1991). It seem likely that a family of apoptotic membrane protein ligands exist each coming into play for different reasons, development, immune selection etc. (Trauth et al., 1989; Itoh et al., 1991). One such protein might exist on all cells, and hold the key to the apoptosis of all cells.

Berke et al., 1993, have recently reported perforin and granzyme independent CTL hybridomas and Rouvier et al., 1993, have shown that Ca^{2+}-independent CTL killing can involve transduction of the lethal hit via Fas. Rahelu et al., 1993, have shown that CTL-CTL killing for $CD4^+$ CTL resembles that described for $CD8^+$ CTL, and Kyburz et al., 1993, have shown that peptide mediated CTL-CTL killing can indeed occur in vivo.

REFERENCES

Barker, E., Wise, J. A., Dray, S., Mokyr, M. B. (1989) Lysis of antigenically unrelated tumor cells mediated by Lyt 2+ spleen T cells from melphalan-cured MOPC-315 tumor bearers. *Cancer Research*, **49**, 5007–5015.

Berke, G. (1991) Lymphocyte-triggered internal target disintegration. *Immunology Today*, **12**, 396–399.

Berke, G., Rosen, D. and Roen, D. (1993) Mechanism of lymphocyte-mediated cytolysis: functional cytolytic T cells lacking perforin and granzymes. *Immunology*, **78**, 105–112.

Burrows, S. R., Fernan, A., Argaet, V. and Suhrbier, A. (In press) Bystander apoptosis induced by CD8+ cytotoxic T-cell (CTL) clones: implications for CTL lytic mechanisms. *International Immunology*.

Burrows, S. R., Suhrbier, A., Khanna, R. and Moss, D. J. (1992) Rapid visual assay of cytotoxic T-cell specificity utilizing synthetic peptide induced T cell-T cell killing. *Immunology*, **76**, 174–175.

Duke, R. C., Peracchini, P. M., Chang, S., Liu, C.-C., Cohen, J. J. and Young, J. D.-E. (1989) Purified perforin induces target cell lysis but not DNA fragmentation. *Journal of Experimental Medicine*, **170**, 1451–1456.

Erb, P., Grogg, D., Troxler, M., Kennedy, M. and Fluri, M. (1990) $CD4^+$ T-cell mediated killing of MHC class II-positive antigen-presenting cells. *Journal of Immunology*, **144**, 790–795.

Fleischer, B. (1986) Lysis of bystander target cells after triggering of human cytotoxic T lymphocytes. *European Journal of Immunology,* 16, 1021–1024.

Golstein, P., Ojcius, D. M., Young, J. D.-E. (1991) Cell death mechanisms and the immune system. *Immunology Review,* 121, 29–36.

Gromkowski, S. H., Brown, T. C., Cerutti, P. A. and Cerottini, J.-C. (1986) DNA of human Raji target cells is damaged upon lymphocyte-mediated lysis. *Journal of Immunology,* 136, 752–756.

Gromkowski, S. H., Hepler, K. M. and Janeway, C. A. Jr. (1988) Low doses of interleukin-2 induce bystander cell lysis by antigen-specific CD4[+] inflammatory T-cell clones in short-term assay. *European Journal of Immunology,* 18, 1385–1389.

Hayes, M. P., Berrebi, G. A. and Henkart, P. A. (1989) Induction of target cell DNA release by the cytotoxic T lymphocyte granule protease Granzyme A. *Journal of Experimental Medicine,* 170, 933–946.

Howell, D. M. and Martz, E. (1987) The degree of CTL-induced DNA solubilization is not determined by the human vs mouse origin of the target cell. *Journal of Immunology,* 138, 3695–3698.

Hudig, D., Allison, N. J., Pickett, T. M., Winkler, U., Kam, C.-M. and Powers, J. C. (1991) The function of lymphocyte proteases. Inhibition and restoration of granule-mediated lysis with isocoumarin serine protease inhibitors. *Journal of Immunology,* 147, 1360–1368.

Itoh, N., Yonehara, S., Ishii, A., Yonehara, M., Mizushima, S.-I., Sameshima, M., Hase, A., Seto Y. and Nagata, S. (1991) The polypeptide encoded by the cDNA for human surface antigen Fas can mediate apoptosis. *Cell,* 66, 233–243.

Jiang, S., Ojcius, D. M., Persehini, P. M, Young, J. D.-E. (1990) Resistance of cytolytic lymphocytes to perforin-mediated killing. Inhibition of perforin binding activity by surface membrane proteins. *Journal of Immunology,* 144, 998–1004.

Ju, S.-T., Ruddle, N. H., Strack, P., Dorf, M. E. and DeKruyff, R. H. (1990) Expression of two distinct cytolytic mechanisms among murine CD4[+] subsets. *Journal of Immunology,* 144, 23–31.

Kolber, M. A., Quinones, R. R. and Henkart, P. A. (1989) Target cell lysis by cytototoxic T lymphocytes redirected by antibody-coated polystyrene beads. *Journal of Immunology,* 143, 1461–1466.

Kupfer, A. and Singer, S. J. (1989) Cell Biology of cytotoxic and helper T-cell functions: Microscopic studies of single cells and cell couples. *Annual Review of Immunology,* 7, 309–337.

Kyburz, D., Speiser, D. E., Aebischer, T., Hengartner, H. and Zingkernagel, R. M. (1993) Virus-specific cytotoxic T-cell mediated lysis of lymphocytes *in vitro* and *in vivo*. *Journal of Immunology,* 150, 5051–5058.

Lancki, D. W., Hsieh, C.-S. and Fitch, F. W. (1991) Mechanisms of lysis by cytotoxic T lymphocyte clones. I. Lytic activity and gene expression in cloned antigen-specific CD4[+] and CD8[+] T lymphocytes. *Journal of Immunology,* 146, 3242–3249.

Lanzavecchia, A. (1986) Is the T-cell receptor involved in T-cell killing? *Nature,* 319, 778–780.

Lenardo, M. J. (1991) Interleukin-2 programs mouse alpha beta T lymphocytes for apoptosis. *Nature,* 353, 858–861.

Liu, C.-C., Jiang, S., Persechini, P. M., Zychlinsky, A., Kaufmann, Y. and Young, J. D-E. (1989) Resistance of cytolytic lymphocytes to perforin-mediated killing. *Journal of Experimental Medicine,* 169, 2211–2225.

Liu, C.-C., Steffen, M., King, F. and Young, J. D.-E. (1987) Identification, isolation, and characterization of novel cytotoxin in murine cytolytic lymphocytes. *Cell,* 51, 393–403.

Liu, Y. and Janeway, C. A. Jr. (1990) Interferon gamma plays a critical role in induced cell death of effector T cells: A possible third mechanism of self-tolerance. *Journal of Experimental Medicine,* 172, 1735–1739.

Luciani, M.-F., Brunet, J.-F., Suzan, M., Denizot, F. and Golstein, P. (1986) Self-sparing of long-term *in vitro*-cloned or uncloned cytotoxic T lymphocytes. *Journal of Experimental Medicine*, **164**, 962–967.

Moss, D. J., Burrows, S. R., Baxter, G. D. and Lavin, M. F. (1991) T cell-T cell killing is induced by specific epitopes: evidence for an apoptotic mechanism. *Journal of Experimental Medicine*, **173**, 681–686.

Newell, M. K., Haughn, L. J., Maroun, C. R. and Julius, M. H. (1990) Death of mature T cells by separate ligation of CD4 and the T-cell receptor for antigen. *Nature*, **347**, 286–289.

Odaka, C., Kizaki, H. and Tadakuma, T. (1990) T-cell receptor-mediated DNA fragmentation and cell death in T-cell hybridomas. *Journal of Immunology*, **144**, 2096–2101.

Ottenhoff, T. H. M. and Mutis, T. (1990) Specific killing of cytotoxic T cells and antigen-presenting cells by $CD4^+$ cytotoxic T-cell clones. I. A novel potentially immunoregulatory T-T-cell interaction in man. *Journal of Experimental Medicine*, **171**, 2011–2024.

Pemberton, R. M., Wraith, D. C. and Askonas, B. A. (1990) Influenza peptide-induced self-lysis and down-regulation of cloned cytotoxic T cells. *Immunology*, **70**, 223–229.

Peters, P., Borst, J., Oorschot, V., Fukuda, M., Kraehenbuehl, O., Tschopp, J., Slot, J. W. and Geuze, H. J. (1991) Cytotoxic T lymphocyte Granules are secretory lysosomes containing both perforin and granzymes. *Journal of Experimental Medicine*, **173**, 1099–1109.

Peters, P. J., Geuze, H. J., van der Donk, H. A. and Borst, J. (1990) A new model for lethal hit delivery by cytotoxic T lymphocytes. *Immunology Today*, **11**, 28–32.

Rahelu, M., Williams, G. T., Kumararatne, D. S., Eaton, G. C. and Gaston, J. S. H. (1993) Human $CD4^+$ cytotoxic T cells kill antigen-pulsed target T cells by induction of apoptosis. *Journal of Immunology*, **150**, 4856–4866.

Robbins, P. A., McMichael, A. J. (1991) Immune recognition of HLA molecules down regulates CD8 expression on cytotoxic T lymphocytes. *Journal of Experimental Medicine*, **173**, 221–230.

Rouvier, E., Luciani, M. F. and Golstein, P. (1993) Fas involvement in Ca^{2+}-independent T-cell mediated cytotoxicity. *Journal of Experimental Medicine*, **177**, 195–200.

Russell, J. H., White, C. L., Loh, D. Y. and Meleedy-Rey, P. (1991) Receptor-stimulated death pathway is opened by antigen in mature T cells. *Proceedings of the National Academy of Science, USA*, **88**, 2151–2155.

Sambhara, S. R., and Miller, R. G. (1991) Programmed cell death of T cells signaled by the T-cell receptor and the alpha 3 domain of class I MHC. *Science*, **252**, 1424–1427.

Schmid, D. S., Tite, J. P. and Ruddle, N. H. (1986) DNA Fragmentation: Manifestation of target cell destruction mediated by cytotoxic T-cell lines, lymphotoxin-secreting helper T-cell clones and cell free lymphotoxin-containing supernatant. *Proceedings of the National Academy of Science, USA*, **83**, 1881–1885.

Shi, L., Kraut, R. P., Aebersold, R. and Greenberg, A. H. (1992) A natural killer cell granule protein that induces DNA fragmentation and apoptosis. *Journal of Experimental Medicine*, **175**, 553–566.

Shi, Y., Sahai, B. M. and Green, D. R. (1989) Cyclosporin A inhibits activation-induced cell death in T-cell hybridomas and thymocytes. *Nature*, **339**, 625–626.

Shi, Y., Szalay, M. G., Paskar, L., Boyer, M., Singh, B. and Green, D. R. (1990) Activation-induced cell death in T-cell hybridomas is due to apoptosis. *Journal of Immunology*, **144**, 3326–3333.

Shiver, J. W. and Henkart, P. A. (1991) A non-cytotoxic mast cell tumor line exhibits potent IgE-dependent cytotoxicity after transfection with the cytolysin-perforin gene. *Cell*, **64**, 1175–1181.

Skinner, M. and Marbrook, J. (1987) The most efficient cytotoxic T lymphocytes are the least susceptible to lysis. *Journal of Immunology*, **139**, 985–987.

Smith, G. J., Bagnall, C. R., Bakewell, W. E., Black, K. A., Bouldin, T. W., Eanhardt, T. S., Hook, G. E. R. and Pryzwansky, K. B. (1991) Application of confocal scanning laser microscopy in experimental pathology. *Journal of Electron Microscopy Techniques*, 18, 38–49.

Suhrbier, A., Burrows, S. R., Fernan, A., Lavin, M. F., Baxter, G. D. and Moss, D. J. (1993) Peptide Epitope-Induced apoptosis of human cytotoxic T lymphocytes: implications for peripheral T-cell deletion and peptide vaccination. *Journal of Immunology*, 150, 2169–2178.

Tian, Q., Streuli, M., Saito, H., Schlossman, S. F. and Anderson, P. (1991) A Polyadenylate binding protein localized to the granules of cytolytic lymphocytes induces DNA fragmentation in target cells. *Cell*, 67, 629–639.

Trauth, B. C., Klas, C., Peters, A. M. J., Matzku, S., Moller, P., Falk, W., Debatin, K.-M. and Krammer, P. H. (1989) Monoclonal antibody-mediated tumor regression by induction of apoptosis. *Science*, 245, 301–305.

Trenn, G., Takayama, H. and Sitkovsky, M. V. (1987) Exocytosis of cytolytic granules may not be required for target cell lysis by cytotoxic T lymphocytes. *Nature*, 330, 72–74.

Tschopp, J. and Nabholz, M. (1990) Perforin-mediated target cell lysis by cytotoxic T lymphocytes. *Annual Review of Immunology*, 8, 279–302.

Vitiello, A., Heath, W. R. and Sherman, L. A. (1989) Consequences of self-presentation of peptide antigen by cytolytic T lymphocytes. *Journal of Immunology*, 148, 1512–1517.

Walden, P. R. and Eisen, H. M. (1990) Cognate peptides induce self-destruction of CD8+ cytolytic T lymphocytes. *Proceedings of the National Academy of Science, USA*, 87, 9015–9019.

Werfel, T., Sonntag, G., Weber, M. H., Goetze, O. (1991) Rapid increase in the membrane expression of neutral endopeptidase (CD10), aminopeptidase N (CD13), Tyrosine phosphatase (CD45), and Fc gamma-RIII (CD16) upon stimulation of human peripheral leukocytes with human C5a. *Journal of Immunology*, 147, 3909–3914.

Young, J. D.-E. and Liu, C.-C. (1988a) How do cytotoxic T lymphocytes avoid self-lysis? *Immunology Today*, 9, 14–15.

Young, J. D.-E. and Liu, C.-C. (1988b) Multiple mechanisms of lymphocyte-mediated killing. *Immunology Today*, 9, 140–144.

Young, J. D.-E., Liu, C.-C., Persechini, P. M. and Cohn, Z. A. (1988) Perforin-Dependent and -Independent pathways of cytotoxicity mediated by lymphocytes. *Immunology Review*, 103, 161–202.

Zychlinsky, A., Zheng, L. M., Liu, C.-C. and Young, J. D.-E. (1991) Cytolytic lymphocytes induce both apoptosis and necrosis in target cells. *Journal of Immunology*, 146, 393–400.

APOPTOSIS AND CANCER

Chapter 22

APOPTOSIS IN IRRADIATED MURINE TUMOURS AND HUMAN TUMOUR XENOGRAFTS

Raymond E. Meyn[1], L. Clifton Stephens[3], K. Kian Ang[2], Luka Milas[1], Nancy R. Hunter[1] and Lester J. Peters[2]

Departments of [1]Experimental Radiotherapy, [2]Clinical Radiotherapy, and [3]Veterinary Medicine and Surgery,
The University of Texas, M. D. Anderson Cancer Center, Houston, Texas 77030, USA

INTRODUCTION

In just the past few years, workers in several fields, notably those in immunology, developmental biology, and endocrinology, have begun to recognize the important role that the mode of cell deletion now known as apoptosis, or programmed cell death, plays in a number of biological processes (Arends and Wyllie, 1991). Radiobiologists, however, have been aware of this particular mode of cell death for many years. Prior to 1972 when the word "apoptosis" was coined (Kerr et al., 1972), it was referred to as "interphase death" and descriptions of its features in irradiated cells, especially in lymphocytes and other cells of lymphoid origin, can be found in literature dating back nearly 40 years. For example, Schrek (1955) characterized the changes observed in the nuclei of lymphocytes irradiated *in vitro* as becoming "crescent-shaped" followed by "pyknotic" and then "fragmented". The particular fragmentation of DNA that we now realize is a feature of apoptosis was also seen in early experiments. Cole and Ellis (1957) noted that large amounts of chromatin fragments could be extracted from lymphoid tissue following irradiation. In his review of interphase death, Okada (1970) made the points that the nuclei of thymocytes become pyknotic within just a few hours of irradiation and that this event precedes the cell's loss of dye exclusion ability confirming that the changes in chromatin conformation represent early steps in interphase death or apoptosis.

Much of what we now know about the biochemistry of apoptosis has come through the study of glucocorticoid-treated thymocytes, and many of the observations have been verified in irradiated cells. For example, the enzymatic fragmentation of the DNA into internucleosomal sized pieces, first shown in rat thymocytes treated with glucocorticoids by Wyllie (1980) was subsequently seen in irradiated thymocytes by Yamada et al. (1981). McConkey et al. (1989) have shown a role for Ca^{2+} in glucocorticoid-induced apoptosis, a requirement that we have recently verified for irradiated thymocytes (Story et al., 1992). Thus, it would appear that apoptosis can be induced in many different situations by many different stimuli but once triggered the actual process is very similar in many of these cases.

APOPTOSIS IN IRRADIATED MURINE TUMOURS

Most of the studies described above relied on model systems whereby the cells examined had been either grown in culture or freshly explanted from animal tissues and then irradiated and analyzed *in vitro*. While much is known about the basic apoptotic process from such studies, less has been learned about the possible importance of this mode of cell death in tissues irradiated *in vivo*. A review on this latter subject by Kerr and Searle (1980), refers to the important observations of Potten (1977) who showed apoptotic bodies in the epithelium of intestinal crypts following exposure to radiation and of Pratt and Sodicoff (1972) in irradiated salivary glands. In that same article, Kerr and Searle (1980) described the production of apoptotic bodies in a transplantable mouse tumour that had received irradiation. Whereas these reports clearly illustrated that cells died by apoptosis in irradiated tissues, the role of this mode of cell death in the pathogenesis of the normal tissue or tumour response could not be adequately ascertained because the magnitude of apoptosis induction was not quantitated.

Our interest in apoptosis was initially stimulated by an observation from the clinic. Irradiation of the major salivary glands is often unavoidable during radiotherapy of head and neck cancer, and the resulting side effects are common and frequently distressing for the patient. The remarkable observation was that the complications become evident during the first week of therapy, i.e. after doses of less than 10 Gy. This clinical problem prompted an investigation of the radiation-induced changes in salivary glands using a primate model (Stephens et al., 1986). Briefly, adult female monkeys were irradiated with single doses of 2.5 to 15 Gy, and acute damage was studied by taking sequential salivary gland biopsy specimens 1 to 72 hours after irradiation. The proportion of apoptotic-appearing cells was quantitated by microscopic examination of histological sections using a scoring system developed as part of this study (Stephens et al., 1989b). Dead cells could be observed as early as 1 hour after irradiation and most of the injury had been expressed by 24 hours after irradiation.

Apoptosis was seen even at the lowest doses used, but the response was extensive at doses of 7.5 Gy and above, reaching levels as high as 60% at 15 Gy. Stephens and co-workers (1989a) in a summary of this work, concluded that cell damage consistent with apoptosis was a primary mechanism of radiation-induced injury to the salivary glands.

The success of this first demonstration, using a quantitative approach, of a definitive role for apoptosis in a tissue response to radiation delivered *in vivo* provided the impetus to evaluate other situations. An obvious case to examine was irradiated tumours. In the initial set of experiments, we chose two mouse tumours whose radiation response with respect to conventional assays were known to be very different (Milas *et al.*, 1987): an ovarian carcinoma, OCa-1, whose TCD_{50} value (i.e. the radiation dose yielding 50% cure) is 52.6 Gy, and a hepatocarcinoma, HCa-1, with a TCD_{50} value of >81 Gy. The OCa-1 tumour is also considerably more responsive to radiation when tumour growth delay is the endpoint, shrinking to <30% of its volume after doses of 30 Gy whereas the HCa-1 tumour shows no measurable shrinkage after a similar radiation dose (Stephens *et al.*, 1991).

Histological sections of these tumours were made at different times following various doses of radiation and scored for apoptotic cells using the same criteria as developed for the analysis of irradiated salivary glands (Stephens *et al.*, 1989b). Photomicrographs depicting the features of radiation-induced apoptosis in histological sections of the OCa-1 tumour are presented in Figures 22.1 and 22.2. The results of this analysis showed that the OCa-1 tumour displayed a dramatic apoptotic response to radiation whereas the HCa-1 tumour essentially showed no apoptotic response

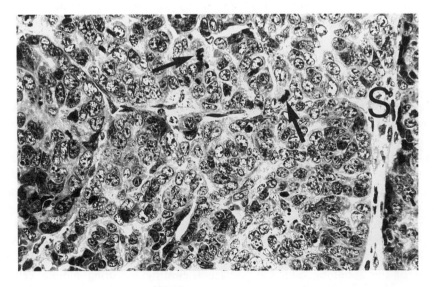

Figure 22.1 (A) An untreated transplanted ovarian carcinoma (OCa-1) composed of discrete clumps of cells separated by fibrovascular stroma (S). Arrows point to mitotic figures.

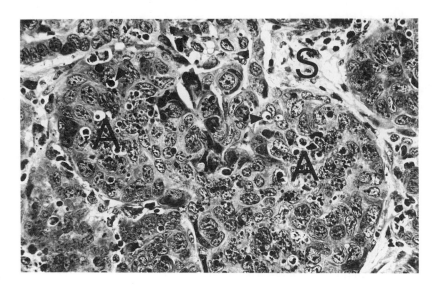

Figure 22.1 (B) An OCa-1 tumour 6 h after irradiation with 35 Gy. Apoptosis is widespread. Apoptotic bodies are the small ovoid structures, usually surrounded by narrow empty spaces, in the vicinity of the letters (A) and arrowheads. Hematoxylin and eosin. × 350. (Stephens et al., 1991, reprinted with permission from *Radiation Research*, © Copyright 1991, Academic Press.)

(Stephens et al., 1991). This difference was surprising, but, in addition, we observed that in the OCa-1 tumour the percent of apoptotic bodies relative to intact nuclei was identical for the three doses tested, 25, 35, and 45 Gy, suggesting that these doses are already on a plateau on the dose-response curve. Moreover, the apoptotic cells appeared within a few hours of irradiation and then slowly disappeared (Figure 22.3).

These initial investigations have been followed up since their publication (Stephens et al., 1991) in a number of more comprehensive studies. First, we re-evaluated the dose response and time-course for apoptosis development in the OCa-1 tumour following single doses of irradiation (Stephens et al., 1993). Our initial impression, that doses of 25 Gy and above are already on a plateau, was confirmed by examining doses between 2.5 Gy and 25 Gy. Indeed, on a per unit dose basis, 2.5 Gy was the most effective dose for inducing apoptosis. The dose-response relationship was shown to plateau at doses above 7.5 Gy. The more detailed time-course was also revealing in that it showed that radiation-induced apoptosis can be visualized starting about 2 hours after irradiation and the percentages peak between 3 and 5 hours reaching levels as high as 30–35%. The numbers of apoptotic cells then drop starting at 6 hours and decline steadily until they reach background levels between 24 and 48 hours after irradiation. By careful examination of the histological sections, we observed that this decline is due to phagocytosis of the apoptotic bodies by adjacent, healthy-appearing tumour cells. By examining the response over time at both 25 and 2.5 Gy, we were able to confirm that the time-course is independent of dose over this range.

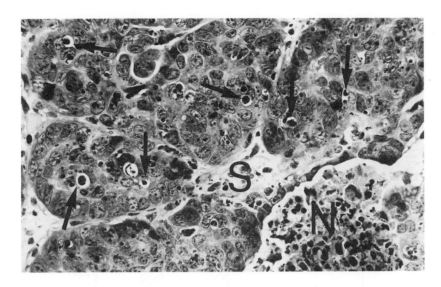

Figure 22.2 An OCa tumour 6 hours after irradiation with 35 Gy. Nuclei that are not lost are hyperchromatic in an area of coagulative necrosis (N). Stroma is identified (S). Tumour above the necrotic area has widely scattered, darkly stained structures, usually surrounded by a narrow empty space, that are cells dying by apoptosis. Arrows point to apoptotic cells, some of which show fragmentation of the condensed nucleus into multiple apoptotic bodies. Note that the nuclei in both apoptotic and necrotic cells can be characterized as pyknotic and karyorrhectic (hematoxylin-eosin, original magnification × 350). (Stephens *et al.*, 1991, reprinted with permission from *Radiation Research*, © Copyright 1991, Academic Press.)

Second, whereas the response of the OCa-1 tumour had provided important insights into the process of radiation-induced apoptosis, it was necessary to evaluate other tumours to see whether apoptosis was limited to this particular tumour. We have recently completed an analysis of 13 other murine tumours and compared their response to the previous data for the OCa-1 and HCa-1 tumours (Meyn *et al.*, in press). For this analysis we chose a limited set of doses and times for study based on the detailed results from the OCa-1 tumour described above. Thus, we examined a total of four mammary adenocarcinomas, five sarcomas, three squamous cell carcinomas and a lymphoma all at 3 and 6 hours following doses of 2.5, 10, and 25 Gy. The results of this survey showed that four additional tumours, three of the four mammary adenocarcinomas and the lymphoma, had a significant apoptotic response to radiation. Considering the limited set of doses and times used for this analysis, these tumours responded in an identical manner to the OCa-1 tumour: i.e. 2.5 Gy was most effective on a per unit dose basis and they all showed an apoptotic response by 3 hours after irradiation.

For many of the 15 tumours, we had previously collected data on tumour response to radiation measured by conventional assays of tumour growth delay and TCD_{50} dose. Thus, we attempted to correlate the apoptotic response determined in the study

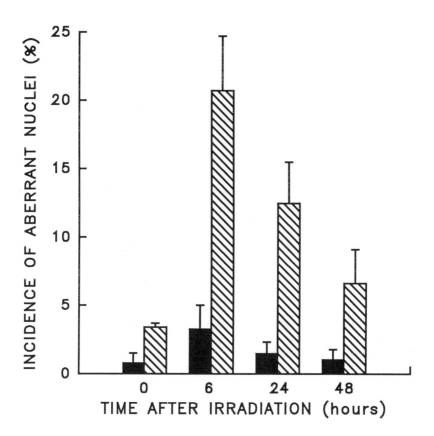

Figure 22.3 Induction of aberrant nuclei as scored in paraffin sections of the hepatocellular (HCa) (solid bars) and ovarian (OCa) (hatched bars) mouse tumours irradiated *in vivo* with 25, 35, and 45 Gy. The given incidence of aberrant nuclei represents the average of the values for the three doses. Tumours were irradiated when they were 8 mm in diameter and examined as a function of time following irradiation. (Stephens *et al.*, 1991, reprinted with permission from Radiation Research, © Copyright 1991, Academic Press.)

above with the known radiation response of these same murine tumours. To summarize briefly, all of the tumours that displayed an apoptotic response had longer specific tumour growth delays than those tumours that did not apoptose following irradiation. Furthermore, the tumours that responded by apoptosis had lower TCD_{50} values than did most of those tumours that did not show an apoptotic response, though this association was weaker than for regrowth delay. Finally, we noticed that there was a significant correlation between the levels of spontaneous apoptosis measured in non-irradiated tumours and their relative apoptotic response when irradiated. In fact, four of the five tumours that responded by apoptosis had spontaneous apoptotic levels of >5% whereas each of the tumours that did not show an apoptotic response to radiation had spontaneous levels of <2%.

Third, in all of the studies described above, single doses of radiation were used; however, if apoptosis as a mode of cell death is to be relevant to tumour response to radiotherapy, it must also occur after repeated fractions of radiation. We, therefore, tested whether apoptosis was repeatedly induced in the OCa-1 tumour by fractionated irradiations using two protocols: two fractions of 12.5 Gy separated by 4–72 hours and two to five fractions of 2.5 Gy separated by 24-hour intervals. There was no evidence of a restoration of the apoptotic subpopulation of cells observed in the 12.5 + 12.5 Gy protocol. However, using five fractions of 2.5 Gy given at 24-hour intervals, a cumulative apoptotic index of over 50% could be achieved, suggesting that the subpopulation of cells susceptible to radiation-induced apoptosis could be at least partially restored in a protocol that closely matches those used in clinical radiotherapy.

APOPTOSIS IN IRRADIATED HUMAN TUMOUR XENOGRAFTS

Considering the dramatic demonstrations in some cases of radiation-induced apoptosis in murine tumours summarized above, the important question became, to what extent does apoptosis occur in different irradiated human tumours? To address this question in a systematic way, we have first analyzed human tumour xenografts as a model system. For these experiments, six different human tumour cell lines growing in culture were used consisting of two ovarian carcinomas, CaOV3 and 2774, two colon carcinomas, Clone A and Clone D, and two melanomas, A-375M and A-375P. Cells were harvested from cultures of these lines and injected into nude mice. After tumours formed, they were irradiated and analyzed using the same protocol and methods that we used above in the survey of fifteen different murine tumours. Preliminary experiments showed that apoptosis was induced in some irradiated human tumour xenografts, including both ovarian carcinomas and both colon carcinomas. Interestingly, apoptosis was induced in the xenograft produced from a primary melanoma cell line, A-375P, but not in the xenograft produced from a cell line derived from the metastasis of that same melanoma. We intend to extend these studies to several other human tumour xenografts since the variety of human tumour cell lines that can be obtained is extensive compared to the rather limited number of transplantable murine tumours available. Once detailed dose-response curves and kinetics of apoptosis development are worked out in these model human tumours it will be important to extend this analysis to biopsy specimens from patients undergoing radiotherapy.

CONCLUSIONS

The results summarized here illustrate that apoptosis is a significant mode of cell death in a substantial proportion of murine and human xenograft tumours exposed to radiation *in vivo*. In those tumours that have an apoptotic response, the apoptosis develops quickly and peaks within a few hours of irradiation. The apoptotic bodies are then rapidly phagocytosed by adjacent tumour cells so that, if tumour specimens are examined as late as 24 hours following irradiation, there may be little or no visible evidence remaining of this particular mode of response.

One of the most intriguing observations to emerge from this work is the overriding presence of heterogeneity in apoptotic response—both with respect to the cells within a given tumour and with respect to tumour type. Heterogeneity in the first instance is quite evident in the dose-response relationships, where the percent apoptosis rises quickly at relatively low doses, plateaus at doses above 7.5 Gy or so, and never reaches levels greater than about 30–35% even after very high doses, i.e. 45 Gy. Such results clearly imply that a subpopulation of tumour cells have the propensity for apoptosis and will do so upon irradiation with relatively low doses whereas the rest of the cells are not subject to apoptosis no matter what dose is delivered. This propensity for radiation-induced apoptosis is predicted from the spontaneous levels of apoptosis measured in the unirradiated tumours, where there was a one-to-one correlation of the spontaneous levels with the percent apoptosis achieved with 25 Gy. Thus, it is possible that irradiation simply accelerates the death of a naturally occurring subpopulation of cells that would have eventually apoptosed spontaneously.

Heterogeneity was most evident among different tumours examined: some had a dramatic apoptotic response and others did not respond by apoptosis at all. This intertumour heterogeneity and the intratumour heterogeneity alluded to above may be explained on the basis of current theories of how apoptosis is regulated. It is now well established that a number of molecular and biochemical factors regulate the ability of cells to apoptose on stimulation including oncogene expression (Wyllie et al., 1987; Hockenbery et al., 1990; Williams, 1991) and exposure to growth factors (Araki et al., 1990; Rotello et al., 1991) and cytokines (McConkey et al., 1990; Sabourin and Hawley, 1990). Arends and Wyllie (1991) have hypothesized that two classes of events, priming and triggering, are required for apoptosis. Priming refers to the cellular expression of certain key proteins (for example the endonuclease responsible for the DNA fragmentation characteristic of apoptosis) that are necessary in order for apoptosis to proceed once triggered. The expression of these proteins is presumably regulated by the molecular and biochemical factors mentioned above. Once primed, apoptosis can then be triggered by an appropriate stimulus, in our case radiation. But since apoptosis is known to be induced by many different types of treatments, other cytotoxins, including chemotherapeutic agents (Eastman, 1990), can also provide the triggering stimulus. Heterogeneity in apoptosis propensity may represent differences that the death induced via these defective receptors was apoptotic) in priming via the regulatory pathways in different types of tumours and within a particular tumour's cell population even though all cells in the tumour receive the same triggering stimulus. Understanding these regulatory pathways is one of the most exciting areas of research in apoptosis because they may offer the possibility for modulating therapeutic response.

Finally, we showed that the apoptotic response correlated with an enhanced tumour response to radiation treatment as measured by two different endpoints—specific tumour growth delay and TCD_{50} dose. These observations imply that reproductive death may not be the only mode of cell death in irradiated tumours, as has generally been believed. Rather apoptosis may also be an important process, at least in some types of tumours. If this can be verified through additional experimentation, the assessment of apoptosis in patient biopsy specimens within a few hours after the initial treatment might be an additional useful predictor of treatment outcome in clinical radiotherapy.

ACKNOWLEDGEMENTS

This work was supported by PHS grants CA06294 and CA16672 awarded by the National Cancer Institute, Department of Health and Human Services and by the Katherine Unsworth Memorial Fund. The skilful technical assistance of Veronica Willingham is gratefully acknowledged. We also thank Tammie Emshoff for typing the manuscript.

REFERENCES

Araki, S., Shimada, Y., Kaji, K. and Hayashi, H. (1990) Apoptosis of vascular endothelial cells by fibroblast growth factor deprivation. *Biochemical and Biophysical Research Communications*, 168, 1194–1200.

Arends, M. J. and Wyllie, A. H. (1991) Apoptosis: Mechanisms and roles in pathology. *International Reviews in Experimental Pathology*, 32, 223–254.

Cole, L. J. and Ellis, M. E. (1957) Radiation-induced changes in tissue nucleic acids: Release of soluble deoxypolynucleotides in the spleen. *Radiation Research*, 7, 508–517.

Eastman, A. (1990) Activation of programmed cell death by anticancer agents: Cisplatin as a model system. In *Cancer Cells*, pp. 275–280.

Hockenbery, D., Nunez, G., Milliman, C., Schreiber, R. D. and Korsmeyer, S. J. (1990) Bcl-2 is an inner mitochondrial membrane protein that blocks programmed cell death. *Nature*, 348, 334–336.

Kerr, J. F. R. and Searle, J. (1980) Apoptosis: Its nature and kinetic role. In *Radiation Biology in Cancer Research*, R. E. Meyn and H. R. Withers (eds.), pp. 367–384. New York: Raven Press.

Kerr, J. F. R., Wyllie, A. H. and Currie, A. R. (1972) Apoptosis: A basic biological phenomenon with wide ranging implications in tissue kinetics. *British Journal of Cancer*, 26, 239–257.

McConkey, D. J., Hartzell, P., Chow, S. C., Orrenius, S. and Jondal, M. (1990) Interleukin-1 inhibits T-cell receptor-mediated apoptosis in immature thymocytes. *Journal of Biological Chemistry*, 265, 3009–3011.

McConkey, D. J., Nicotera, P., Hartzell, P., Bellomo, G., Wyllie, A. H. and Orrenius, S. (1989) Glucocorticoids activate a suicide process in thymocytes through an elevation of cytosolic Ca^{2+} concentration. *Archives of Biochemistry and Biophysics*, 269, 365–370.

Meyn, R. E., Stephens, L. C., Ang, K. K., Hunter, N., Brock, W.A., Milas, L. and Peters, L. J. (In press). Heterogeneity in the development of apoptosis in irradiated murine tumors of different histologies. *International Journal of Radiation Biology*.

Milas, L., Wike, J., Hunter, N., Volpe, J. and Basic, I. (1987) Macrophage content of murine sarcomas and carcinomas: Associations with tumor growth parameters and tumor radiocurability. *Cancer Research,* **47**, 1069–1075.

Okada, S. (1970) Radiation-induced death. In *Radiation Biochemistry,* pp. 247–307. New York, New York: Academic Press.

Potten, C. S. (1977) Extreme sensitivity of some intestinal crypt cells to X and γ irradiation. *Nature,* **269**, 518–521.

Pratt, N. E. and Sodicoff, M. (1972) Ultrastructural injury following X-irradiation of rat parotid gland acinar cells. *Archives of Oral Biology,* **17**, 1177–1186.

Rotello, R. J., Lieberman, R. C., Purchio, A. F. and Gerschenson, L. E. (1991) Coordinated regulation of apoptosis and cell proliferation by transforming growth factor β1 in cultured uterine epithelial cells. In *Proceedings of the National Academy of Science, USA,* pp. 3412–3415.

Sabourin, L. A. and Hawley, R. G. (1990) Suppression of programmed death and G1 arrest in B-cell hybridomas by interleukin-6 is not accompanied by altered expression of immediate early response genes. *Journal of Cellular Physiology,* **145**, 564–574.

Schrek, R. (1955) Cinemicrographic observations and theoretical considerations on reactions of lymphocytes to X-rays. *Radiology,* **65**, 912–919.

Stephens, L. C., Ang, K. K., Schultheiss, T. E., King, G. K., Brock, W. A. and Peters, L. J. (1986) Target cell and mode of radiation injury in Rhesus salivary glands. *Radiotherapy Oncology,* **7**, 165–174.

Stephens, L. C., Ang, K. K., Schultheiss, T. E., Milas, L. and Meyn, R. E. (1991) Apoptosis in irradiated murine tumours. *Radiation Research,* **127**, 308–316.

Stephens, L. C., Hunter, N. R., Ang, K. K., Milas, L., and Meyn, R. E. (1993) Development of apoptosis in irradiated murine tumours as a function of time and dose. *Radiation Research,* **135**, 75–80.

Stephens, L. C., Schultheiss, T. E., Ang, K. K. and Peters, L. J. (1989a) Pathogenesis of radiation injury to the salivary glands and potential methods of protection. *Cancer Bulletin,* **41**, 106–114.

Stephens, L. C., Schultheiss, T. E., Small, S. M., Ang, K. K. and Peters, L. J. (1989b) Response of parotid organ culture to radiation. *Radiation Research,* **120**, 140–153.

Story, M. D., Stephens, L. C., Tomasovic, S. P. and Meyn, R. E. (1992) A role for calcium in regulating apoptosis in rat thymocytes irradiated *in vitro*. *International Journal of Radiation Biology,* **61**, 243–251.

Williams, G. T. (1991) Programmed cell death: Apoptosis and oncogenesis. *Cell,* **65**, 1097–1098.

Wyllie, A. H. (1980) Glucocorticoid-induced thymocyte apoptosis is associated with endogenous endonuclease activation. *Nature,* **284**, 555–556.

Wyllie, A. H., Rose, K. A., Morris, R. G., Steel, C. M., Foster, E. and Spandidos, D. A. (1987) Rodent fibroblast tumours expressing human *myc* and *ras* genes: Growth, metastasis and endogenous oncogene expression. *British Journal of Cancer,* **56**, 215–259.

Yamada, T., Ohyama, H., Kinjo, Y. and Watanabe, M. (1981) Evidence for the internucleosomal breakage of chromatin in rat thymocytes irradiated *in vitro*. *Radiation Research,* **85**, 544–553.

Chapter 23

ANTI-CANCER DRUGS AND APOPTOSIS

L. I. Huschtscha, C. E. Andersson, W. A. Bartier and M. H. N. Tattersall

Department of Cancer Medicine, Blackburn Building, University of Sydney,
Sydney, NSW 2006, Australia

INTRODUCTION

Two biochemically distinct mechanisms of cell death, apoptosis and necrosis have been distinguished based on morphological criteria (Kerr et al., 1972). Cell death by apoptosis is characterised by overall cell shrinkage, margination of chromatin in sharply defined masses adjoining the nuclear envelope and nuclear and cellular fragmentation to produce apoptotic bodies which can be phagocytosed by adjacent cells. However, cells undergoing necrosis are characterised by cell swelling, chromatin flocculation and disruption of membrane integrity followed by cell lysis (Kerr et al., 1972; Wyllie et al., 1980; Wyllie, 1987). Cells dying by apoptosis have been shown to cleave double-stranded DNA at internucleosomal sites giving rise to oligonucleotides which are discrete multiples of about 200 base pairs (Wyllie, 1980; Arends et al., 1990). This degraded DNA produces a ladder-like pattern when run on DNA agarose gels. In contrast, the DNA during necrosis is cleaved randomly, producing a smear pattern on DNA gels.

The onset of DNA fragmentation in apoptotic cells has been shown to require protein and RNA synthesis since the addition of cycloheximide and actinomycin-D respectively, to dexamethasone-treated rat or mouse thymocytes prevented the appearance of nucleosomally cleaved DNA (Cohen and Duke, 1984; Wyllie et al., 1984; Kizaki et al., 1989; McConkey et al., 1988; Yamada and Ohyama, 1988). However, McConkey et al. (1990) suggested an alternative hypothesis based on their results using protease inhibitors which indicated the endonuclease undergoes rapid turnover in the cell rather than synthesis of a new gene product.

More recently apoptotic cells have been distinguished using DNA flow cytometric methods. When glycocorticoid-treated rat thymocytes were stained with ethidium bromide and then analysed by flow cytometry for cell cycle distribution, a new subpopulation of hypo-diploid cells was observed (Afanas'ev et al., 1986; Pechatnikov et al., 1986; Compton et al., 1988). This new population represented apoptotic cells since the addition of inhibitors of translation such as cycloheximide resulted in the disappearance of this population. In other studies, a similar subdiploid peak was seen when propidium iodide was used to stain the DNA (Nicoletti et al., 1991; Telford et al., 1991). Apoptotic cell populations can be distinguished from normal and necrotic when the cells are analysed using forward and side scatter which measure cell size and cellular granularity, respectively (Dive et al., 1992; Huschtscha et al., unpublished data). Populations of viable cells undergoing apoptosis show movement of cells from a high forward light scatter, and low side light scatter compartments, to a low forward light scatter and high side light scatter compartment and are identified as apoptotic.

DNA flow cytometry and morphological studies can distinguish both apoptotic and necrotic cells in heterogeneous populations. This is not possible using DNA gel electrophoresis to assess cell death. It is therefore advisable to use several methods, such as cell morphology studies, DNA gel electrophoresis and DNA flow cytometry when assessing the modes of cell death.

DRUG-INDUCED CELL DEATH

Detailed investigations of the mode of cell death after drug treatment have used *in vivo* and *in vitro* models. Several studies have shown that chemotherapeutic drugs with different modes of action induce apoptosis in mouse and guinea pig colonic crypts (Searle et al., 1974; Ijiri and Potten, 1983; Ronen et al., 1984; Walker et al., 1988; Ijiri, 1989; McConkey and Orrenius, 1989; Anilkumar et al., 1992; Ledda-Columbano et al., 1992). Furthermore, when rat liver was exposed to the carcinogen dimethyl-nitrosamine, an increase in apoptotic bodies was observed (Pritchard and Butler, 1989). However, the administration of CCl_4 to hepatocytes caused death by necrosis.

Several investigators observed that many cytotoxic drugs induced death by apoptosis in various human cell lines, CEM-C7, Daudi, MOLT.4, HL-60, U937, EB1, KG1A and BM13674 (Kaufmann, 1989; Alnemri and Litwack, 1990; Lennon et al., 1990; Johnson et al., 1991; Gunji et al., 1991; Marks and Fox, 1991). Similarly, cytotoxic drug treatment of animal cell lines L1210 and CHO (Sorenson and Eastman, 1988a, b; Barry et al., 1990; Kwok and Tattersall, 1991, 1992), cultured mouse hepatocytes and rat cerebellular granule cells caused apoptosis (Disasquale et al., 1991; Shen et al., 1991).

Chemotherapeutic drugs do not always induce cell death by apoptosis, and necrotic cell death has also been reported (Vedeckis and Bradshaw, 1983; Dyson et al., 1986; Huschtscha et al., unpublished data). Using DNA cytofluorometry and Percoll centrifugation, Dyson et al. (1986) reported necrotic cell death in CEM-C7 cells after methotrexate and adriamycin treatment. Different cell lines were also shown to exhibit different modes of cell death after treatment with the same drug. CCRF-CEM.f2 cells died by necrosis after 5-fluorouracil and mithramycin treatment while other drugs tested induced cell death by apoptosis. Drug treated CCRF-HSB cells showed a similar pattern of cell death after drug treatment to CCRF-CEM.f2 cells except that methotrexate treatment caused cell death by necrosis in the former but not the latter. On the other hand, MOLT.4 cells died by necrosis after all drug treatments investigated when assessed by DNA gel electrophoresis (Huschtscha et al., unpublished data). However, morphological studies on drug treated MOLT.4 cells showed that 2–5% of cells had the morphological characteristics of apoptosis although DNA fragmentation was not observed on DNA gels (McDougall et al., 1991).

The concentration of drug affects the mode of cell death. Exposure to low concentrations of several different drugs in HL-60 cells induced cell death by apoptosis whilst higher concentrations caused necrotic cell death when cells were assessed morphologically and by DNA gel electrophoresis (Lennon et al., 1991). These results suggest that cells which are not acutely damaged may activate a programmed suicide mechanism while acute cytotoxicity may cause necrosis. Three other cell lines also died by apoptosis when they were exposed to low drug concentrations, but the drugs used to induce apoptosis were different for each cell line. Similarly rat thymocytes exposed to low and high concentrations of Bis(tri-n-butyltin)oxide died by different mechanisms (Raffray and Cohen, 1991).

For each drug it seems that cell death mechanisms and time course vary according to drug concentration. For instance, when U937 cells are exposed to Ara-C, for concentrations higher than 10^{-5} M, cell death is induced by apoptosis within 3 h, but the lower concentration of 10^{-6} M Ara-C, caused apoptosis to appear 6 h later (Gunji et al., 1991). In HL-60 cells 3×10^{-6} M Ara-C treatment did not cause apoptosis until 48 h (Kaufmann, 1989). This difference in time course between cell lines of nucleosomally degraded DNA has been noted for other drugs. Fragmented DNA was observed at 18 h when Chinese hamster ovary cells (CHO/AA8) were exposed to 3×10^{-7} M methotrexate, whilst in human myelogenous leukaemic cells, DNA fragmentation of HL-60 cells was not apparent until 48 h of drug treatment (Kaufmann, 1989; Barry et al., 1990). The time of appearance of fragmented DNA ranged from 2–48 h for etoposide-treated cells and is also dependent on the cell line studied (Kaufmann, 1989; Barry et al., 1990; Marks and Fox, 1991). A comprehensive study of etoposide-treated CCRF-CEM cells indicated that the first changes after drug treatment were the appearance of double-stranded DNA breaks followed by alteration in the nucleotide pools. Morphological changes, typical of apoptosis appeared after 24 h and preceded the appearance of fragmented DNA which was not visible until 48 h (Marks and Fox, 1991).

In another study, CCRF-CEM cells were treated with the aminothiol, cysteamine at concentrations of 7×10^{-5} M or higher, and fragmented DNA was observed at 6 h together with a block at the G_1/S boundary of the cell cycle. Cell proliferation was inhibited immediately after drug addition and overall basal phosphorylation was prevented by 30 minutes (Jeitner and Bartier, personal communication).

The mechanisms of cell death induced after drug treatment are dependent on the drug, its concentration, its time-course and the particular cell line used for the study.

DRUG TREATMENT OF CCRF-CEM.f2 CELLS

Four anti-cancer drugs, 5-fluorodeoxy-uridine (FUdR), methotrexate (MTX), 5-fluorouracil (5-FU), and vincristine, over a broad concentration range were used to study the mode of cell death in CCRF-CEM.f2 cells. Cell growth curves were used to monitor drug effects and DNA gel electrophoresis and DNA flow cytometry were used to characterise the mechanism of cell death. At concentrations where FUdR was growth inhibitory ($4 \times 10^{-7} - 4 \times 10^{-6}$ M), DNA fragmentation was first observed at 48 h by DNA gel electrophoresis (Table 23.1). No internucleosomal DNA fragmentation pattern was seen for the lowest studied concentration of FUdR (4×10^{-8} M). However, DNA flow cytometric analysis of FUdR-treated cells showed that by 3.5 h at all concentrations studied, there was a disappearance of the G2 peak and an accumulation of cells in the G1 phase of the cell cycle (Figure 23.1). By 24 h a subpopulation of 'cells' appeared that was smaller in size and exhibited increased cellular granularity, a pattern recognised to correlate with apoptotic cells. The data indicates that changes characteristic of apoptotic cells can be visualised by flow cytometry after FUdR treatment at least 24 h earlier than by DNA gel electrophoresis.

MTX treatment of CCRF-CEM.f2 cells gave similar results to FUdR treatment. At growth inhibitory MTX concentrations ($10^{-7} - 10^{-5}$ M), DNA fragmentation was apparent after 48 h (Table 23.1), but no DNA degradation typical of apoptosis was seen at lower MTX concentrations. At all concentrations studied, flow cytometry showed an accumulation of cells in the G1 phase of the cell cycle at 4 h and an apoptotic profile was visible at 24 h (data not shown).

DNA gel electrophoresis showed DNA degradation typical of necrotic cell death 48 h after treatment with 5-FU at concentrations between $10^{-4} - 10^{-2}$ M, but growth inhibition was apparent only at 10^{-2} M 5-FU (Figure 23.2, Table 23.1). Flow cytometric analysis of 5-FU-treated cells showed concentration dependent changes in S and G2 cell cycle phase distribution by 3.5 h and this was greater by 11 h. By 24 h cellular debris increased with increasing drug concentrations (Figure 23.3). A subpopulation of smaller cells with increased cellular granularity (indicator of apoptosis) was also visible at this time. Preliminary morphological studies of 10^{-3} M 5-FU treatment indicated that both apoptotic and necrotic cells were present at 24 h (data not shown). DNA gel electrophoresis which has limited sensitivity to detect a proportion of nucleosomal DNA in a mixed population of apoptotic and necrotic cells, revealed only a necrotic DNA profile.

($10^{-8} - 10^{-6}$ M) vincristine was growth inhibitory 24 h after drug treatment but internucleosomal DNA fragmentation was observed earlier than with the other drugs

Table 23.1 Summary of drug treatment of CCRF-CEM.f2 cells.

Treatment	Concentration (M)	Growth	DNA Gels[a]		
FUdR			6 h	24 h	48 h
	4×10^{-8}	reduced growth	–	–	–
	4×10^{-7}	cytotoxic	–	–	A
	4×10^{-6}	cytotoxic	–	–	A
MTX			12 h	24 h	48 h
	10^{-8}	reduced growth	–	–	–
	10^{-7}	cytotoxic	–	–	A
	10^{-6}	cytotoxic	–	–	A
	10^{-5}	cytotoxic	–	–	A
5-FU			6 h	24 h	48 h
	10^{-4}	reduced growth	–	–	N
	10^{-3}	reduced growth	–	–	N
	10^{-2}	cytotoxic	–	–	N
VINCRISTINE			6 h	24 h	48 h
	10^{-9}	reduced growth	–	–	–
	10^{-8}	cytotoxic	–	A	A
	10^{-7}	cytotoxic	–	A	A
	10^{-6}	cytotoxic	–	A	A

a. – = no DNA degradation; A = nucleosomal DNA degradation; N = random DNA degradation.

studied, 12 h but not 6 h after drug addition when assessed by DNA gel electrophoresis (Table 23.1). No DNA degradation was seen at lower concentrations, even after 48 h of drug treatment. Analysis by DNA flow cytometry showed that cells accumulated in the G2/M phase of the cell cycle 4 h after drug addition (Figure 23.4). 18 h after drug addition most cells showed an apoptotic flow cytometric profile. The kinetics of vincristine-induced cell death differed from those induced by FUdR and MTX treatment. Vincristine treatment causes apoptotic cell patterns more quickly suggesting triggering of a different pathway to that activated after exposure to the other drugs studied.

Our results indicate that cell death by apoptosis when assessed by DNA gel electrophoresis has a different concentration and time course characteristic depending on the particular drug. Other investigations using the same drugs and similar methods of

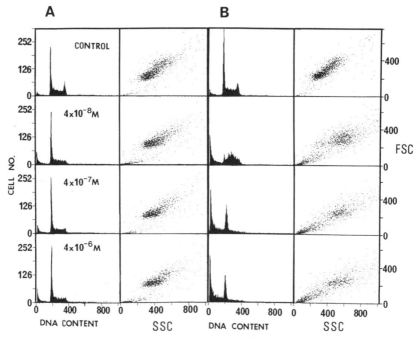

Figure 23.1 Flow cytometric DNA fluorescence profiles and forward light scatter and side light scatter plots of FUdR-treated CCRF-CEM.f2 cells at (A) 3.5 h; (B) 24 h after drug addition.

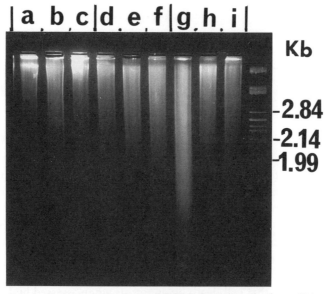

Figure 23.2 Cellular DNA fragmentation of CCRF-CEM.f2 cells after 5-FU treatment. (A) d, g, 10^{-4} M; (B) e, h, 10^{-3} M; (C) f, i, 10^{-2} M 6 h (A,B,C); 24 h (d,e,f) and 48 h (g,h,i) after drug addition.

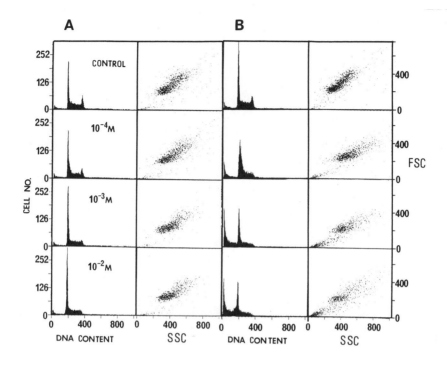

Figure 23.3 Flow cytometric DNA fluorescence profiles and forward light scatter and side light scatter plots of 5-FU-treated CCRF-CEM.f2 cells at (A) 3.5 h; (B) 24 h after drug addition.

detection have shown that different cell lines also have distinct time courses of appearance of fragmented DNA (Kaufmann, 1989; Barry et al., 1990). Changes typical of apoptosis could be visualised earlier when drug-treated cells were studied by DNA flow cytometry than those detected by DNA electrophoresis. However disturbances in the cell cycle phase distribution preceded the appearance of apoptotic cells. DNA flow cytometry after 5-FU treatment of CCRF-CEM.f2 cells revealed different cellular perturbations than those seen using DNA gel electrophoresis to monitor cytotoxicity. We therefore conclude that studies of the mode of cell death should utilise several techniques to monitor this process, namely morphological studies, gel electrophoresis together with flow cytometric analysis of cells treated for differing times over a concentration range.

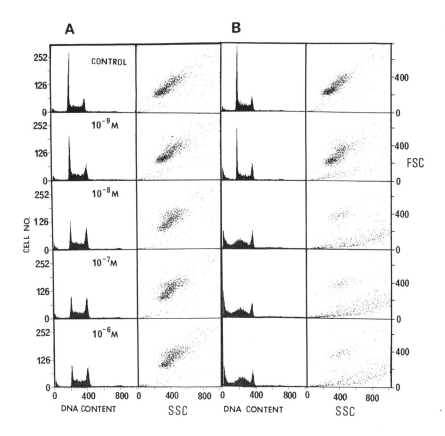

Figure 23.4 Flow cytometric DNA fluorescence profiles and forward light scatter and side light scatter plots of vincristine-treated CCRF-CEM.f2 cells at (A) 4 h; (B) 8 h.

REFERENCES

Afanas'ev, V. N., Korol, B. A., Mantsygin, P. A., Nelipovich, P. A., Pechatnikov, V. A. and Umansky, S. R. (1986) Flow cytometry and biochemical analysis of DNA degradation characteristic of two types of cell death. *FEBS Letters*, **194**, 347–350.

Alnemri, E. S. and Litwack, G. (1990) Activation of internucleosomal DNA cleavage in human CEM lymphocytes by glucocorticoid and novobiocin. *Journal of Biological Chemistry*, **265**, 17323–17333.

Anilkumar, T. V., Sarraf, C. F., Hunt, T. and Alison, M. R. (1992) The nature of cytotoxic drug-induced cell death in murine intestinal crypts. *British Journal of Cancer*, **65**, 552–558.

Arends, M. J., Morns, R. G. and Wyllie, A. H. (1990) Apoptosis: the role of the endonuclease. *American Journal of Pathology*, **136**, 593–608.

Barry, M. A., Behnks, C. A. and Eastman, A. (1990) Activation of programmed cell death/apoptosis by cisplatin, other anticancer drugs, toxins and hyperthermia. *Biochemical Pharmacology*, **40**, 2353–2362.

Cohen, J. A. and Duke, R. C. (1984) Glucocorticoid activation of a calcium-dependent endonuclease in thymocyte nuclei leads to cell death. *Journal of Immunology*, **132**, 38–42.

Compton, M. M., Haskill, J. S. and Cidlowski, J. A. (1988) Analysis of glucocorticoid actions on rat thymocyte deoxyribonucleic acid by fluorescence-activated flow cytometry. *Endocrinology*, **122**, 2158–2164.

Disasquale, B., Marini, A. M. and Youle, R. J. (1991) Apoptosis and DNA degradation induced by 1-methyl-4-phenyl-pyridinium in neurons. *Biochemical and Biophysical Research Communications*, **181**, 1442–1448.

Dive, C., Gregory, C. D., Phipps, D. J., Evans, D. L., Milner, E. M. and Wyllie, A. H. (1992) Analysis and discrimination of necrosis and apoptosis (programmed cell death) by multiparameter flow cytometry. *Biochimica Biophysica Acta*, **1133**, 275–285.

Dyson, J. E. D., Summons, D. M., Daniel, J., McLaughlin, J. M., Quirke, P. and Bird, C. C. (1986) Kinetic and physical studies of cell death induced by chemotherapeutic agents or hyperthermia. *Cell and Tissue Kinetics*, **19**, 311–324.

Gunji, H., Kharbanda, S. and Kufe, D. (1991) Induction of internucleosomal DNA fragmentation in human myeloid leukaemia cells by 1-β-D-Arabinofuranosyl-cytosine. *Cancer Research*, **51**, 741–743.

Ijiri, K., (1988) Apoptosis (cell death) induced in mouse bowel by 1,2-dimethylhydrazine, methyl azoxyl-methanol acetate and X-rays. *Cancer Research*, **49**, 6342–6346.

Johnson, C. A., Forster, T. H. and Allan, D. J. (1991) Hydroxyurea induces apoptosis and regular DNA fragmentation in a human lymphoma cell line. Abstract. *Proc. 35th Annual Conference of the Australian Society for Biochemistry and Molecular Biology*, p.37.

Kaufmann, S. H. (1989) Induction of endonucleolytic DNA cleavage in human acute myelogenous leukaemia cells by etoposide, camptothecin and other cytotoxic anticancer drugs. Cautionary note. *Cancer Research*, **49**, 5870–5878.

Kerr, J. F. R., Wyllie, A. H. and Currie, A. R. (1972) Apoptosis: A basic biological phenomenon with wide-ranging implications in tissue kinetics. *British Journal of Cancer*, **26**, 239–257.

Kizaki, H., Tadakuma, T., Odaka, C., Muramatsu, J. and Ishi, M. (1989) Activation of a suicide process of thymocytes through DNA fragmentation by calcium ionophores and phorbol esters. *Journal of Immunology*, **143**, 1790–1974.

Kwok, J. B. J. and Tattersall, M. H. B. (1991) Inhibition of 2-desamino-2-methyl-10- propagyl-5,8-dideaza folic acid cytotoxicity by 5,10-dide-azatetra hydrofolate in L1210 cells with decrease in DNA fragmentation and deoxyadenosine triphosphate pools. *Biochemical Pharmacology*, **42**, 507–513.

Kwok, J. B. J. and Tattersall, M. H. N. (1992) DNA fragmentation dATP pool elevation and potentiation of antifolate cytotoxicity in L1210 cells by hypoxanthine. *British Journal of Cancer*, **65**, 503–508.

Ledda-Columbano, G. M., Coni, P., Faa, G., Manerti, G. and Columbano, A. (1992) Rapid induction of apoptosis in rat liver by cycloheximide. *American Journal of Pathology*, **140**, 545–549.

Lennon, S. V., Martin, S. J. and Cotter, T. G. (1990) Induction of apoptosis (programmed cell death) in tumour cell lines by widely diverging stimuli. *Biochemical Society Transcriptions*, **18**, 343–345.

Lennon, S. V., Martin, S. J. and Cotter, T. G. (1991) Dose-dependent induction of apoptosis in human tumour cell lines by widely diverging stimuli. *Cell Proliferation*, **24**, 203–214.

Marks, D. I. and Fox, R. M. (1991) DNA damage, poly[ADP-ribosyl]ation and apoptotic cell death as a potential common pathway of cytotoxic drug action. *Biochemical Pharmacology*, **42**, 1859–1867.

McConkey, D. J., Hartzell, P., Duddy, S. K., Haakansson, H. and Orrenius, S. (1988) 2,3,7,8 tetrachlorodibenzo-p-dioxin kills immature thymocytes by Ca^{2+}-mediated endonuclease activation. *Science*, **242**, 256–259.

McConkey, D. J., Hartzell, P. and Orrenius, S. (1990) Rapid turnover of endogenous endonuclease activity in thymocytes: effects of inhibitors of macromolecular synthesis. *Archives of Biochemistry and Biophysics*, 278, 284–287.

McConkey, D. J. and Orrenius, S. (1989) 2,3,7,8-tetrachlorodibenzo-p-dioxin (TCDD) kills glucocorticoid-sensitive thymocytes *in vivo*. *Biochemical and Biophysical Research Communications*, 160, 1003–1008.

McDougall, C. A., Kluck, R. M., Harmon, B. V., Powell, L. W., Kerr, J. F. R. and Halliday, J. W. (1991) Is internucleosomal cleavage part of the final common pathway of cell death by apoptosis. Abstract. *Proc. 35th Annual Conference of the Australian Society for Biochemistry and Molecular Biology*, p.35.

Nicoletti, I., Migliorati, G., Pagliacci, M. C., Grignani, F. and Riccardi, C. (1991) A rapid simple method for measuring thymocyte apoptosis by propidium iodide staining and flow cytometry. *Journal of Immunological Methods*, 139, 271–279.

Pechatnikov, V. A. and Umansky, S. R. (1986) Flow cytometry of DNA degradation in thymocytes of γ-irradiated or hydrocortisone treated rats. *General Physiology and Biophysics*, 5, 273–284.

Pritchard, D. J. and Butler, W. H., (1989) Apoptosis—the mechanisms of cell death in dimethyl nitrosamine-induced hepatotoxicity. *Journal of Pathology*, 158, 253–260.

Raffray, M. and Cohen, G. M. (1991) Bis(tri-n-butyltin)oxide induces programmed cell death (apoptosis) in immature rat thymocytes. *Archives of Toxicology*, 65, 135–139.

Ronen, A. and Heddle, J. A. (1984) Site-specific induction of nuclear anomalies (apoptotic bodies and micronuclei) by carcinogens in mice. *Cancer Research*, 44, 1536–1540.

Searle, J., Lawson, T. A., Abbott, P. J., Harmon, B. and Kerr, J. F. R. (1974) An electron-microscope study of the mode of cell death induced by cancer chemotherapeutic agents in populations of proliferating normal and neoplastic cells. *Journal of Pathology*, 116, 129–138.

Shen, W., Kamendulis, L. M., Ray, S. D. and Corcoran, G. B. (1991) Acetaminophen-induced cytotoxicity in cultured mouse hepatocytes: correlation of nuclear Ca^{2+} accumulation and early DNA fragmentation with cell death. *Toxicology and Applied Pharmacology*, 111, 242–254.

Sorenson, C. M. and Eastman, A. (1988a) Mechanism of cis-diammedichloroplatinum (II)-induced cytotoxicity. Role of G2 arrest and DNA double-strand breaks. *Cancer Research*, 48, 4484–4488.

Sorenson, C. M. and Eastman, A. (1988b) Influence of cis-diammindichloroplatinum (II) on DNA synthesis and cell cycle progression in excision repair proficient and deficient Chinese hamster ovary cells. *Cancer Research*, 48, 6703–6707.

Telford, W. G., King, L. E. and Fraber, P. J. (1991) Evaluation of glucocorticoid- cytometry. *Cell Proliferation*, 24, 447–459.

Vedeckis, W. V. and Bradshaw, H. D. (1983) DNA fragmentation in S49 lymphoma cells killed with glucocorticoids and other agents. *Molecular and Cellular Endocrinology*, 30, 215–227.

Walker, N. I., Bennett, R. E. and Axelsen, R. A. (1988) Melanosis Coli: a consequence of anthraquinone-induced apoptosis of colonic epithelial cells. *American Journal of Pathology*, 131, 465–476.

Wyllie, A. H. (1980) Glucocorticoid-induced thymocyte apoptosis is associated with endogenous endonuclease activation. *Nature*, 284, 555–556.

Wyllie, A. H. (1987) Cell death. *International Reviews in Cytology*, 17(Suppl.), 755–785.

Wyllie, A. H., Kerr, J. F. R. and Currie, A. R. (1980) Cell death: The significance of apoptosis. *International Reviews in Cytology*, 68, 251–306.

Wyllie, A. H., Morris, R. G., Smith, A. L. and Dunlop, D. (1984) Chromatin cleavage in apoptosis; association with condensed chromatin morphology and dependence on macromolecular synthesis. *Journal of Pathology*, 142, 67–77.

Yamada, T. and Ohyama, H. (1988) Radiation-induced interphase death of rat thymocytes is internally programmed (apoptosis). *International Journal of Radiation Biology*, 53, 65–75.

Chapter 24

APOPTOSIS RESULTING FROM ANTI-CANCER AGENT ACTIVITY *IN VIVO* IS ENHANCED BY BIOCHEMICAL MODULATION OF TUMOUR CELL ENERGY

Daniel S. Martin[1,2,5], Robert L. Stolfi[2], Joseph R. Colofiore[2], Jason A. Koutcher[3], Alan Alfieri[3], Stephen Sternberg[4], and L. Dee Nord[2]

[1]Developmental Chemotherapy, Memorial Sloan Kettering Cancer Center (MSKCC), 1275 York Avenue, New York, NY 10021
[2]Cancer Research, Catholic Medical Center, 89–15 Woodhaven Blvd., Woodhaven, NY 11421
[3]Department of Medical Physics and Radiology, MSKCC
[4]Department of Pathology, MSKCC
[5]To whom correspondence should be addressed: 89–15 Woodhaven Blvd., Woodhaven, NY 11421, USA

Abbreviations

FUra, 5-fluorouracil; Adria, adriamycin; CTL, cytotoxic T lymphocyte; NAD, nicotinamide adenine dinucleotide; 6-AN, 6-aminonicotinamide; MMPR, 6-methylmercaptopurine riboside; PALA, N-(phosphonacetyl)-L-aspartic acid; PAM, phenylalanine mustard; NMR, nuclear magnetic resonance; HPLC, high performance liquid chromatography; ATP, adenosine triphosphate; PCr, phosphocreatine; Pi, inorganic phosphate; PR, partial regression of tumour; MTD, maximum tolerated dose.

THE SIGNIFICANCE OF BIOCHEMICAL HETEROGENEITY IN NEOPLASMS

Even effective anti-cancer agents that can induce cancer cell death in a particular type of cancer usually effect only partial tumour response rates rather than cures of advanced tumours. The reason for partial therapeutic failure is the presence of "qualitative" biochemical heterogeneity within the neoplastic cell population of an individual tumour; namely, the presence of biochemically different subsets of cancer cells within an individual tumour. Thus, an individual subset may exist which is not at all sensitive to an agent with therapeutic activity against another subset in the same tumour, and this fact explains the need for administering anti-cancer agents in combination for the treatment of an individual patient with advanced cancer.

Biochemical heterogeneity also usually extends "quantitatively" within each subset of malignant cells. "Quantitative" biochemical heterogeneity means that a population of cancer cells that is sensitive to an individual anti-cancer agent possesses among the individual cells a gradient of responsiveness to that agent. The cancer cells *within* a particular subset differ in that, while all cells of the subset are sensitive to a single effector agent, some cells are more vulnerable than others to a particular dose. Due to this gradient of sensitivity, and the low chemotherapeutic indices of anti-cancer agents, eradication of even slightly resistant cells of the drug-sensitive subset may not be possible at the agent's low *in vivo*-tolerated dose. These apparently resistant cells are actually only "pseudo"-resistant, but, under the described *in vivo* chemotherapeutic treatment conditions, will appear unresponsive to chemotherapy and grow into a treatment failure.

To avoid the latter lethal eventuality, the use of biochemical modulation can be employed to create new intracellular biochemical conditions in the quasi-drug-resistant (but actually drug-sensitive) cells that then will permit their destruction by *in vivo* tolerated doses of the effector agent. Thus, one approach to improving the selectivity of anti-cancer agents is to utilize selective biochemical modulation to increase the therapeutic activity of an individual effector agent, such as FUra or Adria (Martin, 1987a; Martin, 1987b; Martin et al., (in press); Stolfi et al., 1992).

THE BIOLOGICAL RELEVANCE OF APOPTOSIS

Most anti-cancer agents induce cancer cell death by an ordered sequence of events which has been termed apoptosis. Hence, an understanding of the biochemical cascade culminating in apoptosis should offer the opportunity for therapeutic intervention at the biochemical level (i.e., biochemical modulation) to facilitate the process resulting in the death of cancer cells, and thereby to improve chemotherapeutic efficacy. The substantially different morphological features of apoptosis from classic necrosis are discussed elsewhere in this volume (Kerr, Chapter 1). This phenomenon was first outlined (Kerr et al., 1972) and named "apoptosis", in 1972 (Kerr and Searle, 1972). Apoptosis is an important and widely occurring event. It is a type of cell death that occurs under a wide variety of physiological and pathological conditions. For example, it occurs spontaneously in solid cancer (Kerr et al., 1972), in focal elimination of normal tissue in embryogenesis (Hinchliffe, 1981), in withdrawal atrophy of hormone-dependent tissues (Kyprianou and Isaacs, 1988; Kyprianou, et al., 1990), in normal lymphocyte deletion in the thymus (Owen et al., Chapter 18), in CTL-mediated cell kill (Duke, 1991), in withdrawal of growth factor (IL-2) from dependent T cells (Duke and Cohen, 1986), irradiation (Warters, 1992; Meyn et al., Chapter 22), hyperthermia (Harmon et al., Chapter 25), monoclonal antibody anti-APO-1 (Kramer, Chapter 19), autoimmune disease (Pender, Chapter 20), toxic necrosis factor (Rubin, 1988), toxins (Waring, Chapter 7), anti-cancer agents (Barry et al., 1990), and still other stimuli (Gerschenson and Rotello, 1992). Clearly, a great diversity of stimuli can trigger apoptosis. It is therefore likely that multiple biochemical pathways are involved, at least *initially*, and that the apoptotic biochemical cascade can be activated at different points (Duvall and Wyllie, 1986). However, the key initiating biochemical step(s) has (have) not been identified.

BIOCHEMICAL FEATURES OF APOPTOSIS

Apoptosis is considered synonymous with an active (i.e., energy-dependent, at least *initially*) inherent gene-directed program of cell death. This genetic program has not yet been definitively ascertained, although a number of genes that appear to participate in the apoptotic biochemical cascade have been identified (Gerschenson and Rotello, 1992; Cohen et al., Chapter 12; Strasser et al., Chapter 14; Goldstone and Lavin, Chapter 17; Yonish-Rouach et al., Chapter 16). There are studies suggesting, as well as refuting the fact that the endonucleolytic fragmentation results from transcription and translation of a new gene product(s), such as endonuclease.

A consistent feature of apoptosis is the activation of a nuclear endonuclease. Sustained increases in Ca^{2+} have been reported to induce a Ca^{2+}/Mg^{2+}-dependent endonuclease leading to cleavage of nuclear chromatin into oligonucleosome-length DNA fragments, the nucleosome "ladder" pattern considered characteristic and diagnostic of apoptosis (McConkey, et al., Chapter 2). However, other studies report that internucleosomal DNA cleavage is not mediated by a Ca^{2+}-requiring mechanism. And still other studies report that elevated intracellular Ca^{2+} is not universally present during induction of apoptosis, or that an endonuclease is activated by intracellular acidification rather than increases in intracellular Ca^{2+} (Barry et al., 1990; Gerschenson and Rotello, 1992; Alnemri and Litwack, Chapter 8; Iwata et al., Chapter 3; Eastman, 1992; Bertrand et al., 1991; Schanne et al., 1979). Calcium-dependent processes (other than activation of a specific Ca^{2+}-dependent endonuclease) may be required, however, for cytotoxicity. It is to be stressed, however, that there are examples of apoptotic cell killing that do not involve Ca^{2+}-mediated events. Furthermore, although often present, internucleosomal DNA cleavage does not necessarily accompany the appearance of morphological changes of apoptosis, and it is therefore now believed that if the "ladder" pattern is absent, assessment of apoptosis should include morphological studies (Kerr, 1993; Gerschenson and Rotello, 1992). Obviously, however, DNA fragmentation, is a most important feature of the apoptotic biochemical cascade. Although the initial and intermediate steps of apoptosis are likely through different pathways and are essentially uncertain or unknown, programmed cell death is an active cell response to physiological or damaging stimuli with DNA fragmentation the converging step of these preceding processes. The apoptotic biochemical cascade then continues through a common pathway which is largely known, and which is relevant to the final death of the cell.

DNA strand breaks, whether internucleosomal or not, stimulate activation of poly (ADP-ribose) polymerase, a nuclear enzyme (Hayaishi et al., 1977; Juarez-Salinas, 1979; Purnell et al., 1980; Ueda, 1985). The enzyme utilizes NAD^+ as a substrate, cleaving it at the glycosidic bond between the nicotinamide and adenosine diphosphoribose portion of the molecule, and links the poly (ADP-ribose) moieties to cleaved sites of DNA. The result is a marked decrease in intracellular NAD^+ with a consequent fall in ATP until finally there is insufficient ATP to sustain survival of the cell, and apoptosis occurs. Many workers have shown in a variety of cell types that lowering of NAD^+ is due to a greatly enhanced utilization of NAD^+ for the synthesis of poly (ADP-ribose), and that the synthesis of poly (ADP-ribose) is stimulated by molecular damage to DNA caused by a variety of agents (Sims et al., 1983; Rankin et al., 1980; Whish et al., 1975; Skidmore et al., 1979; Berger et al., 1987; Wielckens et al., 1982; Tanizawa et al., 1989; Sorenson et al., 1990; Smith et al., 1992).

An endonuclease potentially involved in apoptosis is itself initially inhibited by direct poly ADP-ribosylation by poly (ADP-ribose) polymerase (Tanaka et al., 1984). The initial blocking of endonucleolytic activity may be an important process to maintain DNA-damaged cells through a period of attempted DNA repair. However, until DNA repair is completed, NAD^+ and ATP levels continue to decrease due to the sustained activity of poly (ADP-ribose) polymerase stimulated by the ongoing presence of DNA strand breaks. Consequent ATP depletion leads to failure of ATP-dependent membrane pumps and results in the disturbance of the cell's ionic (e.g. Ca^{2+}) and pH equilibrium, and activation thereby of intracellular cytotoxic mechanisms involving protease, phospholipase and endonuclease. Proteolysis then ensues and apparently is essential for activation of the endonucleolytic process since apoptosis is prevented by inhibitors of protease (Bruno et al., 1992); the protease inhibitors prevent degradation of certain nuclear proteins (e.g. poly (ADP-ribose) polymerase). The protease degradation of poly (ADP-ribose) polymerase, which has been shown to accompany endonucleolytic DNA degradation (Kaufman, 1989; Smith et al., 1992) stops the inhibition of endonuclease by poly ADP-ribosylation. Other studies have also concluded that inhibition of poly(ADP-ribose) polymerase may be a general mechanism for endonuclease activation during apoptosis (Nelipovich et al., 1988).

ANALYSIS OF ABOVE-DESCRIBED EVENTS OF APOPTOSIS

It appears that apoptosis can be induced by many diverse stimuli and, therefore, may be initiated by a number of different biochemical pathways (i.e. different signalling pathways, different genes, and different gene products), and activated at a number of different points with convergence to the step of primary DNA strand breakage. It can also be activated by DNA strand breaks *per se* (e.g. DNA-damaging anti-cancer agents). At this point, however, the apoptotic biochemical cascade continues on a common final pathway: namely, and in sequential order, the DNA strand breaks activate poly (ADP-ribose) polymerase which lowers NAD^+ pools which in turn leads to ATP depletion, which in turn causes failure of ATP-dependent membrane pumps bringing about a disturbance of intracellular ionic and pH equilibrium that then activates intracellular cytotoxic enzymes (protease, phospholipase, and endonuclease), which produce final loss of cell integrity (death) due to more complete endonucleolysis and destruction of the cytoskeleton. See Figure 24.1.

The "common final pathway" point of view is not new. Gaal *et al.* (1987), have reported that "cellular euthanasia" can be invoked by severe DNA damage leading to overstimulation of poly (ADP-ribose) polymerase, which causes depletion of NAD^+ and ATP which results in cell death. Berger *et al.* (1987a, 1987b), have proposed the same mechanism for death of lymphoid cells stimulated by glucocorticoid. Schraufsatter *et al.* (1986), showed that oxidant injury of cells by H_2O_2 (e.g. as might be brought about by adriamycin) induced DNA strand breaks that activated poly (ADP-ribose) polymerase which caused depletion of NAD^+ sufficient to interfere with ATP synthesis and resultant energy depletion that was incompatible with cell survival.

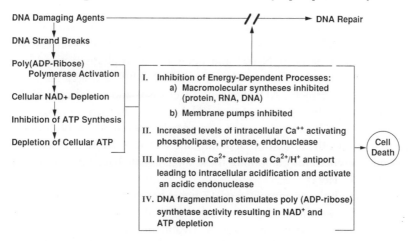

Figure 24.1 The common final pathway of programmed cell death.

Marks and Fox (1990) evaluated a number of anti-cancer agents *in vitro* and found similar patterns of DNA breaks, adenine nucleotide depletion and apoptotic cell death. They suggested that DNA damage, as a primary or secondary effect, associated with poly ADP-ribosylation and apoptotic cell death may be a common pathway of cytotoxic drug action.

THE RELEVANCE OF APOPTOSIS TO CANCER CHEMOTHERAPY

DNA-damaging agents (e.g. anti-cancer agents) provide the opportunity to induce the apoptotic biochemical cascade; indeed, essentially all have been shown to be able to effect tumour cell death by apoptosis. However, as explained in an earlier section on "quantitative" biochemical heterogeneity of neoplastic cell populations, there is a gradient of therapeutic responsiveness to an agent that allows quasi-drug-resistant (i.e. slightly sensitive) cancer cells to appear unresponsive to the chemotherapeutic agent. All that a DNA-damaging agent may do is *initiate* the opportunity for a cell to undergo apoptosis. The outcome of drug therapy to the individual cell will be determined by the response of the cell according to its phenotype, rather than by the primary drug-target interaction alone (Dive and Hickman, 1991). Therefore, concomitant biochemical modulation that effects an augmentation of the final common steps of the biochemical cascade of apoptosis (e.g. ATP depletion) should complement, and, indeed, be complemented by, the ATP-depleting apoptotic cascade induced by DNA-damaging anti-cancer agents.

RATIONALE AND METHODOLOGIC DETAILS FOR STUDIES EMPLOYING A TRIPLE COMBINATION OF AGENTS (PALA + MMPR + 6-AN) TO DEPLETE CANCER CELL ENERGY

The major generation of ATP in mammalian cells occurs during oxidative phosphorylation when NADH is reconverted to NAD^+ with concomitant conversion of ADP to ATP. Limitation of both NAD^+ and adenine diphosphate is the key to ATP depletion. 6-AN is a potent antagonist for NAD^+. MMPR, a *de novo* purine synthesis inhibitor, markedly depletes ATP levels in cancer tissue and can also, in high dosage, decrease pyrimidine levels *in vivo*. Since PALA, in low non-toxic dosage, can lower pyrimidine levels in tumours *in vivo*, it was added to the 6-AN + high dose MMPR therapy to reinforce the inhibition of pyrimidine synthesis. More detailed reasons and background references for combining these three agents have been presented previously (Stolfi *et al.*, 1992).

PALA (100 mg/kg) was administered 17 hours before MMPR (150 mg/kg) plus 6-AN (10 mg/kg). The three drugs were administered i.p. in the described schedule every 10 to 11 days for a total of 3 courses. Treatment was initiated on well-advanced tumours, averaging 260 mg when CD8F1 spontaneous, autochthonous breast cancers were the chemotherapeutic target, and 125 mg when first generation transplants of the CD8F1 spontaneous breast cancer was the tumour model. Mice carrying tumours of approximately equal weight were represented in each treatment group (10–14 mice/group). Therapeutic results were recorded 7–9 days after the third to fourth course of treatment. Differences in the number of partial tumour regressions between treatment groups were compared for statistical significance. Partial tumour regression is defined as a reduction in tumour volume of 50% or greater compared to the tumour volume at the time of initiation of treatment. The data from 3–6 separate experiments were pooled for final analysis. The timing of biochemical measurements is given in relation to the injection of the last chemotherapy (i.e. MMPR + 6-AN) in the triple drug sequence. Students t-test was used for measured biochemical differences between treatment groups, with differences between groups with $P \leq 0.05$ considered to be significant. The CD8F1 spontaneous, autochthonous mammary carcinoma model has been described previously (Stolfi et al., 1971; Martin et al., 1975), and has been included in the murine tumour testing panel of the National Cancer Drug Screening Program (Goldin et al., 1981).

CHEMOTHERAPEUTIC EFFECT OF PALA + MMPR + 6-AN

Table 24.1 records that the triple combination, 6 weeks after the initiation of therapy (i.e. 9 days after the 4th course of treatment), produced a partial tumour regression rate of 38% in mice bearing large (i.e. average = 260 mg), spontaneous, autochthonous breast tumours. There were 24 partial tumour regressions in 64 surviving mice with acceptable toxicity (4% mortality, 3 deaths in 67 treated mice). The data were pooled from six separate, consecutive experiments (Stolfi et al., 1992). In a subsequent series

Table 24.1 Therapeutic comparison of PALA + MMPR + 6-AN chemotherapy with and without the addition of 5-fluorouracil (FUra) in mice bearing spontaneous, autochthonous CD8F1 breast tumours[a].

Treatment[b]	Percent Body Weight Change	Dead/Total	Partial Regressions
1. $PALA_{100}$–17 h->$MMPR_{150}$ + 6-AN_{10}	–10	3/67 (4%)	24 (38%)
2. $PALA_{100}$–17 h ->$MMPR_{150}$ + 6-AN_{10}–2∫h->$FUra_{75}$	–10	5/66 (7%)	41 (67%)

a. Pooled data: Experiments R536; R537; R538; R539; R540; R541. Spontaneous, autochthonous CD8F1 breast tumours averaging 260 mg at the time of initiation of treatment.
b. The indicated treatment was administered at 10–11 day intervals. Subscripts refer to doses in mg/kg. Observations were recorded 6 weeks after initiation of treatment (i.e. approximately 9 days after the fourth course of treatment).

of three separate, consecutive experiments (Table 24.2), totaling 44 mice with spontaneous, autochthonous breast tumours averaging 304 mg at initiation of treatment, the triple chemotherapy produced a partial tumour regression rate of 76% (with a range of 50–92% in the individual experiments and no toxic deaths) (Martin et al., in press). Spontaneous regressions of these tumours has not been observed.

BIOCHEMICAL FINDINGS OF PALA + MMPR + 6-AN

Energy depletion in drug-treated tumours was manifest in decreased ATP pools as measured by HPLC, and is shown in Table 24.3. Compared to saline controls (taken as 100%, group 1), at 6, 24, and 48 hours after administration of MMPR alone (group 2), 6-AN alone (group 3), or the three drug combination, PALA + MMPR + 6-AN (group 4), ATP levels were significantly depressed, reaching levels of 32% and 15% of controls in tumour from mice treated with the 3-drug combination at 24 and 48 hours post treatment. By 72 hours, however, the ATP levels had returned to normal.

NMR spectra, obtained from 1st passage breast tumours pre-and post-treatment with PALA + MMPR + 6-AN demonstrated that both PCr/Pi and NTP/Pi ratios were decreased after treatment with the 3-drug combination, which is clearly indicative of energy depletion (data presented in Stolfi et al., 1992).

It is stressed that the loss of ATP is considered primary to the decrease in cell survival because, although it is not the sole determinant compromising cell viability, a significant drop in ATP concentration has many metabolic consequences, and it is likely that the other below-noted perturbations to cellular biochemistry would act in concert with ATP loss to cause cell death.

Table 24.2 Therapeutic comparison of PALA + MMPR + 6-AN chemotherapy with and without the addition of adriamycin (Adria) in CD8F1 mice bearing spontaneous, autochthonous breast cancers.[a]

Treatment[b]	Percent Body Weight Change	Dead/Total	Regressions/ Survivors Partial Tumour
1. $PALA_{100}$–17h –>$MMPR_{150}$ + 6-AN_{10}	−14	0/42	32/42 (76%)
2. $PALA_{100}$–17h –>$MMPR_{150}$ + 6-AN_{10}–2½h –>$Adria_6$	−16	2/44	42/42 (100%)
3. $Adria_{11}$	−11	1/44	7/43 (16%)

a. Pooled data: Experiments R559, R560, R561. Tumour weight averaged 304 mg at the time of initiation of treatment.
b. Three courses of the indicated treatment were administered with a 10 or 11 day interval between courses. Subscripts refer to doses in mg/kg. Adria was administered i.v. and all other drugs were administered i.p. Observations were recorded 7 days after the third course of treatment.

Table 24.3 Effect of PALA + MMPR + 6-AN on tumour ATP pools in CD8F1 mice[a,d].

Group	Treatment	6 HR[b] ug ATP/mg ± S.E.	% Saline Control	24 HR[a] ug ATP/mg ± S.E.	% Saline Control	48 HR[a] ug ATP/mg ± S.E.	% Saline Control	72 HR[a] ug ATP/mg ± S.E.	% Saline Control
1	Saline	7.1 ± 0.32	—	7.1 ± 0.32	—	7.1 ± 0.32	—	7.1 ± 0.32	—
2	MMPR$_{150}$	3.3[c] ± 0.32	47	3.3[b] ± 0.96	47	1.6[b] ± 0.59	34	9.0[d] ± 2.4	127
3	6AN$_{10}$	5.5[b] ± 0.21	77	5.3[b] ± 0.13	75	4.9[b] ± 0.16	69	12.0[c] ± 3.9	169
4	MMPR$_{150}$ + 6-AN$_{10}$	4.1[b] ± 0.49	58	2.2[b] ± 0.08	31	1.1[b] ± 3.6	15	16.0[c] ± 3.6	225
5	PALA 17 h MMPR$_{150}$ + 6-AN$_{10}$	3.9[b] ± 0.11	55	2.3[b] ± 0.31	32	1.1[b] ± 0.14	15	8.1[c] ± 0.7	114

a. Mean ± S.E. of 10 tumours/group (11 experiments)
b. Statistical comparison to group 1 (saline control); Significant = P value less than or equal to 0.05
c. Mean ± S.E. of 6 tumours/group (4 experiments)
d. Subscripts = mg/kg body weight; i.p. injections; 1st passage tumour transplants of CD8F1 spontaneous breast tumours.

In addition to the primary severe ATP loss, there was a lowering of NAD^+ levels, inhibition of macromolecular synthesis (including protein, DNA and RNA), inhibition of the pentose phosphate shunt (resulting in decreased NADPH levels), marked reduction of ribonucleoside triphosphates, and DNA breaks (3). This pattern of biochemical consequences of treatment with PALA-MMPR-6-AN is remarkably similar to reported biochemical events associated with "programmed cell death", or apoptosis.

Inhibition of DNA repair is a likely mechanism for the anti-cancer effects produced by the triple combination. The PALA-MMPR-6-AN induced reduction in all four individual ribonucleotide pools (Stolfi et al., 1992), which generally correlates with a reduction in the corresponding deoxyribonucleotide triphosphate pools (Hunting et al., 1981), would be expected not only to inhibit DNA synthesis, but to inhibit the potential for DNA repair as well. Repair of DNA damage is an energy-dependent process (Juarez-Salinas, 1979) and the depletion of both "building blocks" (i.e. the depletion of purine and pyrimidine deoxynucleotides), and of cellular energy (i.e. ATP), should result in a deficient ability to repair DNA damage, and enhanced cell death should ensue. Thus, if the PALA-MMPR-6-AN triple combination's major biochemical targets are selectively susceptible in the cancer cells (as the marked *in vivo* tumour PR rate and relative lack of host toxicity indicates), a chemotherapeutically-induced decrease in the rate of DNA repair may contribute to the cell killing by the triple combination.

HISTOLOGICAL EXAMINATION OF TRIPLE CHEMOTHERAPY-TREATED TUMOURS

The morphological hallmarks of apoptosis were found in the histological studies of the treated tumours, and have been previously presented (Stolfi et al., 1992). In the sections studied, some fields had no apoptotic cells, while others did, and the numbers of involved cells varied. Rough estimates of the number of apoptotic cells per high power field were made. However, the *in vivo* quantification of the morphologic incidence and severity of apoptosis is, at best, difficult. The apoptotic bodies are phagocytized and digested rapidly by neighboring cells and/or macrophages. It is estimated that apoptotic bodies remain visible by light microscopy for only a few hours (Bursch et al., 1990); therefore, apoptosis, a discrete phenomenon, is often not particularly conspicuous histologically. Wyllie, Kerr and Currie (1980) note "the involvement of only scattered single cells, the small size of most apoptotic bodies, and the rapid dispersal, phagocytosis, and digestion of bodies in tissues without any inflammatory response all mitigate against detection and even more so against quantification." Nevertheless, the minimal number of apoptotic changes in the controls contrasted sharply with the higher incidence and severity in the treated animals. Since the minor degree of necrosis (less than 10%) was equal in both control and treated groups, it is clear that apoptosis is the mode of cell death that accounts for the observed cell loss (i.e., the partial tumour regressions).

ADDITION OF DNA-DAMAGING AGENTS TO THE TRIPLE COMBINATION

A large number of anti-cancer agents that damage DNA have been demonstrated to induce apoptosis in cancer cells (Stolfi et al., 1992), as has irradiation (Warters, 1992; Meyn et al., Chapter 22; Radford et al., 1993). The spontaneous, autochthonous CD8F1 breast tumour model has demonstrated a remarkable correlation with human breast cancer in terms of both positive and negative sensitivity to individual chemotherapeutic drugs (Stolfi et al., 1988). We, therefore, chose to evaluate the combination of triple chemotherapy with the following apoptosis-inducing agents that are clinically active against human breast cancer: FUra, Adria, PAM and radiation.

FUra + The Triple Combination

Table 24.1 documents a statistically significant enhancement of the PR rate from 38% (group 1, PALA + MMPR + 6-AN) to 67% (with a range of 50–90% in the 6 individual experiments) by the addition of FUra to the triple combination (group 2) in mice bearing spontaneous autochthonous tumours (Stolfi et al., 1992). FUra alone at 75 mg/kg produced less than 5% regressions of spontaneous, autochthonous CD8F1 breast tumours.

Adria + the Triple Combination

Table 24.2 documents a statistically significant enhancement in tumour regressions ($P < 0.01$) over the triple combination (group 1) by the addition of Adria to the three-drug combination (group 2) in mice bearing spontaneous, autochthonous breast tumours (Martin et al., Chapter 24). Remarkably, there were 42 partial tumour regressions in the 42 surviving mice, or 100% (i.e. with no range in the 3 individual experiments), and little toxicity (only 5% mortality). Moreover, 5 of the 42 tumour regressions in group 2 were complete regressions. Adria alone (group 3) at its MTD, 11 mg/kg, produced only 16% partial tumour regressions. The tumour response rate of 100% in this spontaneous and highly heterogenous tumour model is unprecedented, and particularly noteworthy because of the high degree of chemotherapeutic correlation that has been documented between this breast tumour model and human breast cancer (Stolfi et al., 1988).

PAM + the Triple Combination

Table 24.4 reports the pooled results of three experiments with 1st generation tumour transplants of CD8F1 spontaneous, autochthonous breast tumours comparing the PR rates produced by the triple combination to those obtained by the addition of the alkylating agent, PAM, to PALA + MMPR + 6-AN. The first generation tumour transplants are obtained from a tumour cell brei made by pooling 3–4 spontaneously-arising tumours. As in all spontaneous tumours, whether human or murine, each individual spontaneous breast tumour has a biochemically heterogeneous neoplastic cell population. Thus, the individual tumour transplants in each experiment develop from a single brei, and therefore each transplant has a neoplastic cell composition with a number of cells insensitive to one or more of the employed chemotherapeutic agents. Hence, first generation tumour transplants invariably have a lower PR rate than that obtained in mice bearing spontaneous tumours, as for example, in the 14% PR rate obtained with the triple combination of group 1, Table 24.4. All tumour findings, however, are quantitatively relevant within individual first generation tumour transplant experiments, as are similar trends among experiments. The addition of PAM to the triple combination (group 2) markedly enhanced the PR rate to 74% from 14% without PAM, and with acceptable toxicity essentially the same as in the triple combination without PAM. PAM alone in doses as high as 18 mg/kg in the same schedule did not produce tumour regressions in this series of experiments (data not presented).

Radiotherapy + the Triple Combination

Figure 24.2 presents a representative experiment of the triple combination administered concomitantly with radiotherapy to first generation transplants of the CD8F1 spontaneous tumour. To facilitate restricting the administration of the radiotherapy to the tumour, the transplants were placed in one hind limb. All treatments were admin-

Table 24.4 Therapeutic evaluation of phenylalamine mustard (PAM) administered following PALA-MMPR-6-AN.

Treatment[a]	Percent Body Weight Change	Dead/Total	Partial Regressions[b]
1. $PALA_{100}$–17 h–> $MMPR_{150}$ + 6-AN_{10}	–17	2/30 (7%)	4/28 (14%)
2. $PALA_{100}$–17 h–> $MMPR_{150}$ + 6-AN_{10}–2∫h –>PAM_7	–21	3/30 (10%)	20/27 (74%)

a. Pooled results: Exps 2489F, 2541M, 2552M. 1st generation tumour transplants of CD8F1 spontaneous, autochthonous breast tumours averaging 125 mg when treatment initiated. Three courses of the indicated treatment were administered with a 10-11 day interval between courses. Observations were recorded one week after the third course of treatment. Note that PAM alone, in doses from 7 to 18 mg/kg, did not produce any tumour regressions in this series of experiments.
b. PR = partial tumour regressions (No. of regressions/survivors).

istered at the usual 10–11 day schedule for only three treatments (Figure 24.2, arrows), and the animals were observed for tumour response rates without further treatment for an additional 42 days. The initial PR rates were high in all three treated groups of animals but rapidly fell to 0% in the chemotherapy alone (PALA-MMPR-6-AN) group, and to 18% (including 1/10, 10%, complete tumour regressions) in the radiotherapy alone group. In sharp contrast, and despite 42 days without treatment, the combined modality therapy maintained a 90% PR rate (including 7/9 complete tumour regressions).

CLINICAL POTENTIAL FOR EVALUATION OF THE TRIPLE COMBINATION

Two of the three drugs (PALA, MMPR) currently are used clinically. 6-AN was abandoned after clinical trial in the early sixties because of a lack of efficacy as a single agent. However, at that time, the daily administration of anti-cancer agents was in vogue. 6-AN was administered on a daily schedule that resulted in a cumulative toxicity of nicotinamide deficiency that could be reversed merely by stopping treatment with 6-AN, or by administering the antidote, nicotinamide (Herter *et al.*, 1961). Today, the intermittent administration of anti-cancer drugs, has proven practical utility in cancer treatment. As an intermittently administered biochemical modulator (i.e. once every 10–11 days), 6-AN should not produce signs of nicotinamide deficiency. In the present preclinical studies, a low-dose intermittent (every 10–11 day) schedule rather than a daily administration schedule of 6-AN has been

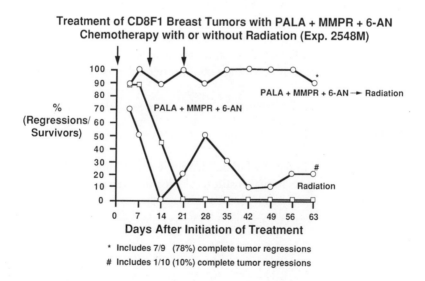

Figure 24.2 Treatment of tumours with the triple drug combination and radiotherapy.

employed successfully. In the early clinical trials it was determined that a dose of 6-AN of 0.2 mg/kg/day for up to 28 consecutive days of treatment was safe (Herter *et al.*, 1961). Using the equivalent surface area dosage conversion factor, this human dose was found to be equivalent to a cumulative dose of 24 mg/kg over 10 days in the mouse. Therefore, our intermittent, low dose schedule (10 mg/kg every 10–11 days) calculates to less than one-half of the safe cumulative human equivalent dose of 6-AN every 10 days. The toxicity identified when 6-AN was administered on a daily schedule in early clinical studies is not expected in the current clinical development of this agent in a therapeutic drug combination proven preclinically safe in an intermittent administration schedule.

CONCLUSION

The term biochemical modulation refers to the pharmacologic manipulation of metabolic pathways by an agent (the modulating agent) to produce the selective enhancement of the anti-tumour effect of a second agent (the "effector" agent) (Martin *et al.*, 1987b). In this context, the triple combination, PALA + MMPR + 6-AN, may be viewed as biochemical modulation employed to establish in tumour cells a wide array of biochemical changes—i.e. a primary diminution of ATP, lowering of all of the nucleoside triphosphates and NAD, inhibition of macromolecular synthesis, and supression of the flux through the pentose phosphate shunt and glycolytic pathway— thereby establishing a level of disrupted metabolism in cancer cells that would complement, and indeed be complemented by, the cascade of similar biochemical derangements induced by the apoptosis effects of DNA-damaging anti-cancer agents (e.g., FUra, Adria, PAM, radiotherapy). This modulation would be expected to result in enhanced cancer cell deaths which reflect, as in the results presented here, in improved tumour regression rates. Greater definition of the critical apoptotic events produced by the combination of these anti-cancer agents should reveal additional metabolic reactions where drug targeting could further enhance tumour cell cytotoxicity.

ACKNOWLEDGEMENT

This work was supported in part by Public Health Service Grant CA 25842 from the National Cancer Institute, National Institute of Health, Department of Health and Human Services.

REFERENCES

Barry, M. A., Behnke, C. A. and Eastman, A. (1990) Activation of programmed cell death (apoptosis) by cisplatin, other anti-cancer drugs, toxins and hyperthermia. *Biochemical Pharmacology*, **40**, 2353–2362.

Berger, N. A., Berger, S. J. and Gerson, S. L. (1987a) DNA repair ADP ribosylation and pyridine nucleotide metabolism as targets for cancer chemother. *Anti-Cancer Drug Design*, **2**, 203–210.

Berger, N. A., Berger, S. J., Sudar, D. C. and Distelhorst, C. W. (1987b) Role of nicotinamide adenine dinucleotide and adenosine triphosphate in glucocorticoid-induced cytotoxicity in susceptible lymphoid cells. *Journal of Clinical Investigation*, **79**, 1558–1563.

Bertrand, R., Kerrigan, D., Sarang, M. and Pommier, Y. (1991) Cell death induced by topoisomerase inhibitors. Role of calcium in mammalian cells. *Biochemical Pharmacology*, **42**, 77–85.

Brune, B., Hartzell, P., Nicotera, P. and Orrenius, S. (1991) Spermine prevents endonuclease activation and apoptosis in thymocytes. *Experimental Cell Research*, **195**, 323–329.

Bruno, S., Lassota, P., Giaretti, W. and Darzynkiewica, Z. (1992) Apoptosis of rat thymocytes triggered by prednisolone, camptothecin or teniposide is selective to G_0 cells and is prevented by inhibitors of proteases. *Oncology Research*, **4**, 29–35.

Bursch, W., Paffe, B., Putz, G. and Schulte-Hermann, R. (1990) Determination of the length of the histological stages of apoptosis in normal liver and in altered hepatic foci of rats. *Carcinogenesis*, **11**, 847–856.

Dive, C. and Hickman, J. A. (1991) Drug-target interactions: only the first step in the commitment to programmed cell death? *British Journal of Cancer*, **64**, 192–196.

Duke, R. C. (1991) Apoptosis in cell-mediated immunity. In: *Apoptosis: The Molecular Basis of Cell Death*, edited by L. D.Tomei and F. O. Cope, pp. 209–226. New York: Cold SpringHarbor Laboratory Press.

Duke, R. C. and Cohen, J. J. (1986) IL-2 addiction: Withdrawal of growth factor activates a suicide program in dependent T cells. *Lymphokine Research*, **5**, 289–299.

Duvall, E. and Wyllie, A. H. (1986) Death and the cell. *Immunology Today*, **7**, 115–119.

Eastman, A. (1992) The pathway of apoptosis activated by anti-cancer agents. *Proceedings of the American Association for Cancer Research*, **33**, 587–588.

Epstein, R. J. (1990) Drug-induced DNA damage and tumour chemosensitivity. *Journal of Clinical Oncology*, **8**, 2062–2084.

Gaal, J. C., Smith, K. R. and Pearson, C. K. (1987) Cellular euthanasia mediated by a nuclear enzyme: A central role for nuclear ADP-ribosylation in cellular metabolism. *Trends in Biochemical Sciences*, **12**, 129–132.

Gerschenson, L. E. and Rotello, R. J. (1992) Apoptosis: a different type of cell death. *FASEB Journal*, **6**, 2450–2455.

Goldin, A., Venditti, J. M., MacDonald, J. S., Muggia, F., Henney, J. and DeVita, V. T., Jr. (1981) Current results of the screening program at the Division of Cancer Treatment, National Cancer Institute. *European Journal of Cancer*, **17**, 129–142.

Hayaishi, O. and Veda, K. (1977) Poly (ADP-ribose) and ADP-ribosylation of proteins. *Annual Review of Biochemistry*, **46**, 95–116.

Herter, F. P., Weissman, S. G., Thompson, H. G., Jr., Hyman, G., and Martin, D. S. (1961) Clinical experience with 6-aminonicotinamide. *Cancer Research*, **21**, 31–37.

Hinchliffe, J. R. (1981) Cell death in embryogenesis. In *Cell Death in Biology and Pathology*, edited by I. D. Bowen, and R. A. Lockshin, pp.35–75. London: Chapman and Hall.

Hunting, D., Hordern, J. and Henderson, J. F. (1981) Effects of altered ribonucleotide concentrations on ribonucleotide reduction in intact Chinese hamster ovary cells. *Canadian Journal of Biochemistry*, **59**, 821–829.

Juarez-Salinas, H., Sims, J. L. and Jacobson, M. K. (1979) Poly (ADP-ribose) levels in carcinogen-treated cells. *Nature*, **282**, 740–741.

Kaufmann, S. H. (1989) Induction of endonucleolytic DNA cleavage in human acute myelogenous leukemia cells by etoposide, camptothecin, and other cytotoxic anti-cancer drugs: A cautionary note. *Cancer Research*, **49**, 5870–5878.

Kerr, J. F. R. and Searle, J. (1972) A suggested explanation for the paradoxically slow growth rate of basal cell carcinomas that contain numerous mitotic figures. *Journal of Pathology*, **107**, 41–44.

Kerr, J. F. R., Wyllie, A. H. and Currie, A. R. (1972) Apoptosis. A basic biological phenomenon with wider implications in tissue kinetics. *British Journal of Cancer*, **26**, 239–257.

Kyprianou, N., English, H. F., Isaacs, J. T. (1990) Programmed cell death during regression of PC-82 human prostate cancer following androgen ablation. *Cancer Research*, **50**, 3748–3753.

Kyprianou, N. and Isaacs, J. T. (1988) Activation of programmed cell death in the rat ventral prostate after castration. *Endocrinology*, **122**, 552–562.

Kyprianou, N. and Isaacs, J. T. (1989) "Thymineless death in androgen-independent prostatic cancer cells. *Biochemical and Biophysical Research Communications*, **165**, 73–81.

Lennon, S. V., Martin, S. J. and Cotter, T. G. (1991) Dose-dependent induction of apoptosis in human tumour cell lines by widely diverging stimuli. *Cell Proliferation*, **24**, 203–214.

Marks, D. I. and Fox, R. M. (1990) DNA damage, poly ADP-ribosylation and apoptotic cell death as a potential common pathway of cytotoxic drug action. *Biochemical Pharmacology*, **42**, 1859–1867.

Martin, D. S. (1987a) Purine and pyrimidine biochemistry, and some relevant clinical and preclinical cancer chemotherapy research. In *Metabolism and Action of Anti-Cancer Drugs*, edited by G. Powis, and R. A. Prough, pp. 91–140. London: Taylor and Francis.

Martin, D. S. (1987b) Biochemical modulation—Perspectives and Objectives. In: *New Avenues in Developmental Cancer Chemotherapy*, edited by K. R. Harrap, and T. Connors, pp.113–162. London: Academic Press.

Martin, D. S., Fugmann, R. A., Stolfi, R. L. and Hayworth, P. E. (1975) Solid tumour animal model therapeutically predictive for human breast cancer. *Cancer Chemotherapy Reports*, Part 2, **5**, 89–109.

Martin, D. S., Stolfi, R. L., Colofiore, J. R., Nord, L. D. and Sternberg, S. (In press) Biochemical modulation of tumour cell energy *in vivo*: II. A lower dose of adriamycin is required and a greater anti-tumour activity is induced when cellular energy is depressed. *Cancer Research*.

Martin, S. J. and Cotter, T. G. (1990) Disruption of microtubules induces an endogenous suicide pathway in human leukemia HL-60 cells. *Cell and Tissue Kinetics*, **23**, 545–559.

Nelipovich, P. S., Nikonova, L. V. and Umansky, S. R. (1988) Inhibition of poly (ADP-ribose) polymerase as a possible reason for activation of Ca^{2+}/Mg^{2+} endonuclease in thymocytes of irradiated rats. *International Journal of Radiation Biology*, **53**, 749–752.

Piacentini, M., Fesus, L., Farrace, M. G., Ghiubelli, L., Piredda, L. and Melino, G. (1991) The expression of "tissue" transglutaminase in two human cancer cell lines is related with the programmed cell death (apoptosis). *European Journal of Cell Biology*, **54**, 246–254.

Purnell, R. M., Stone, P. R. and Whish, W. J. D. (1980) ADP ribosylation of proteins. *Biochemical Society Transactions*, **8**, 215–227.

Rankin, P. W., Jacobson, M. K., Mitchell, V. R. and Busbee, D. L. (1980) Reduction of nicotinamide adenine dinucleotide levels by ultimate carcinogens in human lymphocytes. *Cancer Research*, **40**, 1803–1807.

Rubin, B. Y., Smith, L. J., Hellerman, G. R., Lunn, R. M., Richardson, N. K. and Anderson, S. L. (1988) Correlation between the anticellular and DNA fragmenting activites of tumor necrosis factor. *Cancer Research*, **48**, 6006–6010.

Schanne, F. A. X., Kane, A. B., Young, E. E. and Farber, J. L. (1979) Calcium dependence of toxic cell death: A final common pathway. *Science*, **206**, 700–702.

Schraufstatter, I. U., Hinshaw, D. B., Hyslop, P. A., Spragg, R. G. and Cochrane, C. G. (1986) Oxidant injury of cells: DNA strand breaks activate polyadenosine diphosphate-ribose polymerase and lead to depletion of nicotinamide adenine dinucleotide. *Journal of Clinical Investigation*, **77**, 1312–1322.

Sims, J. L., Berger, S. J. and Berger, N. A. (1983) Poly (ADP-ribose) polymerase inhibitors preserve nicotinamide adenine dinucleotide and adenosine 5'-triphosphate pools in DNA-damaged cells: Mechanism of stimulation of unscheduled DNA synthesis. *Biochemistry*, **22**, 5188–5194.

Skidmore, C. J., Davies, M. I., Goodwin, P. M., Halldorson, H., Lewis, P J., Shall, S. and Zia'ee, A-A. (1979) The involvement of poly (ADP-ribose) polymerase in the degradation of NAD caused by gamma-radiation and N-methyl-N-nitrosourea. *European Journal of Biochemistry*, **101**, 135–142.

Smith, G. K., Duch, D. S., Dev, I. K. and Kaufman, S. H. (1992) Metabolic effects and kill of human T-cell leukemia by 5-deazacyclotetrahydrofolate, a specific inhibitor of glycineamide ribonucleotide transformylase. *Cancer Research*, **52**, 4895–4903.

Sorenson, C., Barry, M. A. and Eastman, A. (1990) Analysis of Events associated with cell cycle arrest at G_2 phase and cell death induced by cisplatin. *Journal of the National Cancer Institute*, **82**, 749–755.

Stolfi, R. L., Colofiore, J. R., Nord, L. D., Koutcher, J. A. and Martin, D. S. (1992) Biochemical modulation of tumour cell energy: I. Regression of advanced, spontaneous murine breast tumours with a 5-fluorouracil-containing drug combination. *Cancer Research*, **52**, 4074–4081.

Stolfi, R. L., Martin, D. S. and Fugmann, R. A. (1971) Spontaneous murine mammary adenocarcinoma: Model system for evaluation of combined methods of therapy. *Cancer Chemotherapy Reports*, **55**, 239–251.

Stolfi, R. L., Stolfi, L. M., Sawyer, R. C. and Martin, D. S. (1988) Chemotherapeutic evaluation using clinical criteria in spontaneous, autochthonous murine breast tumours. *Journal of the National Cancer Institute*, **80**, 52–55.

Tanaka, Y., Yoshihara, K., Itaya, A., Kamiya, T. and Koide, S. S. (1984) Mechanism of the inhibition of Ca^{2+}, Mg^{2+}-dependent endonuclease of bull seminal plasma induced by ADP-ribosylation. *Journal of Biological Chemistry*, **259**, 6579–6585.

Tanizawa, A., Kubota, M., Hashimo, H., Shimizu, T., Takimoto, T., Kitoh, T., Akiyama, Y. and Mikawa, H. (1989) VP-16-induced nucleotide pool changes and Poly (ADP-ribose) synthesis: The role of VP-16 in interphase death. *Experimental Cell Research*, **185**, 237–246.

Tanizawa, A., Kubota, M., Hashimoto, H., *et al.* (1989) VP-16-induced nucleotide pool changes and poly (ADP-ribose) synthesis: The role of VP-16 in interphase death. *Experimental Cell Research*, **185**, 237–246.

Ueda, K. (1985) ADP-ribosylation. *Annual Review in Biochemistry*, **54**, 73–100.

Warters, R. L. (1992) Radiation-induced apoptosis in a murine T-cell hybridoma. *Cancer Research*, **52**, 883–890.

Whish, W. J. D., Davies, M. I. and Sholl, S. (1975) Stimulation of poly (ADP-ribose) polymerase activity by the anti-tumour antibiotic, streptozotocin. *Biochemical and Biophysical Research Communications*, **65**, 722–730.

Wielckins, K. and Delfs, T. (1986) Glucocorticoid-induced cell death and poly adenosine diphosphate ADP-ribosylation: increased toxicity of dexamethasone on mouse S491 lymphoma cells with the poly ADP-ribosylation inhibitor benzamide. *Endocrinology*, **119**, 2383–2392.

Wielckens, K., Schmidt, A., George, E., Bredehorst, R. and Hilz, H. (1982) DNA fragmentation and NAD depletion. Their relationship to the turnover of endogenous mono (ADP-ribosyl) and poly (ADP-ribosyl) proteins. *Journal of Biological Chemistry*, **25**, 12 873–12 877.

Wyllie, A. H., Kerr, J. F. and Currie, A. R. (1980) Cell death: the significance of apoptosis. *International Reviews in Cytology*, **68**, 251–306.

Chapter 25

HYPERTHERMIA-INDUCED APOPTOSIS

Brian V. Harmon[1], Trevor H. Forster[1] and Russell J. Collins[2]
[1]School of Life Science, Queensland University of Technology,
Brisbane, Queensland 4000, Australia
[2]Department of Pathology, Royal Brisbane Hospital,
Brisbane, Queensland 4029, Australia

INTRODUCTION

Hyperthermia is under investigation as a treatment modality for cancer. Whilst there are many reasons why hyperthermia should be a good anti-cancer agent (reviewed in Field, 1983, 1987), in general they fall into two main categories. Firstly, hyperthermia enhances the effectiveness of other treatment modalities and secondly, hyperthermia at temperatures above 42°C kills tumour cells, especially those cells that are normally resistant to other forms of treatment (Overgaard, 1983; Field, 1987). Even though studies carried out some 12 to 15 years ago had shown that hyperthermia did sometimes trigger cell death with the morphological features of apoptosis (Wanner et al., 1976; Schrek et al., 1980), it had been widely accepted, until recently, that hyperthermia always caused necrosis (Buckley, 1972; Fajardo et al., 1980; Overgaard, 1983). It is now clear that this is not the case, a number of studies having shown that hyperthermia induces apoptosis in a range of normal and neoplastic cells and tissues (Allan and Harmon, 1986; Harmon et al., 1989; Harmon et al., 1990; 1991a; Barry et al., 1990; Bruggen et al., 1991; Lennon et al., 1991; Sellins and Cohen, 1991; Takano et al., 1991; Mosser and Martin, 1992). Our studies on murine mastocytoma (Harmon et al., 1990) and a number of human cell lines have shown that heat loads that fall within the therapeutic range ($\approx$ 42 to 44°C for 30 min) induce apoptosis whereas higher heat loads cause necrosis.

Apoptosis induced by hyperthermia occurs extremely rapidly after treatment. This, coupled with the very high levels of apoptosis that can be achieved in some cell lines (almost 100% apoptosis in BM13674 and WW2 Burkitt's lymphoma cell lines; Harmon et al., 1991b) means than an almost synchronous yield of pure apoptotic cells can be obtained. As such, this system lends itself well to utilisation as a model for investigations of the underlying biochemical mechanism of apoptosis.

In this chapter, we will examine:

1. Features of hyperthermia-induced apoptosis including its morphology, DNA fragmentation, kinetics, incidence and the sensitivity of different cell types to apoptotic induction;
2. The importance of accurate delivery and monitoring of heat loads;
3. Results of studies into the mechanism of apoptosis induced by hyperthermia and a proposed molecular pathway for its induction.

FEATURES OF HYPERTHERMIA-INDUCED APOPTOSIS

Morphology

The light microscopic and ultrastructural appearances of apoptosis induced by hyperthermia are identical to those observed in apoptosis occurring in a wide variety of other circumstances (Wyllie et al., 1980; Kerr et al., 1987; Walker et al., 1988; Kerr and Harmon, 1991). As is the case with apoptosis in general, the earliest definitive changes are compaction of chromatin into sharply circumscribed masses of uniformly condensed material abutting on the nuclear envelope (Figure 25.1), and overall cellular condensation. After a series of further changes including convolution of the nuclear and cell outlines, the nucleus fragments and apoptotic bodies containing one or more condensed masses of chromatin are formed. The number of apoptotic bodies formed from each cell is extremely variable and appears to be a function of the cell type involved rather than the initiating stimulus. Thus, whereas each individual HeLa cell undergoing apoptosis after heating, "buds" into hundreds of tiny apoptotic bodies (see Figure 4a in Harmon et al., 1991b), in many cells of the lymphocyte lineage, almost no budding occurs and only a single apoptotic body may be produced. Further information on the morphological features of hyperthermia-induced apoptosis occurring both *in vitro* and *in vivo* can be obtained from previous publications (Allan and Harmon, 1986; Harmon et al., 1990; 1991b; Takano et al., 1991).

DNA Fragmentation

DNA extracted from cells undergoing hyperthermia-induced apoptosis shows the same selective cleavage into oligonucleosomal-sized DNA fragments (Harmon et al., 1990; Kerr and Harmon, 1991; Sellins and Cohen; 1991; Takano et al., 1991) as has been described for apoptosis occurring in a wide variety of other circumstances, both

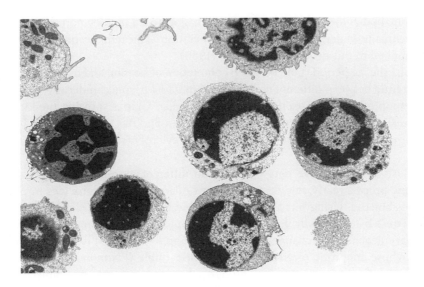

Figure 25.1 Early apoptotic human B-chronic lymphocytic leukaemia (B-CLL) cells 4 h after 43°C heating for 30 min. The amount of apoptosis produced by hyperthermia in B-CLL cells collected from patients was extremely variable, ranging from 2 to 40% apoptosis in the 5 cases studied.

physiological and pathological (for reviews see Wyllie, 1987; Walker *et al.*, 1988; Kerr and Harmon, 1991). The "ladder" pattern resulting from this internucleosomal cleavage can still be observed in DNA extracted from cultures up to 3 days after heating, long after the apoptotic bodies have degenerated into featureless masses of cell debris (Harmon *et al.*, 1990; Collins *et al.*, 1992). Apart from the published reports detailed above, we have found internucleosomal DNA cleavage associated with hyperthermia-induced apoptosis in numerous other murine and human cell lines (data not shown). Currently, the combination of typical morphological appearances with the presence of internucleosomal DNA cleavage is used as a working definition of apoptosis (Kerr and Harmon, 1991). The cell death produced by hyperthermia in the therapeutic range is consistent with this definition.

Kinetics

Induction of apoptosis by hyperthermia occurs very rapidly. In the intestinal crypts of rats and in cultures of murine mastocytoma for example, significantly elevated levels of apoptosis are observed immediately after the completion of a 30 min period of 43 or 44°C heating (Allan and Harmon, 1986; Harmon *et al.*, 1990). Apoptosis peaks in mastocytoma cultures at between 2 and 4 h, and thereafter decreases as the apoptotic bodies undergo secondary necrosis and are degraded to debris. By 24 h, apoptotic counts have returned to levels approaching control values (Harmon *et al.*, 1990).

Apoptotic bodies formed in culture are not phagocytosed (as they normally are *in vivo*) and eventually undergo spontaneous degeneration. The term secondary necrosis is often applied to this process (Wyllie *et al.*, 1980).

Whilst the lag period (time between completion of heating and onset of apoptosis) varies a little between different cell types (from 0 to 4 h), once underway, the kinetics of the apoptotic process itself appears to be similar. Graphs showing the kinetics of apoptosis induced by hyperthermia in two human Burkitt's lymphoma cell lines and the murine mastocytoma have been published previously (see Figure 1 in Takano *et al.*, 1991).

Analysis of quantitative data obtained on different morphological stages of the apoptotic process (early, middle and late; see Allan *et al.*, 1992) provides information about its kinetics. Using this approach we found that in heated cell cultures, it takes only 1 to 2 h for a newly formed (early) apoptotic cell (Figure 25.1) to undergo secondary degenerative changes and a further 6 to 8 h before it is degraded into cell debris whose apoptotic origin is no longer discernible (Harmon *et al.*, 1990). Cells undergoing apoptosis *in vivo* following hyperthermic treatment are rapidly phagocytosed and digested by nearby resident cells or macrophages and all evidence for their occurrence in tissues may be lost within about 8 to 12 h.

The extreme rapidity of the apoptotic process means that unless samples are taken at several early time points after hyperthermia, its occurrence might not be detected. Furthermore, if apoptotic bodies have undergone secondary necrotic changes before samples are taken, the death might be mistakenly identified as necrosis.

Incidence

Cell death with the characteristic morphological features of apoptosis is enhanced by hyperthermia in a range of normal and neoplastic cell populations. For example, it occurs in developing neuroepithelium of the guinea pig foetus (Wanner *et al.*, 1976), in crypt epithelial cells, lymphocytes, eosinophils and plasma cells of rat small intestine (Allan and Harmon, 1986), in spermatogonia and preleptotene primary spermatocytes of rat testis (Allan *et al.*, 1987), in murine thymocytes (Sellins and Cohen, 1991) and macrophages (Bruggen *et al.*, 1991), and in human lymphocytes (Schrek, 1980). Apoptosis is also markedly enhanced by hyperthermia in several murine tumours growing *in vitro* and *in vivo* and in a range of different human tumour cell lines growing *in vitro* (Figure 25.1; Dyson *et al.*, 1986; Harmon *et al.*, 1989; 1990; 1991a; Lennon *et al.*, 1990).

Sensitivity of Cells to Apoptotic Induction

One of the more interesting findings to have come out of studies of hyperthermia-induced apoptosis is that the sensitivity to apoptotic induction varies enormously from cell type to cell type (from < 1.0% in some human melanoma cell lines to almost 100% in some Burkitt's lymphoma cell lines after 43°C heating for 30 min; Harmon *et al.*,

1991b). Understanding why some cell types are sensitive or resistant to apoptotic induction is important as a reduced capacity of cells to undergo apoptosis appears to be associated with development of neoplasia (Hockenberry et al., 1990) and may also be an important factor in cancer treatment failures. What makes a cell sensitive to induction of apoptosis by hyperthermia is not at present clear. However, it appears that most of the extremely heat-sensitive cell lines have a lymphoid origin, grow as suspension cultures, have low levels of constitutively expressed heat shock proteins (see below) and are readily induced to undergo apoptosis by a range of other treatment modalities, especially protein synthesis inhibitors (Takano et al., 1991; Forster et al., 1992; Harmon et al., 1992; Takano and Harmon, 1992).

Liu et al. (1991) found that Burkitt's lymphoma cells that could readily be induced to undergo apoptosis with calcium ionophore did not express bcl-2 oncogene (Hockenberry et al., 1990). Resistant cells on the other hand expressed high levels of bcl-2. Our results in BM13674, an extremely "apoptosis-sensitive" Burkitt's lymphoma cell line we have used extensively in studies of hyperthermia-induced apoptosis, suggest that bcl-2 is expressed only weakly in these cells.

HEAT LOADS AND THE NEED FOR ACCURATE DELIVERY

Thermal Dose

Hyperthermia at temperatures above 43°C kills cells in a time/temperature dependent manner, with an increase in temperature of 1°C being roughly equivalent to changing the heating time by a factor of 2 (e.g. 43°C for 30 min ≈44°C for 15 min ≈ 45°C for 7.5 min; Field, 1983). Therefore, when defining thermal doses in hyperthermic studies, both the time and temperature of heating are equally important determinants. The recent study of Sellins and Cohen (1991) illustrates this point well. They found that whereas heating at 43°C for 30 min produced oligonucleosomal DNA cleavage characteristic of apoptosis, heating at the same temperature for periods greater than 2 h did not cause any DNA fragmentation. The former treatment is within the range of heat loads that produce apoptosis and as such is consistent with the occurrence of DNA fragmentation. The latter (43°C for > 2 h) is equivalent to heating cells at a temperature exceeding 45°C for 30 min, a heat load in the "necrosis-causing" range (Harmon et al., 1990).

Threshold Temperature and the Need for Accurate Heating

Although the sensitivity of different cells is known to be extremely variable (spermatocytes of the testis are the most sensitive; Allan et al., 1987), once the temperature threshold of a particular cell type is reached, an increase in heating time of as little as 20% or an increase in temperature of only 0.5°C may alter the outcome from no cell killing at all to 100% cell killing (Morris et al., 1977; Field, 1983). In general, this threshold temperature occurs somewhere in the vicinity of 43°C (Leith et al., 1977;

Schrek et al., 1980). Above the threshold temperature, a small increase in either the heating time or temperature will lead to a massive increase in cell killing, whereas below this threshold, a small increase will have relatively little effect. Examples of instances where a small increase in temperature leads to a large increase in the amount of apoptosis produced include the murine mastocytoma where a 0.5°C increase from 42.7°C to 43.2°C (for 30 min) increases the rate of apoptosis from < 1.0% to almost 30%, and the BM13674 human Burkitt's lymphoma cell line, where a 0.8°C increase from 42.2°C to 43°C (for 30 min) increases the rate from < 4.0% to almost 100%. As most studies of hyperthermia-induced apoptosis employ heat loads that are close to or just above the threshold temperature of the cell type involved, it is critically important that great attention is paid to the accurate delivery and monitoring of heat loads, if meaningful, repeatable results are to be obtained.

Heating Techniques

When heating cells growing in culture, the usual procedure is to raise the temperature rapidly to the desired level, the period of heating commencing once treatment temperature has been reached. Our experience has shown that an excessively slow rise in temperature can alter the yield of apoptotic cells achieved with a particular heat treatment. Although the most widely used method of raising the temperature rapidly is to heat cells in a small volume of culture medium, the methods we developed sought to avoid this approach, as overcrowding *per se* can cause significant amounts of apoptosis.

Our approach with suspension cultures has been to add freshly centrifuged cells in a small volume of medium (0.05 ml) to flasks containing 10 ml of medium preheated to the treatment temperature. For cells growing as attached monolayers, we discard the medium and replace it with fresh, preheated medium. The heating temperature of suspension cultures or monolayers is then maintained for the period of heating by immersing the flasks containing the cells in a heated waterbath accurate to 0.02°C (Julabo PC). The temperature of cultures during and after heating is monitored by means of a thermistor probe attached to a Harvard digital microprobe thermometer (BAT-12, resolution 0.1°C). All instrumentation is calibrated immediately prior to use by means of a certified, fully immersible thermometer (Dobbie Instruments No. 9945).

Heating of tumour nodules growing in the limbs of mice is achieved by anaesthetising the animals and immersing their tumour-bearing limbs in a heated waterbath for the desired period. The temperature within *in vivo* tumour nodules is monitored during and after treatment by means of sterile microprobes attached to the Harvard digital thermometer. Further information on heating procedures used *in vitro* and *in vivo* are given in Harmon et al. (1990; 1991b). It is important to realise that as the temperature achieved within a tissue culture flask or tumour nodule is usually slightly below that of the waterbath temperature, the temperature of the culture medium or the nodule must be monitored, not just the waterbath.

MECHANISM OF HYPERTHERMIA-INDUCED APOPTOSIS

Hyperthermia-induced apoptosis is produced by heat loads that fall within the therapeutic range (42 to 44°C for 30 min). Once a critical heat load is exceeded (about 45°C for 30 min in murine mastocytoma), the form of death produced changes from apoptosis to necrosis (Harmon et al., 1990). We have used two main cell lines for our studies of the mechanism of hyperthermia-induced apoptosis, the extremely heat-sensitive human Burkitt's lymphoma line, BM13674, and the murine mastocytoma P-815 × 2.1. Heating of the former at 43°C for 30 min produces almost 100% apoptosis, heating the latter for the same period at either 43 or 44°C produces approximately 15 and 50% apoptosis respectively. In this section, we will review the results of these and other studies, and put forward a proposed molecular pathway for the triggering of apoptosis by hyperthermia and by protein synthesis inhibitors.

Inhibition by Zinc and Okadaic Acid

Hyperthermia-induced apoptosis is inhibited by zinc sulphate (Harmon et al., 1988; Takano et al., 1991), a putative inhibitor of the endonuclease thought to be responsible for the internucleosomal DNA cleavage of apoptosis (Cohen et al., 1985; Duvall and Wyllie, 1986). However, whilst prevention of apoptosis by zinc is often ascribed to inhibition of endonuclease, zinc at the concentrations used in many experiments (approximately 1 to 5 mM) does have other damaging effects on cells, not the least being the induction of necrosis (Takano et al., 1991). It is likely that a cell would need to be essentially intact for the apoptotic death program to be effected. Therefore, general disruption to the biochemical machinery of the cell, and not specific endonuclease inhibition alone, might be responsible for the observed zinc inhibition.

Okadaic acid, an inhibitor of protein phosphatases 1 and 2A (Bialojan and Takai, 1988), has been reported (at 500 nM concentration) to markedly reduce the level of apoptosis produced by hyperthermia in the BM13674 human Burkitt's lymphoma cell line (Baxter and Lavin, 1992). However, whilst these workers reported that this inhibitor had no effect on cell viability over the 7 h period of their study, Boe et al. (1991) found that okadaic acid at concentrations between 100 nM and 1 μM induced cell death with the ultrastructural morphology of apoptosis in a range of mammalian cells, including leukaemic cells. These disparate results might simply be a consequence of the cell types involved. However, it is possible that the nuclear condensation observed by light microscopy in the Baxter and Lavin (1992) study and attributed to premature chromosome condensation, was in fact an early apoptotic change. Until further more detailed studies are carried out, it remains to be determined whether inhibition of apoptosis by okadaic acid is a widespread phenomenon in heated cells.

Role of Protein Kinase C

Agents which are known to activate Protein Kinase C (PKC) such as the phorbol esters, TPA and PBDu, and the mitogen Concanavalin A have been reported to prevent apoptosis from occurring in a number of different circumstances (Kizaki et al., 1989; McConkey et al., 1989a; b). Somewhat paradoxically, however, these same agents have been found by others to cause apoptosis (Mercep et al., 1989). We used all three of these agents over a range of concentrations previously shown to inhibit apoptosis in other circumstances (Kizaki et al., 1989; McConkey et al., 1989a; b) in an attempt to block hyperthermia-induced apoptosis in the P-815 × 2.1 mastocytoma and BM13674 Burkitt's lymphoma cell lines. No inhibition of hyperthermia-induced apoptosis was observed (Harmon et al., 1991a). Moreover, PKC levels within mastocytoma (cytoplasmic fraction; Harmon et al., 1991a) and lymphoma cells (both cytoplasmic and membrane fractions; Baxter and Lavin, 1992) appear to be unaffected by heating.

Role of Calcium

A number of studies have shown that an early sustained increase in cytosolic calcium concentration is involved in the activation of thymocyte apoptosis (McConkey et al., 1988; Orrenius et al., 1989; Smith et al., 1989). This finding together with experiments showing that apoptosis can be triggered by calcium ionophore-mediated entry of extracellular calcium into cells (Kizaki et al., 1989; Takahash et al., 1989) and inhibited by the chelators of extracellular or intracellular calcium, EGTA and Quin 2AM respectively, suggest that an increase in calcium levels may be an important trigger of apoptosis. Cells subject to hyperthermia are known to exhibit an increase in cytosolic calcium concentration, with release from internal stores having been implicated in this process (Calderwood et al., 1988; Drummond et al., 1988; Landry et al., 1988). Calderwood et al. (1988) found that a rapid release of the second messenger inositol triphosphate (IP_3) precedes the rise in calcium in heated cells. The above findings together with reports of raised IP_3 levels associated with apoptosis induced by other treatments (Waring, 1990) prompted us to speculate that an IP_3-mediated increase in cytosolic calcium resulting from limited membrane damage might be the critical event responsible for activation of apoptosis by hyperthermia (Takano et al., 1991). However, results subsequently obtained do not support such a view. Firstly, heating cells at 43°C for 30 min in calcium-free medium (DMEM, Gibco) or in the presence of the calcium chelator EGTA at concentrations ranging from 500 μM to 10 mM completely failed to prevent hyperthermia-induced apoptosis in the P-815 × 2.1 mastocytoma and BM13674 Burkitt's lymphoma cell lines (Harmon et al., 1991a). Similarly, preventing a rise in cytosolic calcium levels by addition of the calcium-binding dye Quin 2AM at concentrations ranging from 50 μM to 3 mM had no preventative effect. Moreover, these treatments could *per se* cause apoptosis. Our EGTA results differ from those reported by Lennon et al. (1991) who found that the level of apoptosis induced in HL-60 cells by hyperthermia seemed to be appreciably reduced in calcium-depleted medium. However, the heat load used in their study, 44°C for 1 h, is very severe (equivalent to heating at 45°C for 30 min), and may have caused some necrosis (Harmon et al., 1990). The results of our studies suggest that increased cytosolic calcium levels are not critically involved in the triggering of apoptosis by hyperthermia.

Macromolecular Synthesis Requirement

Inhibitors of transcription and translation such as actinomycin D and cycloheximide suppress apoptosis occurring in a number of circumstances. For example, apoptosis of thymocytes induced by glucocorticoids (Wyllie et al., 1984), dioxin (McConkey et al., 1988) and X-irradiation (Yamada and Ohyama, 1988) is inhibited by one or both of these agents as is apoptosis occurring at the line of fusion of the palatine processes during normal embryonic development (Pratt and Greene, 1976) and in cultures of endothelial cells following withdrawal of fibroblast growth factor (Araki et al., 1990). However, these inhibitors do not prevent hyperthermia-induced apoptosis (Harmon et al., 1989; 1991b; Takano et al., 1991, Sellins and Cohen, 1991). Moreover, our studies with ^{35}S-methionine in BM13674 lymphoma cells show that heat-induced apoptosis is not associated with *de novo* protein synthesis (Figure 25.2). These results taken together suggest that apoptosis induced by hyperthermia occurs independently of protein synthesis. Similar results have been reported for apoptosis induced by cytotoxic T lymphocytes (Duke et al., 1983) or by the fungal metabolite gliotoxin (Waring, 1990). Moreover, gliotoxin, mild hyperthermia, actinomycin D and cycloheximide, which are all inhibitors of protein synthesis *induce* apoptosis in some cell populations (Waring, 1990; Martin et al., 1990; Takano et al., 1991; Collins et al., 1991; Ledda-Columbano et al., 1992). This apparent paradox is poorly understood. However, results of recent studies are beginning to clarify the situation. In thymocytes (the cell type most widely used in studies showing that protein synthesis *is* required for apoptosis to occur), it has been shown that apoptosis induced by gliotoxin (Waring, 1990) and by mild hyperthermia (Sellins and Cohen, 1991) is not prevented by inhibitors of protein synthesis. This suggests that the requirement for *de novo* protein synthesis is not absolute within a particular cell type but varies with the inducing stimulus.

Theoretically, new protein synthesis might be required for priming a cell for apoptosis, for initiation of the process or for its execution. It is attractive to speculate that the requirement for protein synthesis reported in some systems might be linked to early control events involved with the initiation of the apoptotic process. It seems likely however that the final common effector mechanisms leading to the DNA fragmentation and morphological changes of apoptosis are independent of macromolecular synthesis, allowing agents that trigger the process down-stream in the activation pathway to bypass the requirement for protein synthesis. The particularly rapid onset of apoptosis in the case of apoptosis induced by mild hyperthermia (Harmon et al., 1990; see Kinetics section earlier) and CTL-mediated death is consistent with such down-stream activation. It has been suggested by some workers

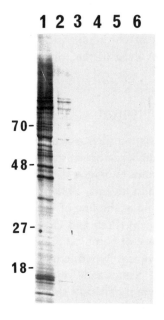

Figure 25.2 Autoradiograph of an SDS-polyacrylamide gel of proteins extracted from control and treated BM13674 cells 4 h after incubation in 15 μCi/ml ^{35}S-methionine. Control cells and cells to be heated were added to 37°C and 43°C medium respectively containing ^{35}S-methionine with or without cycloheximide and left in this medium until sampled. Lane 1: Control 37°C incubation; Lane 2: Incubated at 37°C in 10^{-6} M cycloheximide; Lane 3: Incubated at 37°C in 10^{-4} M cycloheximide; Lane 4: 43°C heating for 30 min; Lane 5: 43°C heating for 30 min with 10^{-6} M cycloheximide; Lane 6: 43°C heating for 30 min with 10^{-4} M cycloheximide.

(Cohen et al., 1992; Chapter 12 this book) that the "death program" within many cell types may be primed but unable to proceed due to concomitant production of short-lived apoptotic "suppressor" protein/s. This concept will be discussed in more detail in the section on a proposed molecular pathway.

Role of Heat Shock Proteins

When cultured cells or whole organisms are subjected to elevated temperatures, they respond by synthesising a small number of highly conserved stress proteins, commonly called heat shock proteins (hsps; Lindquist 1986). These hsps which are also expressed constitutively in cells (albeit to a lesser extent) seem to play an important role in a whole host of cellular functions (normal physiological as well as heat shock). It is widely accepted however, that one of their more important functions is to protect organisms from the toxic effects of heating. Recent evidence suggests that one of the main ways they fulfil this function is by binding to and stabilising unfolded or partially denatured proteins, thus preventing the formation of damaging aggregates (Hahn et al., 1991; Martin et al., 1991). Cosser and Martin (1991; 1992) reported induced thermotolerance to apoptosis in thymocytes collected from newborn mice and also in a human T-lymphocyte cell line heated 10 h after a prior conditioning heat shock. We have observed a similar effect in the BM13674 Burkitt's lymphoma cell line. Normally, virtually 100% of these cells will undergo apoptosis within 8 h of 43°C heating for 30 min (Harmon et al., 1991b; Takano et al., 1991. If however, these cells are subjected to a preliminary sublethal heat-shock (42°C for 30 min) and then incubated at 37°C for several hours, the number of cells that undergo apoptosis when heated at 43°C for 30 min is much reduced (Figure 25.3). The acquirement of thermotolerance by these cells appears to correlate with the synthesis of hsps (Figure 25.4), it does not fully develop until between 6 and 12 h after the conditioning (sublethal) heat shock and largely disappears by around 24 h. Development and decay of thermotolerance in cells has been quantitatively related by others to the absolute levels of hsps in cells (Landry et al., 1982; Li and Werb, 1982). The appearance of markedly enhanced levels of 25 kDa protein in conditioned BM13674 lymphoma cells (Figure 25.4) is interesting as the size corresponds to that of *bcl*-2 oncogene product, a known apoptotic inhibitor (Hockenberry, 1990). Work is currently underway to determine whether this band is in fact the *bcl*-2 oncogene product.

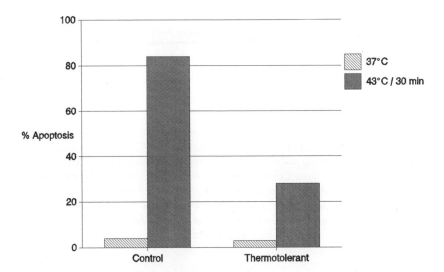

Figure 25.3 BM13674 cells given a sublethal heat-shock (42°C for 30 min) and allowed to recover at 37°C for 12 h resist heat-induced apoptosis (become thermotolerant). Control and thermotolerant cells were heated at 43°C for 30 min and sampled 6 h later. Apoptotic cells are expressed as a percentage of the total number of cells counted in 10 randomly selected high-power fields of 1 μm toluidine blue-stained resin sections (approximately 2000 cells examined).

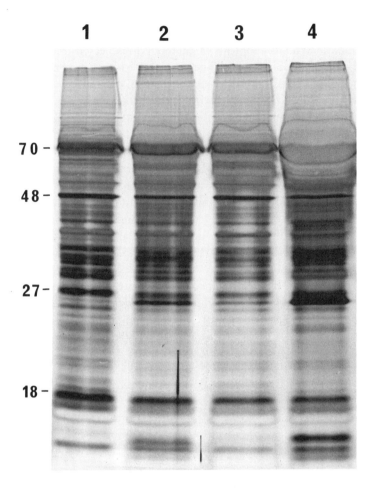

Figure 25.4 SDS polyacrylamide gel (12%) of proteins extracted from control and heated BM13674 cells. Lane 1: Control 37°C incubation; Lane 2: 42°C for 30 min and sampled 2 h later; Lane 3: 42.8°C for 30 min and sampled 6 h later (56% apoptosis produced); Lane 4: 42°C for 30 min and sampled 6 h later. Cells given the sublethal heat shock (42°C for 30 min) show increased staining of a number of protein bands as they acquire thermotolerance, the markedly increased amounts of protein evident in lanes 2 and 4 just below 27 kDa correspond in size to bcl-2, a known apoptotic inhibitor. Loss or reduction of some protein bands (e.g. 27 kDa) is evident in cells undergoing apoptosis after 42.8°C heating (lane 3).

Evidence for Protein Modification/Damage

The nature of the critical event in the triggering of apoptosis by hyperthermia is not yet known. In a recent study, Baxter and Lavin (1992) reported that apoptosis induced in the BM13674 lymphoma cell line by hyperthermia was accompanied by dephosphorylation of a limited number of proteins. These dephosphorylation events might play a central role in induction of apoptosis by hyperthermia. On the other hand, they might simply be one of a large number of secondary events taking place in a cell undergoing apoptosis, or they may even be totally unrelated to the apoptotic process (58% of cells in the study having survived the heat treatment).

A clue to the pathogenesis of hyperthermia-induced apoptosis might be found in the large body of literature on hyperthermic cell killing (killing assessed by clonogenic assays). Although there has been some dispute as to whether membranes or chromatin constitute the principal target for heat damage (Stevenson *et al.*, 1987; Dewey, 1989), it is widely agreed that denaturation of proteins is the critical lesion responsible for hyperthermic cell killing (Streffer and van Beuningen, 1987; Lepock, 1991). Loss or reduction of protein bands is observed in heated BM13674 lymphoma cells undergoing apoptosis after heating (Figure 25.4). No such change occurs in the thermoresistant Burkitt's lymphoma cell line P3HR-1 after the same heat treatment.

A Proposed Molecular Pathway

The absence of a requirement for macromolecular synthesis, the speed of the process and the failure of PKC activators or calcium chelators to prevent hyperthermia-induced apoptosis suggest that hyperthermia bypasses control pathways involved in the production of apoptosis in other systems (e.g. signal transduction, protein synthesis etc.). This concept is illustrated below by means of our schematic representation of apoptosis induced by hyperthermia or by agents that inhibit protein synthesis (Figure 25.5). Cohen *et al.* (1992) have proposed that the putative suppressor proteins that hold in check the primed apoptotic death program within many cells

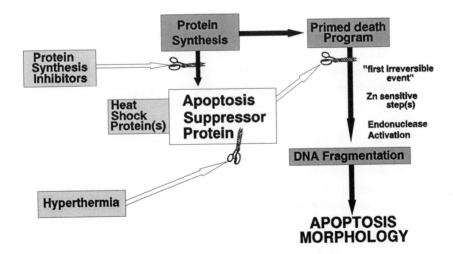

Figure 25.5 Proposed molecular pathway for induction of apoptosis by hyperthermia or by protein synthesis inhibitors.

might have a very short half-life and need to be continuously replenished. In our proposed pathway, inhibition of protein synthesis would cause depletion of the short-lived apoptotic suppressor protein, thereby releasing the primed apoptotic death program within the cell. Hyperthermia, on the other hand, which is known to damage proteins, might trigger apoptosis by directly inactivating the apoptotic suppressor protein, possibly as a result of protein denaturation. The molecular pathway discussed is consistent with most of the known data on hyperthermia-induced apoptosis and with the current thinking on thermal killing of cells. It does not however, explain why some cell types are so sensitive to apoptotic induction by hyperthermia and others so resistant. Our studies and those of Mosser and Martin (1991, 1992) showing that hsps induce resistance to apoptotic induction suggest that variation in the level of hsps in cells might explain, at least in part, the great variability amongst cells in sensitivity to apoptotic induction by hyperthermia. One of the main ways that hsps function is by binding to and stabilising heat labile proteins (Martin et al., 1991). It is possible that hsps might help to protect cells from undergoing hyperthermia-induced apoptosis by binding to and stabilising an apoptotic suppressor protein.

SUMMARY AND CONCLUSIONS

Hyperthermia at doses that fall within the therapeutic range ($\approx$42 to 44°C for 30 minutes) triggers cell death by apoptosis in susceptible cell populations, higher heat loads producing degenerative death or necrosis. Apoptosis occurs rapidly after heating and shows both the ultrastructural features and internucleosomal DNA cleavage characteristic of this mode of cell death. It is prevented by zinc sulphate, a putative inhibitor of the endonuclease presumed to be responsible for DNA cleavage in apoptosis, but is not suppressed by some treatments that have been shown to inhibit apoptosis in other systems. These include activation of protein kinase C (PKC) by phorbol esters or concanavalin A, chelation of extracellular and intracellular calcium with EGTA and Quin 2AM respectively, and inhibition of macromolecular synthesis by cycloheximide and actinomycin D. PKC levels are not significantly altered in cells undergoing hyperthermia-induced apoptosis and studies with ^{35}S-methionine show that the heat-induced apoptosis is not associated with de novo protein synthesis. Heat sensitive cells given a conditioning sublethal heat shock develop resistance to apoptotic induction and this appears to correlate with synthesis of hsps.

The results suggest that hyperthermia triggers apoptosis at a point down-stream of many of the control pathways involved in the initiation of apoptosis in other systems. Studies of the apoptotic mechanism in heated cells therefore, might allow us to gain insight into the "common-core mechanism" of apoptosis. The way in which heating initiates apoptosis still remains to be determined. However, the possibility that denaturation of apoptotic suppressor protein/s might be a critical event in the triggering of apoptosis by hyperthermia warrants further study.

ACKNOWLEDGEMENTS

This work was supported by the Queensland Cancer Fund, the Mayne Bequest Fund of the University of Queensland and the School of Life Science, Queensland University of Technology. The excellent technical assistance of Mr Mark Kenney, Mr Clay Winterford and Ms Karen Kemp is gratefully acknowledged.

REFERENCES

Allan, D. J. and Harmon, B. V. (1986) The morphological categorisation of cell death induced by mild hyperthermia and comparison with death induced by ionizing radiation and cytotoxic drugs. *Scanning Electron Microscopy*, III, 1121– 1133.

Allan, D. J., Harmon, B. V. and Kerr, J. F. R. (1987) Cell death in Spermatogenesis. In *Perspectives on Mammalian Cell Death*, edited by C. S. Potten, pp. 229–258. Oxford: Oxford University Press.

Allan, D. J., Harmon, B. V. and Roberts, S. A. (1992) Apoptosis of spermatogonia has three morphologically recognisable phases and shows no circadian rhythm during normal spermatogenesis in the rat. *Cell Proliferation*, 25, 241–250.

Araki, S., Shimada, Y., Kaji, K. and Hayashi, H. (1990) Apoptosis of vascular endothelial cells by fibroblast growth factor deprivation. *Biochemical and Biophysical Research Communications*, 168, 1194–1200.

Barry, M. A., Behnke, C. A. and Eastman, A. (1990) Activation of programmed cell death (apoptosis) by cisplatin, other anti-cancer drugs, toxins and hyperthermia. *Biochemical Pharmacology*, 40, 2353–2362.

Baxter, G. D. and Lavin, M. F. (1992) Specific protein dephosphorylation in apoptosis induced by ionizing radiation and heat shock in human lymphoid tumor lines. *Journal of Immunology*, 148, 1949–1954.

Bialojan, C. and Takai, A. (1988) Inhibitory effect of a marine-sponge toxin, okadaic acid, on protein phosphatases. *Biochemical Journal*, 256, 283–290.

Boe, R., Gjertsen, B. T., Vintermyr, K. O., Houge, G., Lanotte, M. and Doskeland, S. O. (1991) The protein phosphatase inhibitor okadaic acid induces morphological changes typical of apoptosis in mammalian cells. *Experimental Cell Research*, 195, 237–246.

Bruggen, I. V., Robertson, T. A. and Papadimitriou, J. M. (1991) The effect of mild hyperthermia on the morphology and function of murine resident peritoneal macrophages. *Experimental and Molecular Pathology*, 55, 119–134.

Buckley, I. K. (1972) A light and electron microscopic study of thermally injured cultured cells. *Laboratory Investigation*, 26, 201–209.

Calderwood, S. K., Stevenson, M. A. and Hahn, G. M. (1988) Effects of heat on cell calcium and inositol lipid metabolism. *Radiation Research*, 113, 414–425.

Cohen, J. J., Duke, R. C., Chervenak, R., Sellins, K. S. and Olson, L. K. (1985) DNA fragmentation in targets of CTL: An example of programmed cell death in the immune system. *Advances in Experimental Medicine and Biology*, 184, 493–508.

Cohen, J. J., Duke, R. C., Fadok, V. A. and Sellins, K. S. (1992) Apoptosis and programmed cell death in immunity. *Annuual Review of Immunology*, 10, 267–293.

Collins, R. J., Harmon, B. V., Gobé, G. C. and Kerr, J. F. R. (1992) Internucleosomal DNA cleavage should not be the sole criterion for identifying apoptosis. *International Journal of Radiation Biology*, 61, 451–453.

Collins, R. J., Harmon, B. V., Souvlis, T., Pope, J. H. and Kerr, J.F .R . (1991) Effects of cycloheximide on B-chronic lymphocytic leukaemia and normal lymphocytes *in vitro*: induction of apoptosis. *British Journal of Cancer,* **64**, 518–522.

Dewey, W. C. (1989) Mechanisms of thermal injury and thermal radiosensitisation. In *Hyperthermic Oncology 1988, Vol. 2, Special Plenary Lectures, Plenary Lectures and Symposium and Workshop Summaries (Kyoto, 1988)*, edited by T. Sugahara and M. Saito, pp. 75–80. London: Taylor and Francis.

Drummond, I. A .S., Livingstone, D. and Steinhardt, R. A. (1988) Heat shock protein synthesis and cytoskeletal rearrangements occur independently of intracellular free calcium increases in Drosophila cells and tissues. *Radiation Research,* **113**, 402–413.

Duke, R. C., Chervenak, R. and Cohen, J. J. (1983) Endogenous endonuclease- induced DNA fragmentation: an early event in cell-mediated cytolysis. *Proceedings of the National Academy of Science, USA,* **80**, 6361–6365.

Duvall, E. and Wyllie, A. H. (1986) Death and the cell. *Immunology Today,* **7**, 115–119.

Dyson, J. E. D., Simmons, D. M., Daniel, J., McLaughlin, J. M., Quirke, P. and Bird, C. C. (1986) Kinetic and physical studies of cell death induced by chemotherapeutic agents or hyperthermia. *Cell and Tissue Kinetics,* **19**, 311–324.

Fajardo, L. F., Egbert, B., Marmor, J. and Hahn, G. M. (1980) Effects of hyperthermia in a malignant tumour. *Cancer,* **45**, 613–623.

Field, S. B. (1983). Cellular and tissue effects of hyperthermia and radiation. In *The biological basis of radiotherapy*, edited by G. G. Steel, G. E. Adams and M. J. Peckham, pp. 287–303. Amsterdam: Elsevier Science Publishers.

Field, S. B. (1987) Hyperthermia in the treatment of cancer. *Physics in Medicine and Biology,* **32**, 789–811.

Hahn, B. M., Kim, Y-M. and Lee, K-J. (1991) Proteins, thermotolerance, and heat- induced cell death. In *Radiation Research: A Twentieth Century Perspective, Congress Proceedings (Toronto, 1991)*, edited by W. C. Dewey, M. Edington, R. J. M. Fry, E. J. Hall and G. F. Whitmore, pp. 1010–1015. San Diego: Academic Press.

Harmon, B. V., Allen, J., Gobé, G. C., Collins, R. J., Kerr, J. F. R. and Allan, D. J. (1989) Hyperthermic cell killing in murine tumours with particular reference to apoptosis. In *Hyperthermic Oncology Vol. 1: Summary Papers (Kyoto, 1988)*, edited by T. Sugahara and M. Saito, pp.129–130. London: Taylor and Francis.

Harmon, B. V., Corder, A. M., Collins, R. J., Gobé, G. C., Allen, J., Allan, D. J. and Kerr, J. F. R. (1990) Cell death induced in a murine mastocytoma by 42 to 47°C heating *in vitro*: evidence that the form of death changes from apoptosis to necrosis above a critical heat load. *International Journal of Radiation Biology,* **58**, 845–858.

Harmon, B. V., Forster, T. H., Takano, Y. S., Collins, R. J., Gobé, G. C. and Kerr, J. F. R. (1991a) Cell death induced by hyperthermia: significance of apoptosis. In *Radiation Research: a Twentieth Century Perspective (Toronto, 1991)*, edited by J. D. Chapman, W. C. Dewey and G.F . Whitmore, pp. 377. San Diego: Academic Press.

Harmon, B. V., Takano, Y. S., Winterford, C. M. and Gobé, G. C. (1991b) The role of apoptosis in the response of cells and tumours to mild hyperthermia. *International Journal of Radiation Biology,* **59**, 489–501.

Harmon, B. V., Takano, Y. S., Winterford, C. M. and Potten, C. S. (In press) Vincristine induces cell death by apoptosis in intestinal crypts of mice and in a human Burkitt's lymphoma. *Cell Proliferation*.

Hockenberry, D., Nunez, G. T., Milliman, C., Schreiber, R. D. and Korsmeyer, S. J. (1990) Bcl-2 is an inner mitochondrial membrane protein that blocks programmed cell death. *Nature,* **348**, 334–336.

Kerr, J. F. R. and Harmon, B. V. (1991) Definition and incidence of apoptosis: an historical perspective. In *The molecular basis of cell death*, edited by L. D. Tomei and F. O. Cope, pp. 5–29. New York: Cold Spring Harbor Laboratory Press.

Kerr, J. F. R., Searle, J., Harmon, B. V. and Bishop, C. J. (1987) Apoptosis. In *Perspectives on mammalian cell death*, edited by C. S. Potten. pp. 93–128. Oxford: Oxford University Press.

Kizaki, H., Tadakuma, T., Odaka, C., Muramatsu, J. and Ishimura, Y. (1989) Activation of a suicide process of thymocytes through DNA fragmentation by calcium ionophores and phorbol esters. *Journal of Immunology*, 143, 1790–1794.

Landry, J., Bernier, D., Chrètien, P., Nicole, L. M., Tanguay, R. M. and Marceau, N. (1982) Synthesis and degradation of heat shock proteins during development and decay of thermotolerance. *Cancer Research*, 42, 2457–2461.

Landry, J., Crète P., Lamarche, S. and Chrètien, P. (1988) Activation of Ca^{2+}-dependent processes during heat shock: role in cell thermoresistance. *Radiation Research*, 113, 426–436.

Ledda-Columbano, G. M., Coni, P., Faa, G., Manenti, G. and Columbano, A. (1992) Rapid induction of apoptosis in rat liver by cycloheximide. *American Journal of Pathology*, 140, 545–549.

Leith, J. T., Miller, R. C., Gerner, E. W. and Boone, M. L. (1977) Hyperthermic potentiation: Biological aspects and applications to radiation therapy. *Cancer*, 39, 766–779.

Lennon, S. V., Martin, S. J. and Cotter, T. G. (1991) Dose-dependent induction of apoptosis in human tumour cell lines by widely diverging stimuli. *Cell Proliferation*, 24, 203–214.

Lepock, J. (1991) Protein denaturation: its role in thermal killing. In *Radiation Research: A Twentieth Century Perspective, Congress Proceedings (Toronto, 1991)*, edited by W. C. Dewey, M. Edington, R. J. M. Fry, E. J. Hall, G.F. Whitmore, pp. 992–998. San Diego: Academic Press.

Li, G. C. and Werb, Z. (1982) Correlation between synthesis of heat shock proteins and development of thermotolerance in Chinese hamster fibroblasts. *Proceedings of the National Academy of Science, USA*, 79, 3218–3222.

Lindquist, S. (1986) The heat-shock response. *Annuual Review of Biochemistry*, 55, 1151–1191.

Liu, Y-J., Mason, D. Y., Johnson, G. D., Abbot, S., Gregory, C. D., Hardie, D. L., Gordon, J. and MacLennan, I. C. M. (1991) Germinal center cells express Bcl-2 protein after activation by signals which prevent their entry into apoptosis. *European Journal of Immunology*, 21, 1905–1910.

Martin, J., Langer, T., Boteva, R., Schramel, A., Horwich, A. L. and Hartl, F-U. (1991) Chaperonin-mediated protein folding at the surface of groEL through a 'molten globule'-like intermediate. *Nature*, 352, 36–42.

Martin, S. J., Lennon, S. V., Bonham, A. M. and Cotter, T. G. (1990) Induction of apoptosis (programmed cell death) in human leukaemic HL-60 cells by inhibition of RNA or protein synthesis. *Journal of Immunology*, 145, 1859–1867.

McConkey, D. J., Hartzell, P., Amador-Perez, J. F., Orrenius, S. and Jondal, M. (1989a) Calcium-dependent killing of immature thymocytes by stimulation via the CD3/T-cell receptor complex. *Journal of Immunology*, 143, 1801–1806.

McConkey, D. J., Hartzell, P., Duddy, S. K., Hakansson, H. and Orrenius, S. (1988) 2,3,7,8-Tetracholorodibenzo-p-dioxin kills immature thymocytes by Ca^{2+}-mediated endonuclease activation. *Science*, 242, 256–259.

McConkey, D. J., Hartzell, P., Jondal, M. and Orrenius, S. (1989b) Inhibition of DNA fragmentation in thymocytes and isolated thymocyte nuclei by agents that stimulate Protein Kinase C. *Journal of Biological Chemistry*, 264, 13399–13402.

Mercep, M., Noguchi, P. D. and Ashwell, J. D. (1989) The cell cycle block and lysis of an activated T-cell hybridoma are distinct processes with different Ca^{2+} requirements and sensitivity to cyclosporin A. *Journal of Immunology*, 142, 4085–4092.

Morris, C. C., Myers, R. and Field, S. B. (1977) The response of the rat tail to hyperthermia. *British Journal of Radiology*, **50**, 576–580.

Mosser, D. D. and Martin, L. H. (1991) Heat shock protein synthesis and resistance to heat-induced apoptosis in mouse thymocytes. In *Stress proteins and the heat shock response (Cold Spring Harbor, 1991 Meeting)*, pp. 130, New York: Cold Spring Harbor Laboratory Press.

Mosser, D. D. and Martin, L. H. (1992) Induced thermotolerance to apoptosis in a human T lymphocyte cell line. *Journal of Cellular Physiology*, **151**, 561–570.

Orrenius, S., McConkey, D. J., Bellomo, G. and Nicotera, P. (1989) Role of Ca^{2+} in toxic cell killing. *Trends in Pharmacological Science*, **10**, 281–285.

Overgaard, J. (1983) Histopathologic effects of hyperthermia. In *Hyperthermia in cancer therapy*, edited by F. K. Storm, pp. 163–185. Boston: G. K. Hall Medical Publishers.

Pratt, R. M. and Greene, R. M. (1976) Inhibition of palatal epithelial cell death by altered protein synthesis. *Developmental Biology*, **54**, 135–145.

Smith, C. A., Williams, G. T., Kingston, R., Jenkinson, E. J. and Owen, J. J. T. (1989) Antibodies to CD3/T-cell receptor complex induce death by apoptosis in immature T cells in thymic cultures. *Nature*, **337**, 181–183.

Schrek, R., Chandra, S., Molnar, Z. and Stefani, S. S. (1980) Two types of interphase death of lymphocytes exposed to temperatures of 37–45°C. *Radiation Research*, **82**, 162–170.

Sellins, K. S. and Cohen, J. J. (1991) Hyperthermia induces apoptosis in thymocytes. *Radiation Research*, **126**, 88–95.

Stevenson, M. A., Calderwood, S. K. and Hahn, G. M. (1987) Effect of hyperthermia (45°C) on calcium flux in Chinese hamster ovary HA-1 fibroblasts and its potential role in cytotoxicity and heat resistance. *Cancer Research*, **47**, 3712–3717.

Streffer, C. and van Beuningen, D. (1987) The biological basis for tumour therapy by hyperthermia and radiation. *Recent Results in Cancer Research*, **104**, 24–70.

Takahashi, S., Maecker, H. T. and Levy, R. (1989) DNA fragmentation and cell death mediated by T-cell antigen receptor/CD3 complex on a leukemia T-cell line. *European Journal of Immunology*, **19**, 1911–1919.

Takano, Y. S. and Harmon, B. V. (In press) Apoptosis induced by microtubule disrupting drugs in a human lymphoma cell line: inhibitory effects of a phorbol ester and zinc sulphate but not cycloheximide. *Pathology Research and Practice*.

Takano, Y. S., Harmon, B. V. and Kerr, J. F. R. (1991) Apoptosis induced by mild hyperthermia in human and murine tumour cell lines: a study using electron microscopy and DNA gel electrophoresis. *Journal of Pathology*, **163**, 329–336.

Walker, N. I., Harmon, B. V., Gobé, G. C. and Kerr, J. F. R. (1988) Patterns of cell death. *Methods and Achievements in Experimental Pathology*, **13**, 18–54.

Wanner, R. A., Edwards, M. J. and Wright, R. G. (1976) The effects of hyperthermia on the neuro-epithelium of the 21-day guinea-pig foetus: histologic and ultrastructural study. *Journal of Pathology*, **118**, 235–244.

Waring, P. (1990) DNA fragmentation induced in macrophages by gliotoxin does not require protein synthesis and is preceded by raised IP_3 levels. *Journal of Biological Chemistry*, **265**, 14476–14480.

Wyllie, A. H. (1987) Cell death. *International Review of Cytology Supplement*, **17**, 755–785.

Wyllie, A. H., Kerr, J. F. R. and Currie, A. R. (1980) Cell death: the significance of apoptosis. *International Review of Cytology (New York)*, **68**, 251–306.

Wyllie, A. H., Morris, R. G., Smith, A. L. and Dunlop, D. (1984) Chromatin cleavage in apoptosis: association with condensed chromatin morphology and dependence on macromolecular synthesis. *Journal of Pathology*, **142**, 67–77.

Yamada, T. and Ohyama, H. (1988) Radiation-induced interphase death of rat thymocytes is internally programmed (apoptosis). *International Journal of Radiation Biology*, **53**, 65–75.

Chapter 26

TUBULIN IN APOPTOTIC CELLS

Sally M. Pittman, Melissa Geyp, Steven J. Tynan, Cilinia M. Gramacho, Deborah H. Strickland, Murray J. Fraser and Christine M. Ireland
Children's Leukaemia and Cancer Research Centre, University of New South Wales, Prince of Wales Children's Hospital, Sydney, NSW 2031, Australia

THE PRESENCE OF TUBULIN AND TUBULIN STRUCTURES IN APOPTOTIC CELLS

The distinct morphological features characteristic of apoptosis were first described by Kerr, Wyllie and Currie (1972). We have observed that in apoptotic cells these morphological changes are associated with massive tubulin expression or visible tubulin structures. These are apparent in the cytoplasm or associated with the nucleus, following indirect labelling with a monoclonal antibody to β-tubulin (Amersham), and immunoperoxidase (IPX). Similar tubulin changes were seen in CCRF-CEM T-cell leukaemia, HL-60 promyelocytic leukaemia, IMR-32 neuroblastoma cells and normal peripheral blood lymphocytes undergoing apoptosis. Enhanced tubulin expression and tubulin structures were apparent when apoptosis was induced by treatment with cytotoxic drugs with different modes of action including methotrexate (MTX), vincristine (VCR), VP-16, actinomycin D, and dexamethasone (DEX) in cells of T-lineage. In untreated log-phase cells, the antibody used labelled the centrosome and the mitotic spindle (see Figure 26.1A).

The structures we identified in apoptotic cells were similar in the different cell types and following a variety of treatments. During apoptosis, tubulin structures or enhanced tubulin expression were identified in condensed cells and cells undergoing mild to extreme convolution of the nucleus and cytoplasm as well as cells undergoing blebbing and budding of apoptotic bodies (see Figure 26.1B,C). Certain of the structures appeared to emanate from the centrosome(s) and enclose the nucleus in circular bands and were associated with invagination of the nucleus. In blebbing cells, the structures elongated from the centrosome into the cytoplasmic blebs (see Figure

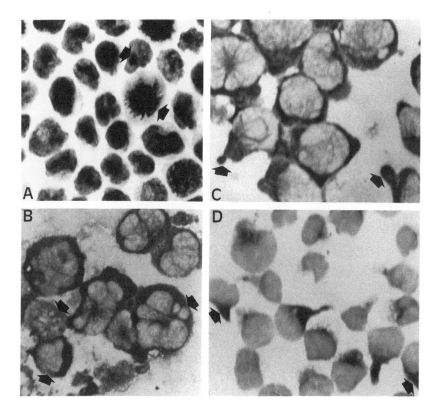

Figure 26.1 Association of tubulin staining with morphological features typical of apoptosis.
A. Control HL-60 cells stained for β-tubulin using immunoperoxidase labelling. Arrows indicate normal mitotic spindle and centrosome.
B. CCRF-CEM cells 24 hours post-treatment with 10^{-8} M VCR. Lower left and right arrows indicate convolution of the nucleus and cytoplasm associated with tubulin staining, top left arrow indicates a mitotic cell with breakdown of the mitotic spindle.
C. CCRF-CEM cells 48 hours post-treatment with 10^{-8} M MTX. Arrows indicate intense tubulin staining in apoptotic bodies.
D. CCRF-CEM cells 6 hours post-treatment with 10^{-8} M VCR. Arrows indicate tubulin structures elongating from the centrosome to cytoplasmic blebs.

26.1D). The nuclear structures in some cells were reminiscent of those associated with the mitotic spindle, with fewer microtubules, while the budding of apoptotic bodies has been likened to the budding of daughter cells at division. Thus it could be suggested that these cells, lacking the condensation of the nuclear material into discrete chromosomes, and frequently without duplication or appropriate migration of the centrosome, are undergoing an aborted mitosis. The intensity of the stain and the tubulin structures varied with changes in cell morphology typical of apoptosis. The most intense tubulin staining appeared in apoptotic bodies and in these and apoptotic cells with dense tubulin positivity it was not possible to determine the degree to which the tubulin had polymerised. However, it was found that indirect labelling with fluorescein (FITC) was a more sensitive detection technique. Increased tubulin monomers were observed with FITC earlier than the changes detected with IPX and in most cells it could be seen that this increased expression was followed by tubulin polymerisation before any change in cell morphology was apparent.

THE TEMPORAL INVOLVEMENT OF TUBULIN IN APOPTOSIS

In order to investigate further these observations of massive increases in tubulin and unique tubulin structures in apoptotic cells, we used molecular and protein analysis. The time-courses of tubulin protein and mRNA expression, formation of tubulin structures, fragmentation of the DNA detected as ladders by agarose gel electrophoresis and the characteristic morphology were determined in CCRF-CEM and HL-60 cells when apoptosis was induced with VP-16, MTX and VCR.

HL-60 or CCRF-CEM cells were exposed to 17 μM VP-16 and cells were harvested every hour for slide preparation and analysis of DNA, RNA and protein. After indirect labelling of β-tubulin with FITC, at 2 hours post-treatment, increased tubulin monomers were observed in 20–30% of the cells. By 3 hours this staining was more intense, with tubulin structures, particularly those associated with the nucleus, clearly apparent and with more cells expressing some increased tubulin positivity. The majority of cells showed massive tubulin staining at 4 hours, with microtubules forming the tubulin stuctures associated with apoptosis. However, at 24 hours most of the cells had undergone secondary necrosis and smeared on the slide preparation. At this time point 90–100% of cells had taken up trypan blue.

In these experiments (summarised in Table 26.1A) the first observable change was enhanced expression of tubulin mRNA at 1 hour post-treatment with VP-16. Northern analysis was performed using the pHTβ-1 probe (kindly provided by Dr P. Gunning, CMRF, Camperdown) which detects all β-tubulin mRNA species, and relative expression was determined by comparison with expression of the glyceraldehyde phosphate dehydrogenase message. Total genomic DNA was isolated, and fragmentation was detected as "ladders" following agarose gel electrophoresis at 3 hours post-treatment. We wished to determine whether the tubulin gene was degraded with the bulk DNA or whether it was protected in some way so as to allow further transcription. Southern transfer of the "laddered" DNA from this experiment and subsequent probing with pHTβ-1 revealed that degradation of the tubulin gene occurred concomitantly with that of the bulk DNA.

Following treatment with 10^{-8} M MTX of HL-60 and CCRF-CEM cells, tubulin expression was detected using indirect staining with IPX (see Table 26.1B). At 24 hours post-treatment, minor blebs of the cytoplasm and nucleus were present in the majority of cells. These changes in morphology were associated with tubulin structures or massively increased tubulin stain. mRNA analysis showed that a relative increase in

Table 26.1 Temporal relationship between tubulin protein and mRNA expression and the observation of DNA fragmentation.

A

DRUG VP16

TIME	CELLULAR TUBULIN	LADDERS	M-RNA's
1hr	No change	0	←
2hrs	←	0	No change
3hrs	⇇	+	No change
4hrs	⇚	++	No change
24hrs	Necrosis	Necrosis	→

B

DRUG MTX

TIME	CELLULAR TUBULIN	LADDERS	M-RNA's
1hr	→	0	No change
5hrs	←	0	No change
24hrs	⇚	0	←
48hrs	⇇	++	No change

C

DRUG VCR

TIME	CELLULAR TUBULIN	LADDERS	M-RNA's
1hr	No change	0	←
5hrs	→	0	←
10hrs	—	+	—
24hrs	⇇	++	→
48hrs	⇇	++	⇉

CCRF-CEM or HL-60 cells were treated with 17 μM VP-16 (A), 10^{-8} M MTX (B) or 10^{-8} M VCR (C). Following treatment, cells were harvested at various times for isolation of DNA, RNA and cellular protein, or for slide preparation. For cellular tubulin arrows indicate: 1, low tubulin staining in a minority of cells; 2, intense tubulin staining in a minority of cells; 3, intense tubulin staining observed in a majority of cells. For ladders: +, laddering just visible in ethidium bromide stained agarose gels; ++, laddering clearly visible with little DNA of high molecular weight present. For mRNA arrow indicates presence of at least 3-4-fold increase in levels observed for tubulin messages as compared to GAPDH. Reversed arrows indicate a corresponding decrease in expression, of both tubulin and GAPDH, indicating degradation.

tubulin message was also apparent. By 48 hours, the cell preparations showed an increased number of apoptotic bodies with intense tubulin expression. Extreme convolutions were observed in the nucleus and cytoplasm of the remaining cells, associated with cytoplasmic and nuclear microtubules. At this stage the DNA fragmentation was apparent. When we investigated the time-course of these events following treatment with 10^{-7} M MTX, the process occurred more rapidly. That is, increased expression of the tubulin mRNA was detected at 5 hours post-treatment, and the ladders were seen at 24 hours associated with the break-up of cells into apoptotic bodies.

As reported by Martin and Cotter (1990), we were able to induce apoptosis by treatment with agents that disrupt microtubules. Despite the fact that the breakdown of microtubules triggers apoptosis, our results showed that in cells treated with these agents, the typical apoptotic morphology was still associated with the formation of microtubules. When HL-60 and CCRF-CEM cells were treated with 10^{-8} M VCR, immunolabelling of dot-blots of cellular protein isolated at various times after treatment revealed an initial decline in the level of tubulin (see Table 26.1C). However, by 24 hours post-treatment, the level of tubulin expression was significantly increased with respect to the untreated control.

This result was consistent with the observation of tubulin staining on slides of VCR treated cells. The cells showed an initial decrease in tubulin, with an increase in staining associated with apoptotic morphology at later time points. Of interest was the identification of at least four morphologically distinct populations of cells up to 24 hours post-treatment. These were: cells arrested in mitosis without the mitotic spindle, apparently normal cells, cells at various stages of apoptosis displaying intense tubulin staining and structures, and necrotic cells. The cell numbers of the two former populations dropped rapidly with time following treatment and the number of apoptotic and necrotic cells increased. In these experiments the level of tubulin mRNA was elevated at five hours post-treatment with VCR and the formation of DNA "ladders" was observed at ten hours post-treatment.

In summary, there seemed to be a consistent temporal relationship observed between the changes in tubulin expression at both the protein and RNA levels, the presence of tubulin staining and structures, and apoptotic morphology. Elevation in the level of tubulin mRNA was usually the first detectable change, followed by the appearance of tubulin structures associated with the morphological features typical of early apoptosis. DNA fragmentation occurred at later time-points, accompanied by the presence of large numbers of apoptotic bodies. It should be noted that cells displaying apoptotic morphology are present even in the control cell populations, about 10% of the total cell population, and to some extent, the timing of the events we have investigated can be affected by the presence of varying numbers of such cells in particular preparations. The other point to be noted is, that, apoptosis is not occurring in synchrony in any given cell population, hence preparations of DNA or RNA reflect this heterogeneity.

CONSEQUENCES OF TREATMENTS THAT DEPOLYMERISE TUBULIN

We wished to investigate whether it was possible to disrupt apoptosis by treating apoptotic cells with drugs or agents known to depolymerise tubulin. Therefore, cells were treated with VP-16 plus VCR and compared to cells treated with VP-16 alone. Identical apoptotic cell morphology, with the presence of apoptotic tubulin structures and similar time of onset of DNA fragmentation was observed regardless of the treatment.

As VCR treatment was unsuccessful in disrupting the tubulin structures associated with apoptosis, cold shock at 4°C was used to depolymerise apoptotic microtubules induced in CCRF-CEM cells following treatment with VP-16. Figure 26.2 demonstrates, at 4 hours post-treatment with VP-16, that cells maintained at 37°C are apoptotic, with intense tubulin staining in the apoptotic bodies, and tubulin structures visible in cells with apoptotic morphology. DNA fragmentation was present at this time point. In contrast, cells which were treated with VP-16 and maintained at 4°C following treatment show break-down of microtubules with neither apoptotic nor mitotic microtubules being observed. It can be seen in Figure 26.2 that the cells do not display apoptotic morphology, with the nuclei remaining intact and no blebbing of the cytoplasm or nucleus occurring. Electrophoresis of DNA isolated from these cells showed that cold shock also inhibited fragmentation.

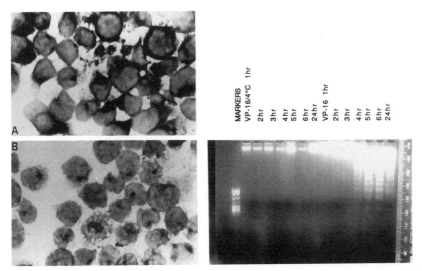

Figure 26.2 Effect of cold shock on apoptosis induced by VP-16 treatment of CCRF-CEM cells. CCRF-CEM cells were treated with 17 μM VP-16 and incubated at 37°C (A) or 4°C (B) for 4 hours. Agarose gel electrophoresis was performed on DNA isolated from cells at the indicated times following treatment.

Further, we used different doses of colchicine (0.01–0.6 μg/ml) to attempt to depolymerise apoptotic microtubules in CCRF-CEM cells following treatment with VP-16. Even at the highest dose used, the presence of colchicine did not affect the formation and maintenance of tubulin structures in apoptotic cells or the time course of DNA fragmentation. However, with both VCR and colchicine in these experiments, breakdown of the mitotic spindle was observed under conditions in which the apoptotic tubulin structures were induced and maintained. This led us to the conclusion that the microtubules in mitotic and apoptotic cells have different properties.

We did not observe breakdown of the apoptotic microtubules following cytotoxic treatments known to depolymerise mitotic microtubules. However, in cells treated with VP-16 and cold shock we observed loss of both apoptotic and mitotic microtubules. This was associated, in non-mitotic cells, with the maintainance of apparently normal morphology, without the blebbing, budding and breakup of cells. It had been reported in a recent paper by Cotter *et al.*, 1992, that cytochalasin B prevented the formation of apoptotic bodies, possibly implicating actin in pinching off these structures. Although the evidence indicates that cytochalasin B interacts with actin, not tubulin, we exposed the CCRF-CEM cells to a pretreatment of 15 μM cytochalasin B, and investigated the tubulin staining following the induction of apoptosis by VP-16. We found that the mitotic spindles in dividing cells were intact, but, like Cotter (1992), saw no cytoplasmic blebbing. In addition, none of the tubulin structures we associate with apoptosis were observed in the cytoplasm of treated cells. In less than 10% of cells some nuclear tubulin structures were maintained and these correlated with invagination or blebbing of the nucleus. In treated cells without the nuclear tubulin structures, the typical nuclear changes characteristic of apoptosis were not observed. Analysis of the DNA showed that at time points when fragmentation was apparent following treatment with VP-16 alone, it could not be detected in cells which were treated with VP-16 and cytochalasin B. Thus, inhibition of actin polymerisation has inhibited the polymerisation of tubulin in apoptotic cells by an unknown mechanism. Further investigations are needed to determine the roles of both actin and tubulin in apoptosis.

SUMMARY

We have observed massive increases in tubulin in apoptotic cells and unique tubulin structures associated with the typical apoptotic morphological changes in the cytoplasm and the nucleus. In addition, our experimental results show that when the tubulin structures in apoptotic cells are prevented from forming or depolymerised, by cold shock, for example, the typical apoptotic morphology is not apparent. These results suggest that tubulin plays a key role in cytoplasmic and nuclear blebbing and in the breakup of apoptotic cells. The increase in tubulin mRNA is an early change in apoptosis followed by the expression and polymerisation of tubulin in cells. Fragmentation of the DNA is usually detected as a late event.

In the presence of VCR or colchicine, mitotic microtubules depolymerised, while apoptotic tubulin structures were formed and maintained. With cytochalasin B, mitotic microtubules were intact, while the polymerisation of apoptotic tubulin structures was inhibited. These results indicate that the microtubular structures observed in apoptotic cells have different properties to those of mitotic cells.

In the normal mammalian cell, microtubules perform diverse and essential functions including the provision of structural support to the cell and involvement in karyokinesis, intracellular transport and cell motility. This functional and structural diversity of tubulin is controlled both at the transcriptional level and by a variety of post-translational modifications, and also by differential binding of microtubular-associated proteins (Joshi and Cleveland, 1990). Six functional human β-tubulin genes have been identified which are expressed in variable quantities in different cell types and during development. The α-tubulins are similar in that four genes which encode distinct isotypic classes are also differentially expressed (Sullivan, 1988). Microtubules are made up of heterodimers consisting of one α- and one β-polypeptide. In the vertebrate cell, interphase and mitotic microtubules are co-polymers of the available isotypes. It will be of interest to determine which of the subtypes of α- and β-tubulin are involved in the formation of the structures we observe in apoptotic cells. Of particular note is the proposed role of γ-tubulin as a requirement for microtubule nucleation throughout the cell cycle (Joshi et al., 1992) as this observation raises the question of whether this type of tubulin is also associated with the unique tubulin structures seen in apoptotic cells.

Microtubule polymerisation/depolymerisation occurring at the interphase-metaphase transition is regulated by cdc2 kinase and related kinases, such as mitogen-activated protein kinase, which has maximum specificity for microtubular associated protein 2 (Gotoh et al., 1991). Changes in cdc2 kinase activity have also been observed under conditions known to induce apoptosis, including exposure to VP-16 or irradiation. cdc2 kinase activity is inhibited with G2 arrest in cells following these treatments, but is enhanced when G2 arrest is released (Lock and Ross, 1990). This enhancement of kinase activity may well correlate with the increased polymerisation of tubulin observed in this laboratory under similar conditions. It would be an obvious extension of our work to compare the nature of mitotic and apoptotic microtubules, specifically investigating post-translational modifications of tubulin and the controlling elements of tubulin polymerisation.

ACKNOWLEDGEMENT

This work was funded by the Children's Leukaemia and Cancer Foundation.

REFERENCES

Cotter, T. G., Lennon, S. V., Glynn, J. M. and Green, D. R. (1992) Microfilament-disrupting agents prevent the formation of apoptotic bodies in tumour cells undergoing apoptosis. *Cancer Research*, **52**, 997–1005.

Gotoh, Y., Nishida, E., Matsuda, S., Shiina, N., Kosako, H., Shiokawa, K., Akiyama, T., Ohta, K. and Sakai, H. (1991) In vitro effects on microtubule dynamics of purified Xenopus M phase-activated MAP kinase. *Nature*, **349**, 251–254.

Joshi, H. C. and Cleveland, D. (1990) Diversity among tubulin subunits: toward what functional end? *Cell Motility and the Cytoskeleton*, **16**, 159–163.

Joshi, H. C., Palacios, N. J., McNamara, L. and Cleveland, D. W. (1992) γ-tubulin is a centrosomal protein required for cell cycle-dependant microtubule nucleation. *Nature*, **356**, 80–82.

Kerr, J. F. R., Wyllie, A. H. and Currie, A. R. (1972) Apoptosis: A basic biological phenomenon with wide ranging implications in tissue kinetics. *British Journal of Cancer*, **26**, 239–257.

Lock, R. B. and Ross, W. E. (1990) Possible role for p34^{cdc2} kinase in etoposide-induced cell death of chinese hamster ovary cells. *Cancer Research*, **50**, 3767–3771.

Martin, S. J. and Cotter, T. G. (1990) Specific loss of microtubules in HL-60 cells leads to programmed cell death (apoptosis). *Biochemical Society Transactions*, **18**, 299–301.

Sullivan, K. F. (1988) Structure and utilization of tubulin isotypes. *Annual Reviews in Cell Biology*, **4**, 687–716.

INDEX

Actinomycin D 124, 156, 269, 305, 315
Adenosine triphosphate (ATP) 103, 279–282
Adenylate kinase 53
Adhesion molecules
  LFA-1 35
  LFA-3 248
  I-CAM-1 35
ADP-ribosylation 114, 118, 282
Adriamycin 279
Agarose gel electrophoresis 109, 112, 128, 140
3-Aminobenzamide 118
Androgen 182
  dependence and independence 183
Anergy 19, 21, 240
Anti-apoptotic signals 162
Anti CD-3 antibodies 26, 34, 90, 157, 168
Antibody-dependent cytotoxicity 231
Anti-cancer drugs 269–276, 279–292 see also methotrexate, vincristine, adriamycin, fluorodeoxyuridine, fluorouracil
Antigen 59
  presenting cells 224, 238

Antisense oligodeoxyribonucleotides
  antisense oligonucleotides 52, 159
  phosphorothioate 52
AP-1 21, 65–69
APO-1 receptor 173
APO-1 229–232 see also Fas antigen
Apoptosis
  common final pathway 162
  CTL-mediated 156, 203, 246, 305
  glucocorticoid-induced 25, 33–6, 40–2, 99–109, 203
  hyperthermia-induced 297–310
  pathway 154
  peptide-epitope induced 245–252
  trigger 266, 281
Apoptotic bodies 2, 129, 155, 269, 300, 315
ATP 104, 279–282
Aurin tricarboxylic acid 74, 115
Autoimmune disease 19, 42, 235–241 see also EAE
Autoreactive T cells 5, 19, 241

B cell 5, 99, 232
  lymphoma 167
  mature 168

B lymphocytes 145, 219
B lymphoid development 145, 173
bcl-2 7, 101, 108, 149, 155, 161–2, 167–174, 179–183, 187, 204, 222, 301
bcr/abl 162
BD/BND-cellulose 134
Biochemical cascade 283
Biochemical mechanisms 19
Bistratene A 50
Blebbing 52, 315, 321
BM13674 48, 270, 298, 303
Burkitt's lymphoma 194, 298

C. elegans 59, 145, 154, 203
Caffeine 137
Calcium 20-1, 38–40, 43, 101, 222, 260, 281, 304
  antagonism 81
  intracellular 38, 304
  ionophore 20, 47, 62, 89, 155, 168
  mobilisation 91
Calcium chelators
  EDTA 116
  EGTA 38, 91–2, 304
Calcium/magnesium dependent endonuclease 20, 81, 103, 111, 124, 205, 281
Calmodulin
  inhibitors 38
Calphostin C 52
Calyculin A 48–9, 54
Camptothecin 106
Castration 5
*cdc*2 kinase 320
cDNA 119, 148, 204, 230
ced genes 150, 212 *see also* death genes
Cells
  B cells 5, 99, 145, 219, 232
  BM13674 48, 270, 305
  CEM 50, 101, 113, 117, 135, 204, 270, 315
  CTL 60, 99, 147, 156, 203, 247–52
  HL-60 77, 119, 270, 315
  MOLT-4 104, 270
  T cells 5, 33, 220, 237, 247
  U-937 51
Cell cycle 54, 59–69, 79, 119, 188, 195, 205, 271 see also G1, G2/M
  S phase 134
Cell deletion 4
Cell proliferation 19
Cell suicide 7
Cell volume 123
Centrosome 315
Chemotherapeutic drugs 6, 266, 270, 284
Chemotherapy 100, 284
Chromatin 2, 81, 111, 259, 281, 309
  associated endonucleases 113
  condensation 2, 48, 51, 87
  fragments 259
  margination 248
  nuclear 2
Clonal expansion 6
CNS 235
Cognate peptide 250
Colchicine 76, 322
Common final pathway 283
Corticosterone 34
Cross-talk 25, 45
CTL 46, 60, 99, 147, 245–52
CTL-induced apoptosis 156
Cyclic AMP 22, 45
Cycloheximide 47, 54, 88, 103, 124, 148, 155–8, 269, 305
Cytochalasin B 321
Cytosine arabinoside (Ara-C) 271
Cyclosporin A 222, 240
Cytolysis 40
Cytosine arabinoside (Ara-C) 155, 271
Cytoskeleton 64 see also microtubules
Cytosolic proteins 92
Cytotoxic T lymphocytes (CTL)60, 235, 245–7, 277, 305
Cytotoxic agents 160, 315
Cytotoxic T cell 59, 236, 245

Cytotoxicity 20, 237

D-loop region 209
Death genes 148, 160, 204, 213
  ced genes 150, 212
Dephosphorylation 49, 308
Development 4, 153
Dexamethasone 36, 63, 88, 101, 117, 124, 148, 203, 269, 315
Diacylglycerol 2
Differential screening 205
Differentiation 5, 20, 59
Dioxin 20, 89, 155, 305
Diptheria toxin 47, 88
DNA
  binding 188
  binding complexes 65
  binding elements 68
  cleavage 4, 105, 281, 283, 299
  damage 87, 90, 188, 196
  damaging agents 284
  degradation 1, 4, 62, 119, 135, 271, 280
  flow cytometry 270
  fragmentation 20, 38, 47, 73, 79, 87, 99, 111, 123, 133, 153, 194, 206, 220, 229, 246, 259, 266, 269, 298, 318
  gel electrophoresis 105, 272, 320
  laddering 115, 125, 189, 205, 220, 266, 269, 299, 317
  repair 134, 282
  replication 134
  structural analysis 133–40
  synthesis 188, 198
  tumour viruses 197
DNase II 113
Drug therapy 284

E1A 198, 210
EAE 235–41
EBV 194
EGTA 302

Electrophoresis
  agarose gel 109, 112, 128, 140
  DNA 271, 320
Embryo 80
Embryonic 41
Endo-exo nucleases 111–20
Endonuclease 74, 81, 101, 111, 114, 140, 156, 269 *see also* calcium/magnesium dependent activation 123 endonuclease
Endoplasmic reticulum 129
*erb*A 42
Etoposide (VP-16) 77, 117, 128, 138, 271, 315
Experimental autoimmune encephalomyelitis EAE 235–41

Fas antigen 230 *see also* APO-1
Fluorescent activated cell sorting
  FACS analysis 26, 93, 124, 129, 169, 190, 221, 274–6
Fluorescent dyes 94
5-Fluorodeoxyuridine 272–6
5-Fluorouracil 272–6, 279
fos 21, 65, 159

G0 phase 59 *see also* cell cycle
G0/G1 arrest 196 *see also* cell cycle
G1 phase 188, 205, 220, 272 *see also* cell cycle
G2/M arrest 51, 115, 191, 273, 322 *see also* cell cycle
G protein 20
Gamma radiation 168, 208
ß-Galactoside binding protein 204, 210 *see also* HL-14
Gel electrophoresis 105 *see also* agarose
  2D 49
Gene expression 203–213
Genes *see also* proto-oncogene, oncogene
  ced
  HL-14
  RP-8 149

RP-2 149
SGP-2 155, 195
TRPM-2, 155, 183
tumour suppressor 8, 187–199 see also retinoblastoma and p53
Gliotoxin 87–94, 305
Glucocorticoid 6, 25, 61, 99, 147, 157, 171, 205, 238, 260, 305
　antagonist 26
　receptor 25, 42, 67, 154, 203
　resistance 64
　response 25
　response elements (GRE) 67
　sensitivity 64
Glucocorticoid receptor
　AP-1 interactions 67–69
Growth factors 161
GTPase 119

Heating techniques 302
Heat shock proteins 306
H-89 52
Heterochromatin 130
HIV infection 7, 225
HL-14 gene 210 see also ß-galactoside binding protein
HL-60 cells 77, 119, 270, 315
Hoechst 33342 128
Homeostasis 6
　lymphoid 168
　T-cell 171
Hormone withdrawal 147
Hybridomas 36, 156, 237
Hyperplasia 6
Hyperthermia 7, 88, 156, 297

I-CAM-1 35
Immune tolerance 5
Immunocytochemistry 238
Immunodeficiency 73
Immunoglobulin 31
　genes 225
Inflammation 2

Inhibitors 52
Interdigital webs 4
Interferon-γ 238
Interleukin-1 22
Interleukin-2(IL-2) 21, 28, 60–69, 238, 247
　receptor 28, 60–69
　signal transduction 62–64
Interleukin-3 161, 193
Interleukin-6 192
Interleukins 80
Inter-nucleosomal DNA fragmentation 73, 92, 189, 194, 272, 282 see also DNA fragmentation
Inter-nucleosomal cleavage 99, 123, 131, 138, 310
Involution 5
Ionizing radiation 147, 205 see also gamma radiation
Ionophore
　A23187 105, 155
　ionomycin 36
Irradiation 128, 155, 260 see also gamma radiation
Ischaemia 6

jun 21, 65

Keratinocytes 5

Lectin 79
Leukaemia 74, 271
LFA-1 35
LFA-3 35
Life-span 59
Ligands 42
Lymphoid 6
Lymphokines 246
Lymphomagenesis 173

Macromolecular synthesis 101, 155, , 292, 305
Macrophage 89

MAP kinase 66
Membrane 62, 250, 309
  blebbing 100, 205
Metamorphosis 4, 42
Methotrexate 271, 315
MHC antigens 33, 224, 237, 247
Microscopy
  electron 128
  light 79
Microtubules 64, 76, 82, 316
Mitochondrial
  D-loop region 209
  DNA 209
  genes 208–9
Mitogenic 27
Mitotic spindle 319
MOLT-4 107, 270
Morphological
  changes 52, 123, 133, 271, 305
  features 280, 298
  studies 270, 275
Morphology 2, 88, 140, 316
mRNA 148, 183, 204, 207, 222, 317
Mutations 6
myc 101, 107, 153–63, 173, 194, 204
Myelin
  basic protein 235
  damage 235

Necrosis 1, 7, 99, 135, 155, 301
Negative selection 19, 59, 146, 171
Neurospora 114
Novobiocin 103–107
Nuclear fragmentation 51
Nuclear chromatin 2
Nucleosomes 4
Nucleus 104, 263, 317
  fragmentation 259

Okadaic acid 46, 303
Oligodendrocyte 236
Oligonucleosomal fragments 4, 154, 298
Oligonucleotides 119

Oncogene 187 *see also* proto-oncogene
  bcr/abl 174
  myc/bcl-2 174
Ontogeny 100
Organelles 123, 129
Ovarian carcinoma 261
Oxidative stress 90

p53 188
p53 knockout mice 196
Perforin 246
Phenanthroline 75
Phorbol ester 21, 36, 47, 60, 67, 168, 304
Phorbol dibutyrate 304
Phosphatases
  PP1, PP2A, 20, 48, 52–55
Phospholipase A2 41
Phospholipase C 20
Phosphorylation 27, 45, 53, 76, 105, 114, 119 *see also* dephosphorylation
Phytohaemagglutinin 80
Poly ADP ribose 77
Poly (ADP ribose) polymerase 81, 282
Polymerase Chain Reaction (PCR) 119, 230
Pre-B cells 101–109, 174
Proliferation 59, 89
Prostaglandins 22
Prostate 5, 148
  cancer 179
Protein kinase A 52
Protein kinase C (PKC) 21, 38, 45, 61–3, 80, 127, 304
  translocation 62
Protein kinases 20, 52, 205
  serine/threonine 20
Protein modification 45–49, 308
Protein phosphatases 45, 303 *see also* phosphatases
Protein synthesis 46, 88
  inhibition 7
  inhibitors 90, 301 see also cycloheximide

Proto-oncogene 8, 161, 222
  see also bcl-2
  erbA 42
  fos 21, 65, 159
  jun 21
  myc 101, 107, 153–162, 173, 194, 204, 222
  ras 23, 161, 195

Radiation 6, 263 see also ionizing radiation
raf-1 64
ras 23, 161, 195
ras-GTPase 23, 119
Receptors 61
  glucocorticoid
  interleukin-2
Repair
  DNA 73
Retinoblastoma (RB) protein 160, 198
Retinoic acid 43
RNA synthesis 46
  inhibitors
RP-8, RP-2 genes 149
RU-486 26

S phase 134 see also cell cycle
Second messengers 20
Self MHC 34, 220
Self-tolerance 239
Serine/threonine/tyrosine protein kinases/phosphatases 20
Serum growth factors 80
SGP-2 195 see also TRPM-2
Shrinkage necrosis 123
Signal transduction 19, 108
  cross talk 25
  pathways 19, 55, 61, 222
Suppressor protein 306
Survival genes 204 see also bcl-2
SV-40 134 see also DNA tumour viruses
  large T antigen 195, 210

T cell 89
  activation 21
  anergy 19, 21, 240
  apoptosis 240
  CD4/CD8 23, 31, 173, 220, 230, 237, 246–7
  development 123, 219
  hybridoma 36–8
  leukaemia 203
  repertoire 33
T-cell receptor (TCR) 21, 31–38, 59, 90, 157–60, 168, 219
  antibodies 20
Target cells 245
Targets 82
Tetrandrine 50
TGF-ß 179, 183
Thermotolerance 307 see also hyperthermia
Thymocyte 21, 33, 60, 79, 89, 99, 124, 146, 241, 259, 269, 304
  apoptosis 19–28, 31–43, 222–4, 307
  immature 21, 220 302
  pre-apoptotic 128
Thymus 19, 219
Tissue regression 59
TNF/NGF receptor superfamily 230
Topoisomerase I 106
Topoisomerase II 105, 118, 126
  inhibitor 126
Toxins 7, 87–9, 266
TPA response elements (TRE) 67
Transcription factors see also AP-1, fos, jun, myc
Transcriptional activation 188
Transforming growth factor see TGF-ß
Transgenic mice 21, 168, 172
Transglutaminase 155
Trigger 88
TRPM-2, 155, 179, 183 see also SGP-2
Tubulin 315–322
  depolymerisation 320

Tumour 8, 262, 280
    breast 284
    cell energy 279–92
    colon carcinoma 265
    heterogeneity 266
    histology 261, 288
    irradiation 259–267, 289
    melanoma 265
    murine 259–67, 299
    ovarian carcinoma 261
    prostate 179
    radiotherapy 260, 290
    xenografts 265
Tumour necrosis factor (TNF) 156, 246
    receptor 230
Tumour suppressor genes 8, 187–99
Tyrosine kinase 22, 63
Tyrphostins 64

Ultrastructure 1–4, 131

Vertebrate development 42
Vincristine 273, 315
VP-16 *see also* etoposide 317

Western blotting 23

X-irradiation 305 *see also* ionizing radiation
Xenografts 259

Zinc 100, 124, 303
    chelators 74
    deficiency 73
    finger proteins 81, 204
    fluorophore 79
    fluxes 79
    intracellular 73–83
    ionophore 77
    targets 80